W9-CMI-230

The Lorette Wilmot Library
Nazareth College of Rochester

REALITY
RULES: II

REALITY RULES: II
Picturing the World in Mathematics
—The Frontier—

JOHN L. CASTI

Technical University of Vienna
Vienna, Austria

and

Santa Fe Institute
Santa Fe, New Mexico

A Wiley-Interscience Publication
JOHN WILEY & SONS, INC.
New York / Chichester / Brisbane / Toronto / Singapore

Copyright ©1992 by John Wiley & Sons, Inc.

All rights reserved. Published simultaneously in Canada.

Reproduction or translation of any part of this work
beyond that permitted by Section 107 or 108 of the
1976 United States Copyright Act without the permission
of the copyright owner is unlawful. Requests for
permission or further information should be addressed to
the Permissions Department, John Wiley & Sons, Inc.

Library of Congress Cataloging in Publication Data:
Casti, J. L.
    Reality rules: picturing the world in mathematics/John L. Casti.

        p.   cm.
    "Wiley-Interscience publication."
    Includes index.
    Contents: v. 1. The fundamentals – v. 2. The frontier.

    1. Mathematical models.   I. Title.
QA401.C3583   1992
511'.8–dc20   92-12213
ISBN 0-471-57798-7 (vol. 2)   CIP
ISBN 0-471-57797-9 (set)

Printed in the United States of America

10 9 8 7 6 5 4 3 2

To the memory of my teacher
RICHARD E. BELLMAN,
who would have been interested

PREFACE

Mathematical modeling is about rules—the rules of reality. What distinguishes a mathematical model from, say, a poem, a song, a portrait or any other kind of "model," is that the mathematical model is an image or picture of reality painted with logical symbols instead of with words, sounds or watercolors. These symbols are then strung together in accordance with a set of rules expressed in a special language, the language of mathematics. A large part of the story told in the 800 pages or so comprising the two volumes of this work is about the grammar of this language. But a piece of the real world encoded into a set of mathematical rules (i.e., a model) is itself an abstraction drawn from the deeper realm of "the real thing." Based as it is upon a choice of what to observe and what to ignore, the real-world starting point of any mathematical model must necessarily throw away aspects of this "real thing" deemed irrelevant for the purposes of the model. So when trying to fathom the meaning of the title of this volume, I invite the reader to regard the word "rule" as either a noun or a verb—or even to switch back and forth between the two—according to taste.

The book you now hold in your hands started its life as a simple revision of my 1989 text-reference *Alternate Realities.* But like Topsy it just sort of grew, until it reached the point where it would have been a misnomer, if not a miscarriage of justice, to call the resulting book a "second" or "revised" or "updated" or even a "new" edition of that earlier work. And, in fact, the project grew to such an extent that sensibility and practical publishing concerns dictated a splitting of the work into two independent, yet complementary, volumes. But before giving an account of the two halves of my message, let me first offer a few words of explanation as to why a three-year old book needed updating in the first place.

Alternate Realities was a mathematical-modeling text devoted to bringing the tools of modern dynamical system theory into the classroom. As such, the focus of the book was on things like chaos, linear system theory, cellular automata, evolutionary game theory, q-analysis and the like. While the book itself was published in early 1989, the actual writing took place during late-1986. As a result, most of the material was based on what was current in the research literature *circa* 1985. In the intervening years research interest and results in dynamical system theory has been nothing short of explosive. The chart below gives some indication of the magnitude of this exponential growth of published research in just the fields of chaos and fractals. And work in the other topical areas addressed in *Alternate Realities* has certainly been no less intense. Hence, the call from both readers and my editor for an update.

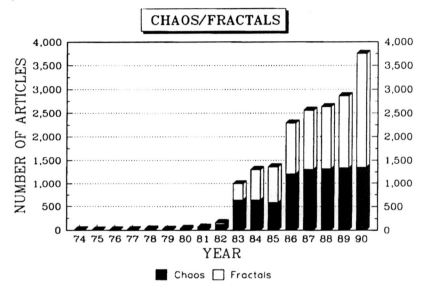

So what are you getting in the two volumes of this book that's new besides some polishing of the reference lists to a little brighter shine and a tidying-up of typos and other literary infelicities in *Alternate Realities?* Briefly, here are the main attractions:

• *New chapter and sections*—Recognizing the fact that for a modern mathematician a computer program has become a legitimate answer to a mathematical question, *The Frontier* contains an entirely new chapter (Chapter 9) on computation and complexity. In particular, this chapter takes up the question of what can actually be done by way of formalizing real-world phenomena within the confines of a computer program, i.e., a set of rules. This chapter offers the student and researcher a point of contact with a host of matters of current intellectual concern running the gamut from Gödel's Incompleteness Theorem and its connection to work in artificial intelligence to the problem of NP-completeness and the complexity of numerical algorithms.

In addition to the new chapter, a number of new sections have been added to the chapters originally constituting *Alternate Realities*. These sections introduce topical areas like artificial life that were only touched on briefly in that earlier work, or treat entirely new areas of concern like the relation between chaos and stock-price fluctuations that were yet to be born in 1985.

• *Exercises*—One of the main comments (and complaints) I heard from readers and users of *Alternate Realities* was that the problems were too hard. Too hard, anyway, to be used as drill exercises in a classroom setting. To remedy this defect, I have added Exercises of the drill-type at the end of almost every section in every chapter of the book (except for the final chapter, which is more a meditation on the philosophy of modeling than on the mathematics). These Exercises make the book much easier for instructors to use as a mathematical modeling text in the classroom, as well as provide drill problems for those using the book for self-study.

• *Examples*—As a further effort toward making the material accessible to students and researchers alike, many new examples have been sprinkled throughout the book at strategic locations. Not only do these examples introduce new applications of the theoretical results, they also serve to illustrate exactly **how** to use the theory in a diverse array of realistic situations.

• *Discussion Questions and Problems*—Several new Discussion Questions have been added to each chapter, both to extend the earlier material and to bring forth ideas nonexistent a few years ago. Moreover, the Problems sections have been augmented with new results from the research literature that for one reason or another didn't fit into the mainstream of the text.

• *Solutions Manual*—From an instructor's point of view, it's always a bit awkward, not to mention annoying, trying to teach from a book containing problems that you yourself can't solve! And I must confess that the problems in *Alternate Realities* were by and large pretty hard, most of them having been taken directly from the primary research literature. So to ease the pain of grappling with these research-type questions, I have prepared a Solutions Manual for the problems appearing in each chapter's *Problems* section. This manual is available at no charge from the publisher. To obtain a copy, ask your Wiley sales representative or write directly to Wiley-Interscience, 605 Third Avenue, New York, NY 10158-0012.

So there you have it. More than 300 pages of new points of contact between the worlds of nature and mathematics—in itself material constituting a fair-sized book. I think that after looking over this new material, readers of *Alternate Realities* and librarians will both agree that a new book was indeed in order. Now let me talk for a moment about the intended audience for this book and how it might be used.

One of the questions that always comes up in deciding on a text for a course is the puzzler: What are the prerequisites? Since this is a book on mathematical modeling, it goes without saying that in general terms the prerequisites for using the book are a *working knowledge* of basic undergraduate mathematics. This means that the reader should have been exposed to *and assimilated* the material typically found in one-semester courses in calculus, linear algebra and matrix theory, ordinary differential equations and, perhaps, elementary probability theory. Moreover, s/he should know about the basic vocabulary and techniques of mathematics. So, for example, things like "sets" and "equivalence relations" should be familiar territory, as should the idea of an inductive argument or a proof by contradiction. The emphasis here, of course, is on the phrase "working knowledge." To illustrate, it is definitely *not* sufficient to have had a course in matrix theory and have just *heard* of the Cayley-Hamilton Theorem. You must actually know what the theorem says and how to use it. Or, at the bare minimum, you should at least know where to go and look up the result. This is what I mean when I say the prerequisites for accessing this book are a working knowledge of basic undergraduate mathematics.

Now let's turn to how the book might be used as a text in a course in mathematical modeling. It's manifestly evident, I think, that there's far more material in these two volumes than can be comfortably addressed in even a one-year course, let alone in a single semester. Hence, one of the prime motivations for splitting the book into two pieces. Since what distinguishes *modern* mathematical modeling from its classical counterpart is its emphasis upon dynamics and nonlinearity, *The Fundamentals* contains the essentials of these matters. By way of illustration, let me briefly indicate how I have made use of this material in my own courses.

First of all, I always include the material from Chapter 1 as it sets the general framework for just about every type of mathematical modeling undertaking. I sometimes follow Chapter 1 with selected material from Chapter 8 on q-analysis. For some reason students seem to find it easier to grasp what's going on in the modeling game by starting with the context of sets and binary relations rather than jumping immediately into the thicket of dynamical systems and its daunting terminology and definitions. Moreover, the generality of the q-analysis idea lends itself to a number of interesting examples in art, literature and life that lie outside the domain of the "hard" sciences. Since just about everyone nowadays wants to know about chaos, my lectures usually continue with a selection of material from Chapters 2, 3 and 4. The second chapter contains the necessary background for dealing with dynamical systems, which is then used in Chapters 3 and 4 within the specific settings of cellular automata and chaotic processes. If there is still time remaining in the term, I conclude the course with a discussion of some of the more general philosophical issues linking science and other reality-generating mechanisms, as outlined in the book's final chapter. Since this program involves mostly material from *The Fundamentals,* it would be remiss of me to close this Preface without saying a word or two about the first half of the book.

As I've already mentioned, the sheer size of this "magnum opus" dictated its division into two volumes. But unlike many multi-volume efforts, this split was not made on the basis of introductory versus advanced material and/or techniques. Quite to the contrary, in fact, as in putting this book together I tried valiantly to ensure that the difficulty level was as uniform as possible throughout. So the line of demarcation between the two volumes lies in a very different direction.

The Fundamentals contains the material that I feel is essential for anyone to consider himself or herself a player in the game of "modern" mathematical modeling. Just as it's difficult to conceive of writing a book without knowing the alphabet, there are certain concepts and results that anyone who wants to get involved with modeling must have at his or her disposal. The first volume provides an account of this irreducible minimum of basic knowledge, material that can form the basis for a one-semester initial encounter with mathematical modeling. For those already familiar with the basics, *The Frontier* introduces a number of application areas and/or associated techniques of modeling that complement the ideas presented in the opening volume. Now let me be a bit more specific about the contents of *The Fundamentals.*

Chapter 1 introduces foundational ideas about modeling like the notions of a state set, system observables and equations of state. This chapter also provides a brief account of the system-theoretic concepts of complexity, self-organization, bifurcation, surprise and error, topics that are developed in more detail throughout the remainder of the book.

A system's equation of state is a function defined on the set of observables used to describe the system. And, generally speaking, the most interesting system behaviors occur at the points where these functions display singularities. Consequently, Chapter 2 is devoted to a recounting of the main ideas in the theory of singularities of smooth functions and mappings, including the famous results of Morse, Whitney, Thom and Arnold. Following development of the principal theoretical tools, a number of applications ranging from animal resource management to electric power generation and on to developmental biology are given to illustrate the theory. The second-half of the chapter is devoted to a presentation of the basics of the theory of dynamical systems, along with an introductory treatment of bifurcation theory for vector fields.

Cellular automata, dynamical systems evolving on discrete state spaces in discrete time, are the objects of attention in Chapter 3. Following a discussion of the basic theoretical and computational aspects of this important class of dynamical systems, the chapter takes up a number of applications arising mostly in biology. These include plant growth, vertebrate skin patterns, the syntactical structure of languages and the process of self-reproduction. The final sections of the chapter treat the problem of artificial life—the question of whether living organisms can be created in a machine.

Chaotic dynamical processes and fractals are the twin theme songs of Chapter 4, the concluding part of *The Fundamentals.* Here consideration is given to the kinds of mathematical structures giving rise to strange attractors, together with a rather extensive discussion of practical ways to test for the presence of chaos in observational data. The second-half of the chapter addresses many applied areas where chaos plays an important role, including stock market price fluctuations, the spread of disease, and turbulent fluid flow. The chapter ends with a discussion of the most complicated object known to humankind—the Mandelbrot set. This leads into a consideration of fractals, in general, together with their interconnections to chaotic processes.

Taken together, it's my hope that the material of the two volumes making up *Reality Rules* will serve as both a text and a reference that students and researchers alike will be able to turn to as a source of inspiration and information as they make their way through the ever-shifting quicksand and minefields of the complex, weird and wonderful world we all inhabit.

JLC
Vienna, Austria
January 1992

CREDITS

Grateful acknowledgment is made to the following sources for permission to reproduce copyrighted material. Every effort has been made to locate the copyright holders for material reproduced here from other sources. Omissions brought to our attention will be corrected in future editions.

Cambridge University Press for Figure 1.2, which is reproduced from Thompson, D'Arcy, *On Growth and Form*, 1942, and for the figures on pages 132 and 149, which are reproduced from Saunders, P., *An Introduction to Catastrophe Theory*, 1980.

E. P. Dutton, Inc. for Figure 2.8, as well as the figures on pages 129 and 136, which are reproduced from Woodcock, T. and M. Davis, *Catastrophe Theory*, 1978.

MIT Press for Figures 2.12 and 2.13 found originally in Arnold, V. I., *Ordinary Differential Equations*, 1973.

American Physical Society for Figures 3.4, 3.5 and 3.8–3.10, which are reproduced from Wolfram, S., "Statistical Mechanics of Cellular Automata," *Reviews of Modern Physics*, 55 (1983), 601–644, and for Figure 4.20 from Eckmann, J. P., "Roads to Turbulence in Dissipative Dynamical Systems," *Reviews of Modern Physics*, 53 (1981), 643–654.

Corgi Books for Figure 3.11 which appeared in Gribbin, J., *In Search of the Double Helix*, 1985.

Basil Blackwell, Ltd. for Figures 3.12, which is taken from Scott, A., *The Creation of Life*, 1986.

Alfred Knopf, Inc. for Figures 3.14, 3.16 and 3.17 from Eigen, M. and R. Winkler, *The Laws of the Game*, 1981.

William Morrow and Co. for Figure 3.18 which appeared originally in Poundstone, W., *The Recursive Universe*, 1985.

Elsevier Publishing Co. for Figure 3.19 which appeared originally in Young, D., "A Local Activator-Inhibitor Model of Vertebrate Skin Patterns," *Mathematical Biosciences*, 72 (1984), 51–58, for Figures 4.17 and 4.18 taken from Swinney, H., "Observations of Order and Chaos in Nonlinear Systems," *Physica D*, 7D (1983), 3–15, and for the figure on page 419, which is reproduced from Barash, D., *Sociobiology and Behavior*, 1977.

Longman Group, Ltd. for Figures 3.20–3.23, which are reproduced from Dawkins, R., *The Blind Watchmaker*, 1986.

Springer Verlag, Inc. for Figure 4.3 taken from Arnold, V. I., *Mathematical Methods in Classical Mechanics*, 1978, and for Figure 4.5 found in Lichtenberg, A., and M. Lieberman, *Regular and Stochastic Motion*, 1983.

Mathematical Association of America for Figure 4.14 from Oster, G., "The Dynamics of Nonlinear Models with Age Structure," in *Studies in Mathematical Biology, Part II,* S. Levin, ed., 1978.

Academic Press, Inc. for Figure 5.1 from Vincent, T., and J. Brown, "Stability in an Evolutionary Game," *Theoretical Population Biology,* 26 (1984), 408–427.

Heinemann Publishing, Ltd. for Figure 8.10 from Atkin, R., *Mathematical Structure in Human Affairs,* 1974.

Houghton Mifflin Company for Figures 9.2 and 9.4, which are reproduced from Rucker, R., *Mind Tools,* 1987.

Professors E. C. Zeeman for Figures 2.17 and 2.18, as well as for the figures on pages 134 and *II–77;* Jeff Johnson for Figures 8.6 and 8.9, as well as the figures on pages *II–272* and *II–279;* Robert Rosen for Figures 1.7, 1.8 and 7.8; Thomas C. Schelling for Figures 3.2 and 3.3; Helen Couclelis for Figures 3.6 and 3.7; Christopher Langton for the figures on pages 239 and 240; Erik Mosekilde for Figures 4.10–4.13; Otto Rössler for Figures 4.24 and 4.25, as well as the figure on page 350; Douglas Hofstadter for Figure 9.5.

CONTENTS

Chapter Ten How Do We Know?: Myths, Models and Paradigms in the Creation of Beliefs

CONTENTS OF
THE FUNDAMENTALS

REALITY
RULES: II

CHAPTER FIVE

STRATEGIES FOR SURVIVAL: COMPETITION, GAMES AND THE THEORY OF EVOLUTION

1. Evolutionary Epistemology

In the Darwinian battle for survival, Fortune's formula may be compactly expressed as

$$\text{heredity} + \text{variation} + \text{selection} = \text{adaptation},$$

encapsulating in four everyday words over a century's worth of scholarly and public debate about the nature of change in living organisms, and the degree to which that process, whatever it may be, is homologous to the flow of human affairs. The intensity of the debate as to the precise interpretation of the terms in this formula has often been raised (or lowered?) to levels of hysteria and metaphysics quite far beyond the bounds of mere science, even spilling over occasionally from the professional journals and popular press into the courts and pulpits. We have two principal goals in this chapter: (1) to avoid (like the plague) entering into any discussion of the metaphysical or theological aspects of Darwinian theories and, what's more important, (2) to provide an account of some current system-theoretic and mathematical approaches to the formalization of the basic ingredients of evolutionary theory expressed in the above "evolutionary equation."

In the preceding chapters we have been primarily occupied with matters of dynamics, showing the way in which modern mathematics attempts to formalize the processes of change. Now we shift our emphasis a bit, considering less matters of analytic technique and more how to actually use our tools to address that most quintessential of dynamical processes: (r)evolutionary change. The mathematical formalization of evolutionary processes offers a nearly unlimited playpen for modeling, about which are strewn many of the favorite toys of the system scientist. Here we will find bifurcation theory, catastrophes, complexity, randomness, stability, self-organization and self-reference all making much more than a cameo appearance. We find as well many other tricks and subterfuges of the modern applied mathematician. All of these weapons will be brought to bear upon a handful of basic questions:

- What is the relationship between a change at the microlevel of an organism's genotype, and the resultant change at the macrolevel of its physical and/or behavioral phenotype?

• What does it mean to say that a particular phenotypic property is "selected for" in the competition for survival, i.e., what is the relationship between selection criteria and the idea of adaptive fitness?

• What are the causal paths linking environmental, genetic and phenotypic change?

• Where do we draw the dividing line between evolutionary and revolutionary change?

• To what degree, if any, are the concepts and results of *biological* evolution relevant to the changes in social and behavioral systems?

Despite the virtually limitless variations possible on the above themes, we can compactly summarize the problem of evolution as: "What changes what, how does it happen and why?" When stated in such a direct and crude manner, it's clear that a procedure for the formalization of evolutionary phenomena is tantamount to a procedure for the formalization of *any* dynamical process. In this sense the study of evolution serves as a universal metaphor for change in *all* processes. With these thoughts in mind, let's take a more detailed look at the major components in the modern view of the world according to Darwin.

Exercises

1. We have described the study of evolution as a universal metaphor for change. Yet as the term "evolution" is generally employed, it refers to a process satisfying Fortune's formula given in the text. But not every dynamical process obeys this formula, suggesting that evolution cannot be a universal metaphor after all. Discuss how you might reconcile these two seemingly contradictory claims.

2. It's often claimed that social groups like a city or organizations like corporations also adapt to their environment by an evolutionary process. Taking these claims at face value, consider what the genotype and phenotype of such organizations might be. Also, what kind of selection principle is at work for evolutionary processes of a social nature?

3. The theory of "punctuated" evolution argues that evolutionary change is not smooth and gradual but rather proceeds in fits and spurts, with long periods of stasis interrupted by relatively short periods of rapid change. How does this idea fit in with the notion of *revolutionary* change? Can you think of situations in the social realm where punctuated evolution might be a better metaphor than the more conventional slow, plodding, continuous Darwinian view?

4. Is there any necessary connection between an evolutionary process of adaptation and an optimization process? That is, is the endpoint of

every evolutionary process an organism that is "optimal" by some measure of optimality?

2. The Neo-Darwinian Paradigm

Darwinian evolution is based upon a population of organisms that:

1) Vary in heritable traits influencing their chances (fitness) for reproduction and survival;

2) Have offspring resembling their parents more than they resemble randomly chosen members of the population.

3) Produce more offspring, on the average, than are needed to replace members removed from the population by emigration and death.

We can refer to these properties of the population as *variation, inheritance* and *reproduction.*

On intuitive grounds, it seems evident that populations producing more offspring that survive until reaching reproductive age than other populations will be favored in an environment having limited resources, since the other groups will not be able to establish themselves in competition with such a population. This thesis lies at the heart of what Darwin called *natural selection.* According to Darwinian theory, all aspects of the biological structure and behavior of living organisms are molded by this process of natural selection. The current *neodarwinian* paradigm for species change adds to the classical Darwinian picture a theory of inheritance based on a genetic mechanism, as well as a theory of how genes spread in a population (population genetics).

In the neodarwinian set-up, it's important to note the controversial point that selection may occur at several levels of organization. The unit of selection may be the gene, the chromosome, the individual, the population or even the society. There is some evidence to support the claim that selection at levels of aggregation higher than the classical Darwinian level of the individual organism are far less efficient. But it's still of considerable research interest to understand how these various levels of selection are integrated into a single biological community.

Probably the most mysterious component of the entire Darwinian setup is the concept of *fitness.* To understand the nature of this a little better, let's suppose we have a population of asexually-reproducing individuals. Assume that each individual gives rise to α offspring per unit time, and that these offspring mature within one time unit. Let q_o represent the probability that each such offspring survives one time unit, and let q_p be the corresponding survival probability for an adult. Then if N_t represents the expected population level at time t, we have

$$N_{t+1} = (\alpha q_o + q_p)N_t.$$

We can define the quantity $\lambda \doteq \alpha q_o + q_p$ to be the mathematical representation of *individual fitness*. Thus, λ is the net rate of population growth and, consequently, represents a quantitative formalization of the adaptation component of Darwin's world. In the case of asexual reproduction, natural selection acts to single out those strategies (determined by α, q_o and q_p) that maximize λ. In common parlance, we use the term "fittest" to characterize those individual strategies that are so selected. Similar arguments involving generalized notions of fitness can be given for sexually-reproducing populations, as well as for inhomogeneous populations.

Darwin never really had a proper understanding of the processes of variation and inheritance, primarily because he had no knowledge of the concept of a *gene*, nor any notion of how new, inheritable traits arise and spread through a population. It was left to the German biologist August Weissman to formulate with his theory of "germ plasm" the basis of our current understanding of these two key components of the Darwinian picture. Basically, Weissman's argument was that any fertilized egg gives rise at an early stage of its development to two independent populations of cells: the *germ line*, which is composed of the sex cells, and the *soma line*, which makes up all the other cells in the body. Weissman then hypothesized that genetic changes occur only in the germ line, and that these changes are independent of changes taking place in the soma line—at least insofar as characteristics acquired by the organism during its life are not passed along to subsequent generations.

The distinction between the germ line and the soma line corresponds to the distinction between what's now termed the *genotype* and the *phenotype*. An organism's genotype consists of its total complement of genes (i.e., it's *genome*); the phenotype is the organism's total assemblage of physical and behavioral characteristics. The phenotype is determined by the individual's environment as well as its genotype, so the reproductive fitness of an individual is a phenotypic property. As a result, conventional wisdom has it that natural selection operates on the phenotype. However, by this action the average genotypic properties of the population are changed over generations in response to selective pressures posed by the environment. Thus there is an interplay between the intrinsic features of the individual (its genotype) and the features of the environment that ultimately determine the survival of the species. The reader will note that Weissman's idea, first presented in 1866, corresponds directly to what we have already called in Chapter 3 the Central Dogma of Molecular Biology: "DNA → RNA → Protein." Only now the germ line corresponds to the DNA/RNA, while the phenotype matches up with the proteins. So genetic information, once it has passed from the genotype into the phenotype, cannot get back into the genotype again. Lamarckian inheritance, in which acquired phenotypic traits can be passed on to subsequent generations, is directly opposed to the Weissmanian view, and thus contravenes the Central Dogma. There is strong empirical evidence supporting Weissman's claim, at least for biological inheritance; however, Lamarck's position is exactly what we observe when it comes to learning and cultural inheritance. Thus, it cannot be neglected when it comes to the consideration of evolutionary processes in the social and behavioral areas. But more of this later.

It's important to emphasize the point that the overall morphology (form) M of an organism is determined *directly* by gene products p_i (the proteins) and the environment E, i.e.,

$$M = f(p_1, \ldots, p_n; E).$$

But by definition each protein is determined by exactly one gene and the environment $p_i = h(g_i; E)$. Here, while almost nothing is really known about the form of the functions f and h, they are both assuredly highly nonlinear. Because of this nonlinearity, small changes in the genotype and/or environment and/or proteins *may* result in large or small changes in the resulting phenotype. However, if evolution through natural selection brings about adaptation, we would expect that organisms will be close to their adaptive peak ($\lambda \approx \lambda_{max}$). So forms deviating greatly from those in the current population will, in general, be less fit. For this reason, evolutionary biologists have traditionally not been much concerned with these sorts of macromutations. However, this is now becoming an active area of research interest, as recent empirical evidence suggests that such "hopeful monsters" may play a more important role than previously thought in evolutionary processes. With the above bit of Darwiniana now under our belts, let's move on to a consideration of how we might formulate in more precise terms some of the questions surrounding the processes of evolutionary change.

Exercises

1. Give examples of situations in which selection would seem to most naturally operate at the level of the genotype or the group, rather than at the conventional Darwinian level of the phenotype. Can you think of any circumstances in which selection might be acting at all three levels simultaneously?

2. Distinguish the germ line from the soma line in: (a) a living cell, (b) a corporation, (c) a national economy, (d) a society.

3. Consider situations in which the Central Dogma of Molecular Biology might be false.

3. Genetic vs. Phenotypic Change

Organisms can change at either the level of genes, in which case our interest usually focuses upon how such changes (mutations) affect phenotypic traits like size, shape or behavior, or they can change at the behavioral level, leading to a consideration of how such changes diffuse throughout the population. These are qualitatively different types of evolutionary change and, consequently, require quite different types of modeling frameworks for their mathematical representation. To fix ideas for future reference, let's look at a typical example of each type of situation.

Example 1: Mendelian Genetic Inheritance

The modern theory of inheritance is based upon principles set down by Mendel governing the transmission of genes from parents to offspring. A particular gene may occur in different forms termed *alleles,* with the resultant genetic pattern of the child determined by the pairing of various alleles contributed by each parent to the fertilized egg. The simplest case is that of a gene with two alleles A and B.

The three possible pairs AA, AB and BB determine the possible genotypes of the organism (Note: Since there is no way to identify the father's genetic contribution from the mother's, the combination AB has the same effect as BA). Since the sex cells of the parents have only a single copy of the gene on each chromosome, the genetic makeup of the child is fixed by the pairing of a chromosome from each parent. So if the parents are *homozygous,* i.e., either both of type AA or BB, then the child must also be homozygous; however, if the parents are *heterozygous,* i.e., AB, then the child's genotype is not fixed and can be any of the pairs AA, BB or AB with probabilities $\frac{1}{4}, \frac{1}{4}$ and $\frac{1}{2}$, respectively. Finally, there is the case when one parent is AA while the other is BB. In this situation, the child is *always* heterozygous. The allele A is said to be *dominant* if the two genotypes AA and AB have the same effect on the phenotype. In this event, the allele B is called *recessive* if the genotype BB is observably different from AA and AB.

Now suppose we have a large population with the alleles A and B present in proportions p and $q = 1 - p$, respectively. Further, assume there is random mating in the population and that the genotypes AA, AB and BB are equally viable. What will be the proportion of the population bearing the three different genotypes in the first and subsequent generations? To answer this question, note that an individual of the first generation will be of type AA only if both parents contribute the allele A to its genotype. Since this allele occurs with likelihood p, the probability of the genotype AA equals p^2. Similarly, the genotypes BB and AB occur in the first generation with proportions q^2 and $2pq$, respectively. Consequently, the proportion of allele A in the first generation is

$$p_1 = \Pr(AA) + \tfrac{1}{2}\Pr(AB) = p^2 + \tfrac{1}{2}(2pq) = p(p+q) = p,$$

while the proportion of allele B is $q_1 = 1 - p_1 = q$. Thus, under the random mating and equal viability hypotheses, the proportions of the two alleles remains fixed from generation to generation.

Now let's change the situation by allowing a kind of non-Darwinian evolutionary effect so that one allele may displace the other, leading to a genetic drift in the population. Suppose we have a population of N reproducing individuals. This means that our gene with alleles A and B has $2N$ representatives in each generation of this population. Let allele A occur α times in generation m, whereas allele B occurs $2N - \alpha$ times. What is the likelihood that A occurs μ times in generation $m+1$ for $\mu = 0, 1, \ldots, 2N$? Again assuming random mating and equal viability of the three genotypes, this question involves carrying out 2N Bernoulli trials having probability of "success" (the appearance of allele A) $p = \alpha/2N$. Thus, the probability of exactly μ appearances of the allele A in generation $m+1$ is

$$p_\mu = \binom{2N}{\mu}\left(\frac{\alpha}{2N}\right)^\mu \left(1 - \frac{\alpha}{2N}\right)^{2N-\mu}.$$

As a corollary, we find the probability that allele A will disappear in the next generation is given by

$$p_0 = \left(1 - \frac{\alpha}{2N}\right)^{2N},$$

while the chance of disappearance of allele B is

$$p_{2N} = \left(\frac{\alpha}{2N}\right)^{2N}.$$

These expressions come from standard results in probability theory, and show the overwhelming likelihood that the population will fix on one genotype or the other. Note also the crucial importance of the population size N. If N is small, as it often is in real-life situations, then convergence to one or the other genotypes is not assured, and we face a much more difficult situation.

Thus far we have assumed that neither allele confers a selective advantage, thus increasing its chances of appearing in the next generation. To model this situation, we must modify the probability of occurrence of alleles A and B in a particular trial. So instead of the earlier probability of occurrence $p = \alpha/2N$ for allele A, suppose we take the probability of A to be $(\alpha/2N)^\psi$, where $\psi \geq 0$. So A has a differential advantage if $\psi < 1$. But if $\psi > 1$, allele B has the edge. Thus, the parameter ψ is a kind of "fitness" index.

The fitness index gives a selective advantage to one of the alleles, showing that the process of genetic transmission has the character of a "game" in which the rules and the outcome are fixed by nature without any consideration of the phenotypes (individuals) at all! We will return to these matters of genetic change in a later section when we discuss replicator dynamics. But now as motivation for the next section, let's follow up this idea of an evolutionary game within the context of changes of behavior at the phenotypic level.

Example 2: Hawks and Doves

In a famous conflict resolution game, John Maynard Smith considered a population consisting of animals competing for a resource (e.g., food, territory, or mates). Each animal can choose one of two strategies: HAWK, an aggressive and potentially dangerous line of action, or DOVE, an unaggressive, relatively safe strategy. Given these strategies, HAWK will do well against DOVE, but badly against another HAWK because of the risk of injury. The results of any contest between two animals will be measured in some units of Darwinian fitness (e.g., expected number of offspring). Maynard Smith postulated payoffs of the following sort:

$$
\begin{array}{cc}
 & \begin{array}{cc} \text{HAWK} & \text{DOVE} \end{array} \\
\begin{array}{c} \text{HAWK} \\ \text{DOVE} \end{array} & \left(\begin{array}{cc} 0 & 3 \\ 1 & 2 \end{array} \right).
\end{array}
$$

Here, by convention, we assume that the payoffs are to the animal using the strategy on the left if the opponent employs the strategy at the top. The actual numbers appearing in the payoff matrix are not too important, as it's only their relative differences that are of interest for the arguments that follow.

Suppose a large population of animals plays this game repeatedly, pairing off randomly. Further, assume that a given animal always uses either the strategy HAWK or DOVE when engaged in a contest, and that the offspring of that animal inherit this strategic choice. Finally, let the number of offspring produced equal the payoff in the matrix plus, perhaps, some constant value. Under these conditions, we have a model of evolution by natural selection involving asexual reproduction. Our interest is with how strategies evolve in such a population, i.e., we want to know about the steady-state proportions of HAWKS and DOVES in the population.

To understand how the strategic balance in the population shifts, consider a strategy that is "uninvadable" in the sense that if a large population adopts this strategy, then any mutation causing individuals to adopt some other strategy will be eliminated from the population by natural selection. Such an uninvadable strategy is termed an *evolutionary stable strategy* (ESS). Considering the payoff matrix above, it's clear that HAWK is not ESS since DOVE does better against HAWK than HAWK does against itself. A similar argument shows that DOVE is also not an ESS. So there is no "pure" ESS for this game. But now suppose that in a given encounter an animal adopts the strategy HAWK with probability p, and DOVE with probability $q = 1 - p$. Such a strategy is termed "mixed." Later we will show that a mixed strategy with $p = \frac{1}{2}$ is an ESS for this particular game. One way of interpreting this result is to say that if individuals are not allowed to play mixed strategies, then over the course of time the population would evolve to a distribution in which half of its members always play HAWK, while the other half always play DOVE.

There are a number of ways to extend the above game to bring it into closer contact with reality, incorporating features like sexual reproduction, more strategies, and asymmetric contests. Some of these refinements will be taken up later. For now, the important point to note is that the evolutionary trend of the population is defined in terms of the phenotypic *behaviors*, implicitly assuming that these behaviors are observable indicators of underlying genetic variations. Thus, in contrast to the Mendelian inheritance situation described earlier, the Hawk-Dove game makes no explicit reference to genes, but only to phenotypic behavioral patterns. What is common to the two situations is the notion of the competition between behavioral patterns or gene pools being formalized in terms of a game. This is an important, rather recent development in the mathematical treatment of evolutionary change, and forms the basis for some of the most exciting contemporary applications of game theory. So as a prelude to further discussion of evolutionary matters, let's have a brief *intermezzo* to review the basics of the theory of games of strategy. With these concepts in hand, we will then be in a position to return to a more detailed look at evolution as a dynamical process.

Exercises

1. Consider a population in which each genotype comes in n alleles, $A_1, A_2, \ldots, A_n$. Assume that allele A_i is present in proportion p_i in the population, that each allele is equally viable and that there is random mating. (a) Show that the proportion of each allele type remains fixed in the population through all generations. (b) Generalize the genetic drift extension of the text to this n-allele case. In particular, calculate the likelihood that allele A_i disappears from the population in the first generation. (c) How would you introduce the idea of "fitness" into this n-allele situation?

2. In the Hawk-Dove game, one way to interpret an ESS is that it specifies the fraction of the population that should *always* play HAWK and the fraction that should *always* play DOVE, so that the population as whole will be in a stable strategic balance. An alternate interpretation is that the ESS specifies the *fraction* of the time that an individual should play HAWK versus the fraction of the time that it should play DOVE. Consider the differences, biologically speaking, between these two interpretations.

4. Game Theory—The Basics

The distilled essence of the classical theory of games is captured in the following example involving Sherlock Holmes and his archenemy, Professor Moriarty. Holmes wants to take a train from London to Dover, at which point he can take the boat to the Continent and make a safe escape from Moriarty. He boards the train and then sees Moriarty on the platform. Holmes can safely assume that Moriarty has managed to get on the train as well, and is now faced with one of two choices: continue with the train to Dover, or get off at Canterbury which is the only stop. Moriarty has the same two choices. It can be assumed that if they meet in either Canterbury or Dover, Holmes' chances of survival are zero. But if Holmes gets off at Dover and Moriarty gets off at Canterbury, his chances of survival are 100%, since he can then make good his escape to the Continent; on the other hand, if Holmes departs at Canterbury and Moriarty stays on till Dover, then Holmes' survival odds are only 50% since the chase could then continue. Both Holmes and Moriarty are ultra rational men and consider their options accordingly. So where should they get off the train?

To analyze this situation, consider the payoff matrix

		Moriarty	
		Canterbury	Dover
	Canterbury	0	50
Holmes			
	Dover	100	0

Here it's reasonable to assume that Holmes wants to choose his strategy to maximize his return, while Moriarty's interests are diametrically opposed and he wants to select his action to make the payoff to Holmes as small as possible. This is an example of a *zero-sum* game, in which the payoff to one player equals the loss to the other. Thus, in such games the sum of the payoffs to both players is zero. Since Holmes wants to maximize while Moriarty wants to minimize, it's clear that both will be happy if there is an entry in the payoff matrix that is simultaneously the largest element in its row (to satisfy Holmes), and the smallest element in its column (to make Moriarty happy). Such an entry in a payoff matrix corresponds to what's termed a *saddle-point,* and represents an obvious solution to the game. It's easy to see that the Holmes-Moriarty dilemma has no saddle-point, implying that there is no clear-cut choice for either of them. That is, there is no choice that a rational player will *always* make in the absence of information about his opponent's action, other than that his opponent is rational too. Thus, the game has no solution in what are termed pure strategies for either Holmes or Moriarty. To decide what to do, they are going to have to *mix* their actions, the relative weight attached to getting off in Dover or Canterbury being determined by the relative payoffs for the two alternatives.

Suppose Holmes encounters Moriarty many times in this kind of situation, and that in each encounter there is a probability p that Holmes gets off in Canterbury and a probability $1 - p$ that he disembarks in Dover. Then the *expected* payoff to Holmes for using such a strategy is

$$\mathcal{E}(\text{Holmes}) = p \cdot 0 + (1 - p) \cdot 100 = (1 - p) \cdot 100$$

if Moriarty gets off in Canterbury, and it is

$$\mathcal{E}(\text{Holmes}) = p \cdot 50 + (1 - p) \cdot 0 = p \cdot 50$$

if Moriarty leaves the train in Dover. Of course, Moriarty has his own set of probabilities for leaving the train. So let's assume that he decides to get off in Canterbury with likelihood r, and in Dover with probability $1 - r$. Thus the overall expected payoff to Holmes is

$$\mathcal{E}^*(\text{Holmes}) = 100r(1 - p) + 50p(1 - r) = 100r + 50p - 150pr.$$

Holmes wants to select p to maximize this quantity. Arguing in a similar fashion, we compute Moriarty's overall expected return as

$$\mathcal{E}^*(\text{Moriarty}) = 100r(1 - p) + 50p(1 - r) = 100r + 50p - 150pr,$$

which is exactly the same as the expected return to Holmes since this is a zero-sum game. Moriarty wants to minimize this quantity while, as we saw above, Holmes wants to maximize it. Equating the above expressions for the returns to Holmes and Moriarty and doing the resulting algebra, we see that the antagonists will achieve their objectives if $p = \frac{2}{3}$ and $r = \frac{1}{3}$. So two-thirds of the time Holmes should alight in Canterbury, while he should leave the train in Dover only one-third of the time. Similarly, Moriarty should get off in Canterbury one-third of the time, and leave in Dover two-thirds of the time. With these strategies, the expected payoff to Holmes is

$$\frac{1}{3}(0) + \frac{2}{3}(50) = 33\frac{1}{3}.$$

Thus, Holmes can guarantee himself a one-third chance of survival (on the average) no matter what Moriarty chooses to do, while Moriarty can ensure himself at least a one-third chance of doing Holmes in, regardless of where Holmes decides to leave the train. It's a simple matter to verify that weighing their actions according to these probabilities gives both Holmes and Moriarty at least as great a return (on the average) as by playing a pure strategy involving either always getting off at Canterbury or at Dover. These values of p and r represent *mixed strategies* for Holmes and Moriarty. It's amusing to note that the most likely outcome of a single play of this game will have Holmes getting off in Canterbury with Moriarty continuing on to Dover— exactly what happened in the actual story by Conan Doyle. Not only is this the most likely outcome, but it's a choice that has the inestimable virtue of enabling the chase to continue!

The above game shows that the maximal expected return to Holmes equals the minimal expected return to Moriarty, under the assumption that both players are rational and play optimally. The existence of mixed strategies ensuring such an outcome for any zero-sum, two-person game, regardless of the number of pure strategies available to each player, forms the essence of the celebrated Minimax Theorem, which was first proved by John von Neumann in 1928.

Let the array

$$A = (a_{ij}), \qquad i = 1, 2, \ldots n; \, j = 1, 2, \ldots, m,$$

be the payoff matrix for a zero-sum, two-person game, where Player I chooses the action i with probability p_i. Let $p = (p_1, \ldots, p_n)$ be the probability vector representing the mixed strategy of Player I, while $q = (q_1, q_2, \ldots, q_m)$ denotes the corresponding probability vector for Player II. Then we can state von Neumann's result more formally as the

MINIMAX THEOREM. *For a two-person, zero-sum game with payoff matrix A, there exist probability vectors p^* and q^* such that*

$$V(p^*, q^*) \doteq \max_p \min_q V(p, q) = \min_q \max_p V(p, q),$$

where $V(p, q) \doteq (p, Aq)$(i.e., V is the payoff to Player I using strategy p, which equals the negative of the payoff to Player II using strategy q). The number $V(p^*, q^*)$ is termed the value of the game.

Unfortunately, the Minimax Theorem provides no information on how to actually *find* the optimal mixed strategies. But it turns out that the computation of the optimal strategy vectors p^* and q^* can always be reduced to a standard linear programming problem (see Problem 1), whose solution can be obtained by numerical methods based upon the Simplex Method or one of its many variants. There are a number of ways in which we can extend the foregoing, admittedly artificial, situation to encompass more realistic conflicts. One of the most important is to allow for more than two players, an extension leading to the thorny difficulties surrounding "n-person game theory." For now, let's steer away from these deep waters, introducing instead an equally important generalization in which the interests of the two players are not diametrically opposed. This is the territory of the *mixed-motive* or, as it is sometimes termed, *nonzero-sum* game.

In mixed-motive games, the interests of the two players are neither in direct opposition nor directly parallel, leading to a nonzero-sum outcome for any play of the game. For this reason, each entry in the payoff matrix consists of two numbers, the first representing the payoff to Player I, the second to Player II under the corresponding actions. Suppose, for simplicity, that each player has two actions at his disposal which, for reasons that will become apparent later, we'll label "C" and "D." Further, let's assume that the game is *symmetric* so that the outcome is the same if we reverse the roles of the two players. Under these hypotheses, the general form of the payoff matrix will be

Player II

		C	D
	C	(R, R)	(S, T)
Player I			
	D	(T, S)	(P, P)

where P, R, S and T are real numbers. Since the two-person, mixed-motive games play an extremely important role in the modeling of evolutionary processes, we examine their structure in more detail in the next section.

Exercises

1. How would the payoffs have to be rearranged in the Holmes-Moriarty game in order to make Holmes's optimal strategy be to get off in Canterbury and Dover with equal likelihood?

2. In February 1943 the Allies received an intelligence report stating that the Japanese were planning a troop and supply convoy to resupply their army in New Guinea. The convoy could sail either north of the island of New Britain, where the weather could be bad, or they could sail south, where they would probably have fair weather. By either route, the trip would take three days. The Allied commander, General Kenney, was ordered by General MacArthur, the supreme commander, to attack the convoy and inflict maximum damage. General Kenney had to decide whether to concentrate the bulk of his reconnaissance aircraft on the northern or the southern route.

Kenney's advisors felt that if the aircraft were concentrated on the northern route, the convoy would probably be sighted after one day, regardless of whether it sailed north or south, and would therefore be subjected to two days of bombing. On the other hand, if the aircraft went south, then either one or three days of bombing would be possible, depending upon whether the Japanese sailed north or south. (a) Formulate this Battle of Bismarck Sea in terms of a two-person, zero-sum game, with the options for each player being either the northern or the southern route. Assume the payoffs are the number of days of bombing resulting from each combination of choices, where Kenney wants to maximize the days, while the Japanese commander wants to minimize. (b) Compute the solution of this game. (Hint: In this game there is a pure strategy for both Kenney and the Japanese commander.)

3. Suppose we have a game with two players A and B, whose choices together give a point on the unit square $0 \leq x, y \leq 1$. Assume that A's action is to select a number x in the unit interval $0 \leq x \leq 1$, while B chooses a number y between 0 and 1. The players make their choices independently and simultaneously, with the payoff to A being

$$P(x, y) = -x^2 - y^2 - 4xy + 4x + 2y - 2.$$

Assume that A wishes to maximize P, while B wants to make this quantity as small as possible. (a) Show that the solution to this game lies at the point $(x^*, y^*) = (0.8, 0.6)$. (b) What is the value of this game? (Remark: This Exercise shows that the set of allowable choices for a player need not necessarily be finite—or even countable.)

5. A Typology of Mixed-Motive Games

There are 24 different ways to order the four numbers P, R, S and T, each of which gives rise to a different type of mixed-motive game. But in view of the symmetry hypothesis, only 12 of these games are qualitatively different. And of these, there are eight that possess optimal pure strategies for both players. Such games are conceptually uninteresting, so we focus attention on the remaining four two-person, mixed-motive games without pure strategies. Each of these games has been extensively studied in the literature, and can be represented by a prototypical situation capturing the concepts peculiar to that particular type of game. These qualitatively different games can be classified by the relative magnitudes of the numbers P, R, S and T:

- *Leader*—$(T > S > R > P)$.
- *The Battle of the Sexes*—$(S > T > R > P)$.
- *Chicken*—$(T > R > S > P)$.
- *The Prisoner's Dilemma*—$(T > R > P > S)$.

To see the differences in character and strategy for each of these games, let's look at the prototypical example from each category.

• *Leader*—Consider the case of two drivers attempting to enter a busy stream of traffic from opposite sides of an intersection. When the cross traffic clears, each driver must decide whether to concede the right-of-way to the other (C), or drive into the gap (D). If both concede, they will each be delayed, whereas if both drive out together there may be a collision. However, if one drives out while the other waits, the "leader" will be able to carry on with his trip, while the "follower" may still be able to squeeze into the gap left behind by the leader before it closes again. A typical payoff matrix for this Leader game

Driver I

		C	D
	C	(2, 2)	(3, 4)
Driver II			
	D	(4, 3)	(1, 1)

Clearly, there is no dominant strategy in the game of Leader. According to the minimax principle, to avoid the worst possible outcome each driver should choose strategy C, thereby ensuring that neither will receive a payoff less than 2 units. However, the minimax strategies are not in equilibrium as each driver would have reason to regret his choice when he sees what the other has done. This simple observation shows that the minimax principle cannot be used as a basis for prescribing rational choices in mixed-motive games.

In fact, there are *two* equilibrium strategies in Leader: the strategies (C, D) and (D, C), which appear at the off-diagonal corners of the payoff matrix. If Driver I chooses D, the second driver can do no better than to choose C, and vice versa. In other words, neither can do better by deviating from such an equilibrium outcome. However, in contrast to zero-sum games in which such equilibrium points are always equivalent, in the Leader game Driver I prefers the (D, C) equilibrium, while Driver II prefers (C, D). There is no mathematical way of settling this difference. But in real-world situations of this type, the impasse is often resolved by the fact that one of the equilibrium points is more *visible* to the players than the other. So, for example, in the simple traffic example above, cultural and/or psychological factors may enter to break the deadlock, with various tacit rules like "first come, first served," or signaling schemes like blinking of lights used to single out one of the equilibria. This kind of signaling, incidentally, is in sharp contrast to the situation in zero-sum games, where such signals would definitely not be to a driver's advantage.

• *Battle of the Sexes*—In this game a married couple has to choose between two options for their evening entertainment. The husband prefers one type of entertainment (e.g., a movie), whereas his wife prefers another, say, going out for a pizza. The problem is that they would both rather go out together than alone. If they each opt for their first choice (call it action D), they end up going out alone and each receives a payoff of 2 units. If they each make a sacrifice and go to the activity they don't like (action C, say), each suffers and they receive a payoff of only 1 unit apiece. But if one sacrifices while the other gets their first choice, then they still go out together, but the "hero" who sacrifices receives 3 units of "reward" while the other party gets 4. The payoff matrix for this game is shown below.

Wife

		C	D
Husband	C	(1, 1)	(3, 4)
	D	(4, 3)	(2, 2)

There are a number of features in common between the Battle of the Sexes and the Leader games: (1) neither the Husband nor the Wife has a dominant strategy, (2) the minimax strategies intersect in the nonequilibrium result (D, D), and (3) both strategies (C, D) and (D, C) are in equilibrium. But in contrast to the Leader game, in Battle of the Sexes the player who deviates unilaterally from the minimax strategy rewards the other player more than himself or herself. This is just the opposite of what happens in Leader, where the deviator receives a greater reward than his opponent. But just as in Leader, here also a player can gain by communicating with the other player in order to obtain some level of commitment to strategy C. So, for instance, the husband might announce that he is irrevocably committed to his first choice of going to the movies, in which case this will work to his advantage if his wife then acts so as to maximize her own return. The only difficulty is in convincing her that he's serious. The main point to note here is that some kind of commitment is needed in order for both parties to achieve the best possible joint outcome in the Battle of the Sexes game.

• *Chicken*—A well-known game, whose underlying principles date back at least as far as the Homeric era, involves two motorists driving toward each other on a collision course. Each has the option of being a "chicken" by swerving to avoid the collision (C), or of continuing on the deadly course (D). If both drivers are chicken, they each survive and receive a payoff of 3 units. But if one "chickens out" while the other drives straight on, the chicken loses face (but not his life) and the "macho man" wins a prestige victory. In this case the chicken receives 2 units, whereas the opponent receives 4. Finally, if they both carry on to a fatal collision, they each receive Death's reward of 1 unit. The payoff matrix for the Chicken game is given by

		Driver I	
		C	D
	C	(3, 3)	(2, 4)
Driver II			
	D	(4, 2)	(1, 1)

Again, in Chicken there is no dominant strategy. The minimax strategies intersect in the outcome involving both drivers "chickening out." Also as before, an exploiter who deviates from a minimax strategy can gain an advantage for himself. But this time he *invariably* affects the other player adversely by such a deviation. So not only does the deviator harm the other player, he also puts himself and the other player in a position where they may have a disastrous outcome.

Chicken also has the peculiar feature that it is impossible to avoid playing the game with someone who is insistent, since to refuse to play is effectively the same as playing and losing. In addition, the player who succeeds in making his commitment to the dangerous D option appear convincing will *always* win at the expense of the other player, assuming the other player is rational. Thus, a player who has a deserved reputation for hot-headed recklessness enjoys a decided advantage in Chicken over one who is merely rational. Perhaps this accounts for the aversion of the typical academic to this kind of irrational winning strategy, as most academics seem to pride themselves on being both risk averse and ultra rational, an unhappy losing combination for professors engaging in the game of Chicken. This becomes especially true if Chicken is played a number of times, because the player who gains an early advantage usually maintains (or even increases) that advantage later on. For once he has successfully exploited the other player, he gains confidence in his ability to get away with the risky strategy in the future while making his opponent all the more fearful of deviating from the cautious minimax alternative.

• *The Prisoner's Dilemma*—The last basic type of nonzero-sum game, and by far the most interesting, is the famous game of two prisoners who are accused of a crime. Each of them has the option of concealing information from the police (C) or disclosing it (D). If they each conceal the information (i.e., they cooperate), they will both be acquitted with a payoff of 3 units to each. If one conceals and the other "squeals" to the police, the squealer receives a reward of 4 units, while the payoff to the "martyr" is only 1 unit, reflecting his role in the obstruction of justice. Finally, if they both talk, they will each be convicted thereby receiving a payoff of only 2 units apiece.

The appropriate payoff matrix for the Prisoner's Dilemma game is

Prisoner I

		C	D
	C	(3, 3)	(1, 4)
Prisoner II			
	D	(4, 1)	(2, 2)

Incidentally, our use of the symbols C and D to represent the possible actions by the players in all of these games is motivated by the usual interpretation of the actions in the Prisoner's Dilemma game. Here C represents "cooperating" with your pal and not confessing, whereas D signifies "defecting" to the police and giving the information needed for a conviction.

The Prisoner's Dilemma is a real paradox. The minimax strategies intersect in the choice of mutual defection, which is also the only equilibrium point in the game. So neither prisoner has any reason to regret a minimax choice if the other also plays minimax. The minimax strategies are also dominant for both prisoners, since each receives a larger payoff by defecting than by cooperating when playing against either of the other player's strategies. Thus, it appears to be in the best interests of each prisoner to defect to the police—*regardless* of what the other player decides to do. But if both prisoners choose this individually rational action of defecting, the 2 units they each receive is less than the 3 units could have gotten if they had chosen to remain silent.

The essence of the paradox in the Prisoner's Dilemma lies in the conflict between individual and collective rationality. According to individual rationality, it's clearly better for a prisoner to defect and give information to the police. Yet, paradoxically, if both try to be "martyrs" and remain silent, they each wind up being better off. What's needed to ensure this better outcome for both players is some kind of selection principle based upon their collective interests. Perhaps the oldest and most well-known such principle is the Golden Rule of Confucius: "Do unto others as you would have them do unto you." Note, however, that the Golden Rule can be disastrous in other kinds of games. For example, if both the husband and wife adopt the Golden Rule in the Battle of the Sexes, the outcome is the worst possible and leads to each of them going out alone to an entertainment they don't like.

The foregoing archetypal two-person, mixed-motive games have many features in common, especially Leader, Battle of the Sexes and Chicken. Here are a few of the most important common properties of all the games except the Prisoner's Dilemma:

1) There exists a "natural" outcome if each party plays minimax, i.e., each player choosing option C.

2) The outcome (C, C) is a nonequilibrium point that's vulnerable in the sense that both players are tempted to deviate from it.

3) Each of the games possesses two asymmetric equilibrium points. But neither of these points is stable, since the players are not in agreement as to which of the two equilibria is preferable.

4) None of the games possesses a dominant strategy for either player.

5) The worst possible outcome for both players results if both choose their non-minimax strategies, i.e., option D.

By way of contrast, the Prisoner's Dilemma situation possesses none of these features.

From the standpoint modeling evolutionary processes, the Prisoner's Dilemma is far and away the most interesting of these games and the one that we shall focus considerable attention upon later. But for now, let's return to the issue of dynamics and spend some time looking at how we can superimpose a notion of dynamical change on the essentially static character of a two-person game.

Exercises

1. Compute the equilibrium point of the following game:

Player I

		A	B
	A	(4, 4)	(2, 3)
Player II			
	B	(3, 2)	(1, 1)

2. Consider the following payoff matrix representing a vastly over-simplified version of the nuclear arms race:

USSR

		Limit production	Increase production
	Limit production	(3, 3) *(Status quo)*	(1, 4) *(USSR advantage)*
USA			
	Increase production	(4, 1) *(US advantage)*	(2, 2) *(possible nuclear war)*

(a) Is this a mixed-motive game? (b) If so, what type is it? (c) Does the game have an equilibrium solution?

3. The theory of games rests on the assumption that the players act *rationally*. Consider various possibilities for what this might mean within the context of the four basic types of mixed-motive games considered in this section.

6. Replicators, Hypercycles and Selection

We have seen that evolutionary models come in two basic flavors: genetic and phenotypic. Despite the radically different physical interpretation that we attach to these two classes, it turns out to be possible to give a common mathematical framework suitable for investigating many of the important properties of both types of models. This formalism goes under the rubric *replicator dynamics,* and involves postulating an abstract unit of selection termed a *replicator.* Roughly speaking, the properties characterizing a replicator system consist of the replicators being able to give rise to an unlimited number of copies of themselves (at least in principle), and that the replicators occur in many variants whose properties influence the number of copies of each variant that will be produced. Thus, genes can be replicators. But so can chemical molecules and behavioral patterns. We shall consider these examples later. For now, let's just assume that we have at hand an abstract replicator system composed of entities that occur in one of n different forms or variants, which we label $i = 1, 2, \ldots, n$. Assume, further, that the fraction of each variant present is given by the vector $x = (x_1, x_2, \ldots, x_n)$.

To develop the general form of the replicator dynamics, assume that there exists a function $f(x) \doteq (f_1, f_2, \ldots, f_n)$ expressing the fitness of the different variants of the replicator. Let's normalize things so that the flow describing the change of the variant frequencies takes place on the unit simplex S^n in R^n. This entails scaling x so that

$$0 \leq x_i \leq 1, \qquad \sum_{i=1}^{n} x_i = 1.$$

If we assume that the rate of increase of variant i is directly proportional to the current level of x_i, as well as to the fitness of variant i as compared with the fitness of the other variants, we are led to the dynamics for x_i as

$$\dot{x}_i = x_i \left[f_i(x) - \Phi(x) \right], \qquad i = 1, 2, \ldots, n. \tag{$\dagger$}$$

Here the term $\Phi(x)$ represents the average fitness of the population and is given by the expression

$$\Phi(x) = \sum_{j=1}^{n} x_j f_j(x).$$

The quantity Φ is introduced to ensure that the system trajectory remains within the unit simplex S^n. It's clear that the nature of the change in the replicator frequencies is determined solely by the interaction function $f(x)$.

To make contact with our earlier examples, assume that the fitness function $f(x)$ involves only first-order terms, i.e., there are only linear interactions. Then we have

$$f_i(x) = (Ax)_i = \sum_{j=1}^{n} a_{ij}x_j, \qquad i = 1, 2, \ldots, n,$$

where $A = [a_{ij}]$ is a matrix whose terms specify the nature and degree of interaction between variants i and j. Under this linearity assumption, the replicator dynamics (†) can be used to describe equally well the effect of selection on allele frequencies in a gene pool or the distribution of behavioral phenotypes in a species population. The difference is not in the mathematics, but in the interpretation we attach to the replicator variants. In the first case, x_i represents the frequency of allele A_i in the gene pool, whereas in the second case we regard x_i as the frequency of behavioral phenotype B_i in the species population. In both cases, the elements $\{a_{ij}\}$ represent the manner in which the variants interact and are "selected for" in the next generation. To see how this set-up works in a familiar setting, let's revisit the animal conflict problem considered earlier.

Let $x_1, x_2, \ldots, x_n$ be the respective frequencies of the behavioral phenotypes $B_1, B_2, \ldots, B_n$ within a population. Further, let a_{ij} be the expected payoff to an animal using strategy B_i in competition with one employing strategy B_j, $i, j = 1, 2, \ldots, n$. Assuming random encounters as before, it's easy to see that the quantity $(Ax)_i$ represents the average payoff to a contestant playing strategy B_i when the population is in the state x, with the average payoff for the entire population then being

$$\Phi(x) = (x, Ax) = \sum_{i=1}^{n} x_i(Ax)_i.$$

Again assuming asexual reproduction, the relative rate of increase of phenotype B_i is

$$\frac{\dot{x}_i}{x_i} = (Ax)_i - (x, Ax) = (Ax)_i - \Phi(x),$$

which are the original dynamics (†) for the special case of linear interactions. A similar argument can be used to derive *exactly* the same equations for the case of genetic selection. Rather than going through this boring and repetitious calculation, let's look at another quite different physical setting in which the same replicator dynamics describe chemical processes of interest in the study of the origin of life. This is the case of *prebiotic molecular evolution*.

Let $C_1, C_2, \ldots, C_n$ be self-replicating polynucleotides (primitive RNA or DNA strands) swimming about in a primordial "soup." The concentration of nucleotide C_i is given by x_i, $i = 1, 2, \ldots, n$, and we assume there is a dilution flow $\Phi(x)$ that keeps the overall nucleotide concentration constant, normalized to be equal to 1. Under these circumstances, independent replication of the nucleotide strands leads to a situation in which all but one of the molecular species vanishes, with loss of the corresponding information encoded in the other species. To prevent this information loss, Manfred Eigen and Peter Schuster have considered *networks* of catalytically interacting polynucleotides. Of special interest are those closed feedback loops in which each molecular species is catalyzed by its predecessor. These loops are termed *hypercycles*. They evolve according to the dynamics

$$\dot{x}_i = x_i[x_{i-1}H_i(x) - \Phi(x)], \qquad i = 1, 2, \ldots, n,$$

where the indices are taken modulo n and the functions $H_i(x)$ are strictly positive on S^n. The simplest case of a hypercycle is when each $H_i(x)$ is a constant $k_i > 0$. This case results from the general linear replicator dynamics if we take A to be the permutation matrix

$$A = \begin{pmatrix} 0 & 0 & \ldots & 0 & k_1 \\ k_2 & 0 & \ldots & 0 & 0 \\ 0 & k_3 & \ldots & 0 & \\ \vdots & \vdots & & \vdots & \vdots \\ 0 & 0 & \ldots & k_n & 0 \end{pmatrix}$$

The above discussion illustrates the ubiquitous nature of the replicator equations (†) for the description of a wide variety of processes involving the selection of replicating variants in a population of genes, animal behaviors or molecules. With these physical situations in mind, we turn now to a more detailed look at the mathematical properties of such systems.

Exercises

1. In the text we have formulated the replicator equations in continuous time. What is the discrete-time analogue?

2. Express the genetic inheritance problem and the Hawk-Dove game of Section 3 as replicator systems.

3. Consider a first-order replicator system with an anti-symmetric interaction matrix, i.e., $A = -A'$. Show that in this case the replicator system reduces to

$$\dot{x}_i = x_i(Ax)_i, \qquad i = 1, 2, \ldots, n.$$

7. Equilibrium Properties of Replicator Systems

There are two questions of fundamental concern surrounding the dynamical behavior of replicator systems: (1) Given an initial distribution of replicator variants, what is the equilibrium distribution as $t \to \infty$? (2) In the steady state, do any species (behavior patterns, molecular types) die out? Clearly, the second question is a special case of the first. So we begin by investigating the equilibrium behavior of the replicator system

$$\dot{x}_i = x_i[f_i(x) - \Phi(x)], \qquad i = 1, 2, \ldots, n.$$

First of all, note that the manifold on which the replicator dynamics unfold is the n-simplex

$$S^n = \left\{ x \in R^n : 0 \le x_i \le 1, \sum x_i = 1 \right\}.$$

Since $\sum \dot{x}_i = 0$ on S^n, each face of the simplex forms an invariant set for the replicator dynamics (Recall: the faces consist of subsimplices characterized by $x_i = 0$ for indices i contained in a nonempty proper subset of $\{1, 2, \ldots, n\}$). In particular, this means that the corners e_i are equilibria. Let's ignore these "trivial" equilibria for the moment, focusing our attention upon the nontrivial equilibria located in the interior of S^n.

The interior equilibria are given by the positive solutions of the equations

$$f_1(x) = f_2(x) = \cdots = f_n(x), \qquad x_1 + x_2 + \cdots + x_n = 1.$$

Any solution x^* of the above system must be such that $f_i(x^*) = \Phi$. In the special case of a first-order replicator system, the interior equilibria satisfy the equations

$$\sum a_{1j} x_j^* = \sum a_{2j} x_j^* = \cdots = \sum a_{nj} x_j^*.$$

Such solutions form an affine space, and it has been shown that generically for first-order replicator systems there is at most a single interior solution. That is, there is an open, dense subset of matrices $A \in R^{n \times n}$ such that the corresponding first-order replicator system has at most one equilibrium in the interior of S^n and in the interior of each face. A specific test for the absence of such an interior equilibrium is provided by considering $\tilde{A}$, the adjoint matrix of A. If $\tilde{A}u$ has both positive and negative row sums, then there is no interior equilibrium and no interior periodic orbits for the first-order replicator system with matrix A.

We should note also that the genericity of the interior equilibrium, when it exists, also rules out the appearance of elementary catastrophes. If such bifurcations could occur in the dynamic, there would then be a perturbation of the system leading to more than a single interior equilibrium. So for first-order replicator systems there is at most a single candidate for an equilibrium distribution that preserves *all* the species or behaviors in the system. This fact leads us to speculate that nature's way is to let some perish so that others may survive. We shall explore this point in more detail in a moment.

Suppose that the general replicator equations have an equilibrium point $p \in S^n$ (not necessarily an interior point). Then an easy calculation shows that the Jacobian matrix of the system has $\Phi(p)$ as one of its characteristic values, with a corresponding characteristic vector that does not belong to the tangent space to S^n at p. Thus, this characteristic value is irrelevant as far as determination of the stability properties of the equilibrium at p and thus can be ignored. So, for instance, the relevant characteristic values for the first-order system at a corner e_i are the $n-1$ values $a_{ij} - a_{ii}, j \neq i$. Now let's look at a couple of examples.

Example 1: The Hypercycle

For the hypercycle system discussed in the last section, the matrix A is a permutation matrix with entries $\{k_i\}$. It's straightforward to show that in this case there always exists an equilibrium point $p \in$ int S^n. The coordinates of this point are given by

$$p_i = \frac{k_{i+1}^{-1}}{\sum k_j^{-1}} \ , \qquad i = 1, 2, \ldots, n \quad \mathrm{mod}\ n.$$

Here the characteristic values are (up to a positive multiplier) the n roots of unity, with the irrelevant root being 1 itself. Thus, the point p is locally asymptotically stable for $n \leq 3$, but is always unstable if $n \geq 5$. The case $n = 4$ is more delicate, since we then have two characteristic values lying on the imaginary axis. But by use of the Lyapunov function $\prod x_i$, it can be shown that in this case p is actually globally asymptotically stable. It's also interesting to note that numerical results point to the existence of a stable periodic attractor when $n \geq 5$, although there is as yet no rigorous proof for the existence of this attractor.

Example 2: Molecular Replication and Mutation

The replicator equations can be used to model the simplest kind of molecular mechanism of catalyzed replication *and* mutation. The basic chemical relations are of the form

$$R + I_j + I_i \ \xrightarrow{Q_{kj} \cdot A_{ji}} \ I_j + I_k + I_i, \qquad i, j, k = 1, 2, \ldots, n.$$

Here I_j is the template that is replicated and I_i is the catalyst, while R represents the various substrates needed for replication. The mutation frequencies depend on the template I_j and the target species I_k, but not on the catalyst. The matrices R and Q represent the processes of replication and mutation, respectively. Thus, the rate constants of the individual reactions are the products of a replication and a mutation factor. In particular, the product $Q_{kj} \cdot R_{ji}$ represents the following sequence of processes: I_i catalyzes the replication of I_j, which then yields I_k as a possibly erroneous copy of I_j. Thus, error-free replication (the case $I_j = I_k$) occurs with frequency Q_{jj}, while a mutation from I_j to I_k happens with frequency Q_{kj}. The concentration of the individual molecular species k is denoted by c_k. So using the normalized (i.e., relative) concentrations $x_k = c_k / \sum_{j=1}^{n} c_j$, we are led to the replicator equations

$$\dot{x}_k = \sum_{i,j=1}^{n} Q_{ki} R_{ij} x_i x_j - x_k \Phi, \qquad k = 1, 2, \ldots, n,$$

where, as before, $\Phi = \sum A_{ij} x_i x_j$.

In this formulation of the mutation process, we can measure the average mutation rate ϵ by the equation

$$\epsilon = \frac{1}{n(n-1)} \sum_{\substack{i,j=1 \\ i \neq j}}^{n} Q_{ij}.$$

As special cases of this setup, we obtain the well-known *house of cards* model of mutation if we set

$$Q_{ij} = \begin{cases} 1 - (n-1)\epsilon, & i = j, \\ \epsilon, & i \neq j. \end{cases}$$

Another important special case arises when we come to consider *error-free* replication. Then we set $Q_{jj} = 1$, $Q_{ij} = 0, j \neq i$. It has been shown that at least in this case there is the possibility for the replicator system to display chaotic behavior. The literature cited in the Notes and References section spells out the details.

By the above discussion, if the replicator system has an interior equilibrium we can employ standard techniques of stability analysis to determine the local—and sometimes the global—stability characteristics of the equilibrium point. If there is no interior equilibrium, then it can be shown there is a constant vector $c \in R^n$, with $\sum_i c_i = 0$, such that the function $V(x) = \prod x_i^{c_i}$ increases along the orbits of the replicator system. Hence, by Lyapunov's Theorem, it follows that each orbit of the system has its ω-limit contained in the boundary of S^n. Thus, if there is no interior equilibrium point, there are no periodic, recurrent or even nonwandering points in the interior of S^n. However, this does **not** mean that $\lim_{t\to\infty} x_i(t) = 0$ for some i. There exist examples of systems whose ω-limit of every interior orbit is a cycle consisting of the corners and edges of the simplex S^n. Such a limiting trajectory implies only a temporary die-off of a given species or behavior, with the species re-emerging Phoenix-like out of the ashes infinitely often. Since the persistence of species is of great importance in many biological settings, let's consider what kind of conditions on the replicator dynamics would ensure the survivability of all species initially present in the system.

Exercises

1. Consider a first-order replicator system with anti-symmetric interaction matrix A, and let Δ^r be an r-dimensional face of S^n. Prove that the following property is generic for such systems: If r is even, then the system has at most one equilibrium in the interior of Δ^r, while if r is odd there is no equilibrium in Δ^r. (Hint: Consider the fact that every such system is completely characterized by the interaction matrix A. Then use the generic properties of subsets of such matrices.)

2. Consider the first-order replicator system with interaction matrix

$$\begin{pmatrix} 0 & 6 & -4 \\ -3 & 0 & 5 \\ -1 & 3 & 0 \end{pmatrix}.$$

Show that this system has an equilibrium at the point $p = (\frac{1}{3}, \frac{1}{3}, \frac{1}{3})$. Is this point asymptotically stable?

3. For a first-order replicator system, show that the barycentric transformation $x_i \to y_i = c_i x_i / (\sum_j c_j x_j)$, $c_i > 0$ maps an interior equilibrium $p \in S^n$ to the central point $m = \frac{1}{n}(1, 1, \ldots, 1)$. (Note: This transformation is often useful for computational purposes.)

8. Perseverance of Species in Replicator Systems

As emphasized above, the issue of persistence centers about the question of when a species i with initial fraction $x_i(0) > 0$ *perseveres* for all $t > 0$. More formally, we say that the system is *perseverant* if there exists a $\delta > 0$ such that $\lim_{t \to \infty} \inf x_i(t) \geq \delta$ for all i whenever $x_i(0) > 0$. In short, a system is perseverant when the population levels of all species remains uniformly bounded away from zero. Systems having this property are clearly of great practical importance since small fluctuations in the population of any species cannot result in a species being wiped out. Moreover, if the system starts on the boundary with one or more components absent, mutations that generate these components will spread, resulting in a system that is safely cushioned from extinction.

Remarks

1) Perseverance is *not* a structurally stable property.

2) Nonperseverance doesn't necessarily imply that some component of the system is driven to extinction. For example, there exist systems with attractors on the boundary, as well as in the interior of S^n. It can also happen that each orbit of the system remains bounded away from ∂S^n, but the bound is not uniform and depends upon the particular orbit. For perseverance, the same bound must work for all orbits.

Since there appear to be no useful conditions for perseverance that are both necessary and sufficient, we must make do with the following separate conditions:

• *Sufficiency*—A replicator system is perseverant if there exists a function $P \colon S^n \to R$ such that: (1) $P(x) > 0$ for $x \in \mathrm{int}\, S^n$ and $P(x) = 0$ for $x \in \partial S^n$, and (2) $\dot{P} = P\Psi$, where Ψ is a continuous function such that there exists a $T > 0$ making the integral

$$\frac{1}{T} \int_0^T \Psi[x(t)]\, dt > 0$$

for all $x(t) \in \partial S^n$. In other words, the function P is increasing on the average near the boundary of S^n. So P acts as a kind of average Lyapunov function.

An example of the use of this result arises for the hypercycle system in which the function $P(x) = x_1 x_2 \cdots x_n$ satisfies the condition for an average Lyapunov function. Thus, we conclude that the hypercycle equations are perseverant, even though we have already seen that the interior equilibrium is, in general, unstable for $n \geq 5$. Now let's look at some simple necessary conditions.

• *Necessity*—The Brouwer Fixed Point Theorem implies that a necessary condition for perseverance is the existence of an equilibrium in the interior of S^n. In the case of first-order replicator systems, this equilibrium is necessarily unique, and for perseverance it can be shown that it's necessary that the trace of the system Jacobian matrix be strictly negative when evaluated at this interior equilibrium point.

For the special (but practically important) case of first-order systems having matrices of the form

$$A = \begin{pmatrix} 0 & - & - & \cdots & + \\ + & 0 & - & \cdots & - \\ - & + & 0 & \cdots & - \\ \vdots & \vdots & \vdots & & \vdots \\ - & - & \cdots & + & 0 \end{pmatrix},$$

where "+" indicates a positive entry and "−" denotes an entry that is either negative or zero, the following result of Josef Hofbauer and Robert Amann characterizes the perseverance of the corresponding replicator system.

PERSEVERANCE THEOREM. *For matrices A of the special form above, the following conditions are equivalent for first-order replicator systems:*

    i) The system is perseverant.

    ii) There is a unique interior equilibrium p such that $\Phi(p) > 0$.

    iii) There is a vector $z \in R^n$, with $z_i > 0$ for all i, such that all components of zA are positive.

    iv) The matrix C, whose components are $c_{ij} = a_{i+1,j}$ (with the indices taken mod n), has only positive principal minors.

Note that the matrix C described in part (iv) of the theorem is obtained from A by moving the first row of A to the bottom. Such matrices C having strictly positive diagonal elements, with all other elements nonpositive, play an important role in mathematical economics, where $-C$ is termed a *Metzler matrix*.

As an easy application of the above theorem, we note that the hyper-cycle equation with interaction matrix

$$A = \begin{pmatrix} 0 & 0 & \cdots & 0 & k_1 \\ k_2 & 0 & \cdots & 0 & 0 \\ 0 & k_3 & \cdots & 0 & 0 \\ \vdots & \vdots & & \vdots & \vdots \\ 0 & 0 & \cdots & k_n & 0 \end{pmatrix}$$

leads to a matrix C that is diagonal with positive entries; hence, the system is perseverant by part (iv) of the Perseverance Theorem.

There exist a variety of additional conditions for the perseverance of replicator systems, some of which are explored in the Problems section. But for now we take a different tack and look at another important question surrounding replicator systems, the matter of classification. In Chapter 2, we saw that it was possible to classify all smooth functions of codimension no greater than five, but that a similar classification for smooth vector fields presented difficulties. Here we look at the same question, but in the restricted case of vector fields specified by first-order replicator systems.

Exercises

    1. Consider a first-order replicator system whose interaction matrix A has a zero diagonal, i.e., $a_{ii} = 0$, $i = 1, 2, \ldots, n$. Assume that the system

is perseverant, with an interior equilibrium point p. Prove the following inequalities: (a) $(p, Ap) > 0$, and (b) $(-1)^{n-1} \det A > 0$.

2. With any first-order replicator system having interaction matrix A containing only nonnegative entries, we can associate a directed graph G in the following way: Place an arc from node j to node i in G if $a_{ij} > 0$. The graph G is called *irreducible* if for any two nodes i and j, there is an oriented path leading from i to j. (Note: This does **not** mean that there need be an arc directly connecting i and j.) Prove that if the replicator system is perseverant, then its associated graph G must be irreducible.

3. Consider the replicator system matrix

$$A = \begin{pmatrix} 0 & - & + \\ + & 0 & - \\ - & + & 0 \end{pmatrix}.$$

(a) Prove that the corresponding first-order replicator system is perseverant if and only if it has a global stable interior equilibrium. (b) Show that this will hold if and only if $\det A > 0$.

9. Classification of Replicator Systems

The basic idea underlying the classification of first-order replicator systems is to find conditions under which two matrices A and B generate the same flow on S^n. For instance, if a constant is added to every element of A, the flow remains unaltered. So given any matrix, we can reduce its diagonal to zero without altering the flow by subtracting an appropriate constant from each column. Let K_n be the set of matrices in $R^{n \times n}$, all of whose columns are scalar multiples of $u = (1, 1, \cdots, 1)'$, and let Z_n be the set of matrices whose diagonals are zero. It's clear that any matrix $M \in R^{n \times n}$ can be written as the direct sum

$$M = Z \oplus K, \qquad Z \in Z_n, \qquad K \in K_n.$$

The following result due to Christopher Zeeman provides a simple test for when two matrices yield the same flow on S^n.

EQUIVALENCE LEMMA. *The matrices A and B yield the same flow for the first-order replicator system if and only if $A - B \in K_n$.*

As a consequence of the above lemma, we see that every equivalence class is of the form $E \oplus K_n$, where E is an equivalence class in Z_n. Thus, it suffices to classify matrices $A \in R^{n \times n}$ in the smaller subset Z_n.

For the case $n = 2$, the above remarks lead to the consideration of matrices of the form

$$A = \begin{pmatrix} 0 & a \\ b & 0 \end{pmatrix}.$$

There are four distinct cases to consider, depending upon the zero/nonzero pattern of the two elements a and b.

1) $a, b > 0$. In this case, the point $(\frac{a}{a+b}, \frac{b}{a+b})$ is an interior equilibrium that is an attractor of the flow.

2) $a, b < 0$. This situation is the reverse of case (1), except that the interior equilibrium is now a repellor; hence, it is unstable.

3) $a \geq 0 \geq b$, but not both zero. Here the corner point $(1, 0)$ is an attractor. This case is also equivalent to the case $a \leq 0 \leq b$, but not both zero, except that now the point $(0, 1)$ is an attractor.

4) $a = b = 0$. Here all points are equilibria.

We call the matrix A *stable* if there is a neighborhood of A such that every matrix in this neighborhood generates a flow that is equivalent to that of A. The above analysis shows that there are three stable classes for two-dimensional replicator dynamics. These classes are characterized by

$$I: a, b > 0, \qquad II: a, b < 0, \qquad III: a > 0 > b \text{ or } a < 0 < b.$$

It should be noted that if we allow time reversal, cases I and II are the same, leading to only two stable classes for $n = 2$. Similar arguments lead to the strongly supported conjecture that for $n = 3$ there are 19 stable classes.

For higher-dimensional replicator systems, there appears to be no hope for a simple classification since the number of qualitatively distinct possibilities grows geometrically with the system dimension n. The only general result that can be given is that stability implies that all equilibria of the system are hyperbolic; i.e., if p is such an equilibrium, the system Jacobian matrix at p has no characteristic values lying on the imaginary axis.

With the foregoing excursion into the dynamical behavior of replicator systems completed, let's now return to the connection between these dynamical systems and the game-theoretic concepts underlying advantageous survival strategies.

Exercises

1. Prove that the stable matrices are dense in $R^{n \times n}$ if and only if they are dense in Z_n.

2. Let M be the set of anti-symmetric matrices whose sign pattern is either

$$\begin{pmatrix} 0 & + & - \\ - & 0 & + \\ + & - & 0 \end{pmatrix} \qquad \text{or} \qquad \begin{pmatrix} 0 & - & + \\ + & 0 & - \\ - & + & 0 \end{pmatrix}.$$

(a) Prove that the associated replicator system has a single equilibrium in the simplex S^3. (b) Show that the flow pattern is that shown in the diagram below. (c) If A is anti-symmetric with a different sign pattern, prove that all trajectories converge to the single corner equilibrium at $e = (1, 0, 0)$.

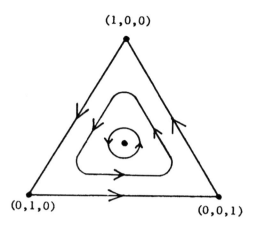

Flow Pattern for an Anti-Symmetric Replicator System

10. Game Theory and Replicator Systems

In our earlier discussions using game-theoretic concepts for the determination of competitively advantageous strategies, we made no explicit mention of dynamics. However, the first-order replicator systems considered above offer a natural way to attach a dynamical process to the earlier static competitive situation. For simplicity, we consider the case when both players have the same strategy set and the payoffs are symmetric; i.e., the payoff matrix A is the same for both competitors. By the arguments already given, the rate of growth of those individuals playing strategy i is proportional to the advantage of that strategy. But the advantage of strategy i is measured by the average payoff against *all* strategies. If we let x_i be the fraction of the population playing i, this observation leads to the replicator dynamics

$$\dot{x}_i = x_i[(Ax)_i - (x,\,Ax)], \qquad i = 1, 2, \dots, n.$$

We have considered earlier the concept of an evolutionary stable strategy (ESS) in an informal way. Now we define the concept formally as a strategy p possessing the following equilibrium and stability properties: (1) p is at least as good a reply against itself as any other strategy x, and (2) if x is a best reply against p, then p is a better reply against x than x itself. Mathematically, we can state these two conditions succinctly as

   1. $(p,\,Ap) \geq (x,\,Ap)$ for all $x \in S^n$ *(Equilibrium)*

   2. If $(p,\,Ap) = (x,\,Ap)$ for $x \neq p$, then $(p,\,Ax) > (x,\,Ax)$ *(Stability)*

The question that arises most naturally at this point is to ask what the relationship is, if any, between an asymptotically stable equilibrium of the replicator system associated with the payoff matrix A and the ESS p. The following result goes most of the way toward answering this query.

ESS THEOREM. *The following statements are equivalent:*

  i) The point p is an ESS.

  ii) For all sufficiently small $\epsilon > 0$ and for all $q \in S^n$, $q \neq p$, we have

$$(p, A[(1 - \epsilon)p + \epsilon q]) > (q, A[(1 - \epsilon)p + \epsilon q]).$$

  iii) For all $x \neq p$ in some neighborhood of p, we have

$$(p, Ax) > (x, Ax).$$

  iv) The function $V = \prod x_i^{p_i}$ is a Lyapunov function at p for the replicator dynamics.

Remarks

1) The last condition of the ESS Theorem shows that an ESS must be an asymptotically stable attractor of the replicator dynamics. However, the converse is not true; there exist asymptotically stable attractors that are not ESS.

2) The second part of the theorem can be interpreted as saying that if the state of the population is p and a subpopulation (mutation) arises in state q, such a subpopulation will become extinct since the p population does better than q against the mixture $(1-\epsilon)p+\epsilon q$ as long as the fluctuation q in the original population is sufficiently small. In other words, a population using the strategy p is uninvadable.

We have now explored a number of theoretical aspects of static games, replicator dynamics and evolutionary processes, as well as several of their interconnections. In the next few sections we put much of this machinery to work by examining in some detail two evolutionary processes involving plants that suggest an interpretation of nature's strategy for survival in terms of ESS. Our first set of examples of the uses of ESS in an ecological context involves the behavioral strategies observed in plants as they attempt to maximize their ability to propagate seeds into future generations.

Exercises

1. Consider the Hawk-Dove game with the following payoff matrix:

$$\begin{array}{cc} & \begin{array}{cc} \text{HAWK} & \text{DOVE} \end{array} \\ \begin{array}{c} \text{HAWK} \\ \text{DOVE} \end{array} & \begin{pmatrix} (V - C)/2 & V \\ 0 & V/2 \end{pmatrix} \end{array}.$$

(a) Use this payoff matrix to convert the problem to a scalar replicator system defined on the unit interval $0 \leq x \leq 1$, where x denotes the fraction of the population that should play HAWK. (b) Show that if $V < C$, the replicator system has a global attractor at $x = V/C$. (c) From first principles or otherwise, show that this attractor is an ESS for the game. (d) What happens if $V \geq C$?

2. Consider the game with payoff matrix

$$A = \begin{pmatrix} 0 & a_1 & -b_3 \\ -b_1 & 0 & a_2 \\ a_3 & -b_2 & 0 \end{pmatrix},$$

where $a_i, b_i > 0$. (a) Show that the system has an interior equilibrium $p \in S^3$. (b) Prove that p is an ESS if and only if $a_i > b_i$, where the three numbers $c_i = a_i - b_i$ correspond to the lengths of the sides of a triangle.

3. Consider the "generalized" Hawk-Dove game, where in addition to pure HAWK and DOVE strategists, there is now a third phenotype HAVE that plays HAWK with probability V/C and DOVE with likelihood $1 - V/C$. (a) Formulate the extended payoff matrix for this game. (b) Show why a population of HAVE's cannot be invaded by a group of HAWK's alone or DOVE's alone. (c) Is the corner equilibrium $e_3 = (0, 0, 1)$, representing a population consisting entirely of HAVE's, an ESS? Why? (d) Can a simultaneous invasion of HAWK's and DOVE's displace a population of HAVE's?

4. Let the payoff matrix of a three-strategy game be given by

$$A = \begin{pmatrix} 0 & 6 & -4 \\ -3 & 0 & 5 \\ -1 & 3 & 0 \end{pmatrix}.$$

(a) Show that the corresponding replicator system has an attractor at the barycenter $e = (\frac{1}{3}, \frac{1}{3}, \frac{1}{3})$. (b) Is e stable or unstable? (c) Prove that e is **not** an ESS. (d) This game does have an ESS, however. Find it.

11. Desert Plant Root Systems and ESS

A good illustration of how ESS arise in nature is provided by the way desert plants organize their root structures to acquire enough water to ensure their survival and reproduction in an arid environment. Such plants can develop two types of root systems: a *lateral* system that spreads out horizontally utilizing water sources near the surface, and a *tap root* system that uses deeper sources of water. Most desert plants specialize in one or the other of these two systems; however, some shrubs have the capacity to develop either a lateral or a tap root system. Generally, surface water is limited, so there is considerable competition between neighboring plants for this source. But underground water is usually quite plentiful. So there is little competition between plants using tap root systems. But underground water is only available in certain spots, so the development of a tap root system entails some amount of risk associated with failing to find an underground water supply.

Now consider the situation of two neighboring plants, each having the option of developing either a lateral or a tap root system. Assume that the evolutionary utility, in the sense of Darwinian fitness, is directly proportional to the rate of water intake. Let S be the utility of a lateral root system in the absence of competition, while U denotes the utility of a tap root system. Finally, assume that if both plants develop a lateral root system, they divide equally the utility associated with a lateral root system in the absence of competition. Under the foregoing hypotheses, the payoff matrix in the evolutionary game between these two plants is given by

		Plant I	
		L	T
	L	$(S/2, S/2)$	(S, U)
Plant II			
	T	(U, S)	(U, U)

Here L indicates the lateral-root system strategy, while T represents the tap root option. Let us also emphasize that the payoff U is an *expected payoff*, obtained by multiplying the payoff associated with the presence of underground water by the probability of actually finding water at the site where the tap root system is developed.

The equilibria of the above game depend upon the relative magnitudes of the quantities $S/2$ and U. The case $S/2 > U$ leads to a single equilibrium associated with both plants developing lateral root systems. In fact, in this case L is a dominant strategy for each plant, as it's the best reply against both strategies of the other plant. The more interesting case comes about when we have $S/2 < U \le S$.

If $S/2 < U$, the game has two equilibria: (T, L) and (L, T). In other words, whatever one plant does, the other does the opposite. But since the game is symmetric, neither the pure strategy "always build lateral" nor the pure strategy "always build tap root" can be ESS. This is because neither of these strategies is the best reply to itself. Thus, the ESS is a mixed strategy. Let x be the fraction of the time the plant "plays" L, with $(1 - x)$ being the fraction of the time it plays T. Then for $(x, 1 - x)$ to be an ESS, we must have

$$\left(\frac{S}{2}\right) x + (1 - x)S = U,$$

which leads to the expression for the ESS solution as

$$x = 2\left(1 - \frac{U}{S}\right).$$

Thus, the strategy play L a fraction $2(1 - U/S)$ of the time and T a fraction $2U/S - 1$ of the time is uninvadable.

It's of some interest to derive the above result from first principles using the replicator dynamics discussed in earlier sections. If we let x_1 be the fraction of the plants playing the strategy L, with x_2 being the fraction playing T, the interaction matrix is

$$A = \begin{pmatrix} S/2 & S \\ U & U \end{pmatrix}.$$

This leads to the first-order replicator system

$$\dot{x}_1 = x_1 \left[\frac{S}{2}(x_1 + x_2) - \frac{S}{2}x_1^2 - Sx_1x_2 - Ux_2\right],$$

$$\dot{x}_2 = x_2 \left[U - \frac{S}{2}x_1^2 - Sx_1x_2 - Ux_2\right].$$

Using the fact that $x_1 + x_2 = 1$, we can eliminate x_2 from the foregoing system arriving at the final system for $x \doteq x_1$ as

$$\dot{x} = x \left[(S - U) - x(S - U) + \frac{S}{2}(x^2 - x)\right].$$

It's easy to verify that the point $x^* = 2(1 - U/S)$ is an equilibrium point of the above system. Furthermore, it's a straightforward matter to see that x^* is a globally stable attractor, thereby satisfying both the equilibrium and stability conditions for an ESS.

The assumption that a plant has only a single competing neighbor is somewhat unrealistic, so let's consider a situation in which a given plant is equidistant from a set of neighbors. This means that each plant in the region is at the center of a hexagon whose vertices are occupied by similar plants. Thus, each shrub has six neighbors, and the competition for surface water is with those of the six that choose the lateral root strategy. Competition for underground water is global, with the total amount of underground water available assumed to be $U(1 - uq)$, where U is the amount that would be available if there were no competitors for underground water, q is the proportion of tap root players in the population and u is a parameter denoting the relationship between the number of competitors and water availability in the underground reservoir.

In this many-player game, we cannot analyze the strategies in payoff matrix form. However, if the game consists of a large number of players, we can replace the payoff matrix with the payoff function $W(I, J)$, which represents the expected change in fitness received by a plant playing the strategy J in a population of I players. Using this idea, it can be shown that the strategy $(x, 1 - x)$, with

$$x = \frac{(S/U) + u - 1}{u + (2S/U)} ,$$

is ESS. As expected, this strategy reduces to the earlier one when $u = 0$.

Exercises

1. If the payoffs in the Desert Plant game are such that $U < S/2$, which type of mixed-motive game are the plants playing? Does the type of game shift if $U > S/2$?

2. Consider how to formulate the Desert Plant game when there are two species of plants that differ slightly in the efficiency with which they can exploit the two water sources. This means that if species A is more efficient than species B, the payoff parameters would satisfy the relation $S_A/U_A > S_B/U_B$. In your formulation, you may also assume that the difference in the surface and underground water exploitation efficiencies of A and B are not strong enough for either of them to favor a given root system, independent of what the rest of the community is doing. How would you translate this condition into mathematical inequalities involving the quantities S_A, S_B, U_A and U_B?

12. ESS vs. Optimality

It's often asserted in ecological analyses that natural selection has arranged matters so that a population is driven to a state in which the resources are optimally exploited. Here we want to indicate briefly how this argument can break down in situations in which the selection mechanism is frequency-dependent, as in the preceding desert plant example.

In the sense of species optimality, the goal of a shrub population is to adopt a strategy J permitting the maximum water uptake per individual. If p represents the fraction of the population adopting the lateral root strategy, the total water uptake arising from considering the four possible interactions between tap and lateral root players is

$$f(p) = U - pU + pS - \tfrac{1}{2}p^2 S.$$

The shrubs should choose p in an attempt to maximize this function. It's easily seen that the maximum is attained at the value

$$p_{\text{max}} = 1 - \frac{U}{S}.$$

Thus, maximum water uptake per shrub occurs if the probability of adopting a lateral root strategy is $1 - U/S$. On the other hand, the ESS probability of adopting a lateral root system is 1 if $U < S/2$ and $2(1 - U/S)$ if $U > S/2$. So ESS analysis predicts that *twice* as many plants will adopt a lateral root strategy than would be suggested by optimality considerations.

The two criteria—ESS and optimality—also differ considerably in their predictions about the actual amount of water uptake achieved by the two strategies. In the case $U < S/2$, the ESS strategy predicts a water uptake of $S/2$; on the other hand, using the optimality strategy we find a water uptake of $(S^2 + U^2)/2S$. Thus, the water uptake per individual as predicted by the optimality criterion solution is consistently higher than that from the ESS solution, and can be as much as 25 percent greater depending upon the particular value of the ratio U/S.

The main point to note about these optimality versus ESS results is that there can be a significant difference between the predictions based upon one criterion or the other. Which criterion is "correct" cannot really be settled by armchair analysis alone, but is dictated by the considerations of the problem and, most importantly, by the experimental data. So, for instance, in the desert shrub example above, the difference between the two criteria in predicted numbers of shrubs using a lateral root system is so large that it should be a relatively straightforward matter to construct a laboratory experiment to test whether nature uses an ESS or an optimality selection criterion. Now let's turn to another type of plant growth problem illustrating some of the same features seen above but in a context where there can be a *continuum* of strategies.

Exercises

1. It's common in ecological analyses to assume that natural selection has driven a population to a state in which resources are optimally exploited. Yet in the desert plant example we find that the optimal strategy and the ESS differ considerably. (a) Can you offer any explanation for why plants might prefer to forego optimality in favor of stability? (Hint: Consider the differences between frequency-independent and frequency-dependent selection.) (b) In the plant example, how would the payoffs have to be arranged in order for the optimal strategy to also be ESS?

2. Discuss the general issue of trading off stability versus optimality in other contexts, such as managing a corporation or building a bridge. (Chapter 7 takes up this kind of tradeoff in some detail.)

13. *Flowering Times for Plants under Competition*

Consider an annual flowering plant having a growing season of length T. During the growing season the plant can devote energy to either vegetative growth or to seed production. Under the hypotheses of a single plant in the absence of competition, it has been shown that the optimal strategy for maximizing seed production is to grow only vegetatively up to a time $u < T$, then shift over and devote the remainder of the time $T - u$ solely to seed production. Under plausible assumptions, this leads to the optimal crossover point

$$u^* = T - \frac{1}{RL},$$

where R is the net photosynthetic production per unit leaf mass and L is the ratio of leaf mass to remaining vegetative mass. The total seed production in this case will be

$$H(u) = (T - u)RA_0 \exp(RLu),$$

where A_0 is the initial leaf biomass of the plant.

Here we are interested in the more general case where there are many such plants in competition. Let S_i be the seed production of the ith plant in the absence of competitors, with u_i being the time when the ith plant would switch from vegetative to reproductive growth if there were no competitors. In this situation we have

$$S_i = (T - u_i)f(u_i),$$

where $f(\cdot)$ is a function that measures the plant's ability to produce seeds given that flowering starts at time u_i. Since $(T - u_i)$ represents the time remaining in the growing season to produce seeds, if $f(0)$ is small and $f(u_i)$ increases rapidly with u_i, the maximum value for S_i will occur between 0 and T. Further, it's reasonable to suppose that $f'(u_i) > 0$, since $f(u_i)$ is a measure of the plant's size and photosynthetic capability. Note that we assume the plant population to be homogeneous; i.e., the function f is the same for all plants in the population.

Now let's assume there are two phenotypes in the population: Type 0 with a flowering time U_0, and Type 1 with a flowering time U_1. Let's postulate that the generating function for the average fitness of the two phenotypes is given by

$$G = \frac{(T - u_i)f(u_i)}{1 + W(U_0, U_1, u_i, P_0, P_1, n)},$$

where P_0 and P_1 are the initial populations of the two phenotypes, and n is the total number of plants (players) in the population. The function W

represents the effect of competition and is zero when $n = 1$. Furthermore, it should be possible for one phenotype to diminish the competitive effect of the other by flowering earlier or to intensify the competitive effect by postponing its flowering time. These considerations lead to the conditions

$$\frac{\partial W(s)}{\partial u_i} < 0, \qquad \frac{\partial W(s)}{\partial s} > 0.$$

In the first expression, the partial derivative is evaluated at $s = U_0 = u_i$, but in the second expression s is substituted for U_0, U_1 and u_i before taking the partial derivative. In what follows, it's convenient to introduce the function $E(s)$ as

$$E(s) = (T - s)\frac{df(s)}{du_i} - f(s).$$

We now examine flowering times using the average fitness function G under the conditions of maximizing seed production when there is no competition, and finding an ESS when there is competition. Finally, we compare both results with the solution obtained for a community of plants experiencing competition, in which *community* seed production is maximized.

In the case of no competition, $W = 0$. If we set $\partial G/\partial s = 0$, we obtain the condition $E(s) = 0$. Or equivalently,

$$\frac{df(s)}{ds} = \frac{f(s)}{T - s}.$$

This condition for optimal seed production is similar to a well-known result in economics associated with the cost of production. By producing more of a product, the average cost of production goes down, while the marginal cost goes up. Cost is minimized at the point where marginal and average costs are equal. For plants, seed production is maximized when the marginal rate of increase in seed production equals the average rate of seed production taken with respect to the time remaining in the growing season. This relationship is expressed graphically in Fig. 5.1, where the average and marginal rates of production are displayed. Their crossover point is the optimal time of flowering s^*.

Now let's assume competition ($n > 1$) and calculate the ESS strategy. The necessary condition for an ESS strategy is that

$$\frac{\partial G}{\partial u_i} = 0.$$

Computing this quantity yields

$$\frac{\partial G}{\partial u_i} = \frac{(1 + W)[-f + (T - u_i)(\partial f/\partial u_i)] - (T - u_i)(\partial W/\partial u_i)f}{(1 + W)^2}.$$

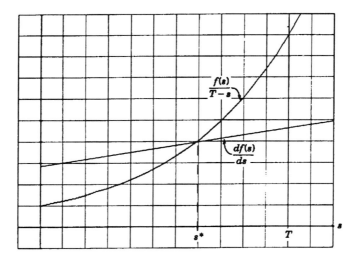

Figure 5.1. The Average and Marginal Rates of Seed Production

Substituting s for U_0, U_1 and u_i in the above terms and setting the resulting expression to zero, we find

$$E(s) = \left[\frac{(T-s)f(s)}{1+W(s)}\right]\frac{\partial W(s)}{\partial u_i} = G(s)\frac{\partial W(s)}{\partial u_i} \ .$$

From the assumptions on W, this means $E(s) < 0$ at an ESS solution. And from the definition of $E(s)$ this requires that

$$\frac{df(s)}{ds} < \frac{f(s)}{T-s} \ .$$

Comparing this result with that obtained under no competition, we find that the above inequality is satisfied only if the ESS flowering time is *greater* than s^*. That is, when faced with competition the plants should begin to flower later. Just as in our desert shrub example, it can be seen that the number of seeds produced per plant under competition will always be less than the number produced in a noncompetitive environment.

In order to obtain the optimal *community* seed production level, we must have

$$\frac{\partial G(s)}{\partial s} = 0.$$

We compute this expression using the function G as above, obtaining

$$E(s) = \frac{\partial W(s)}{\partial s}\left[\frac{(T-s)f(s)}{1+W(s)}\right].$$

From the assumptions on W, this implies that

$$\frac{df(s)}{ds} > \frac{f(s)}{(T-s)} \, .$$

This inequality will be satisfied only if the flowering time is *earlier* than s^*, the optimal flowering time in the absence of competition. This solution will yield more seeds per plant than the ESS solution, which is clearly desirable from an agricultural point of view. The catch is that this cooperative solution is unstable and, hence, cannot be maintained in nature.

The foregoing plant examples have shown some of the simpler interconnections between the concepts of ESS, optimality, and the properties of replicator dynamics. At this point it's tempting to engage in a bit of rash speculation, conjecturing that perhaps *human* behavioral patterns are also determined in somewhat the same manner as for plants and animals. This leads to the core hypothesis of the field now termed *sociobiology*. The next section examines the pros and cons of this hypothesis by way of considering the role of evolutionary concepts within the broader context of cultural and social change.

Exercise

1. Suppose there is no competition, i.e., there is just a single plant species in the region. In this case $W = 0$. (a) Show that in this case a necessary condition for the optimal flowering date is given by $E(s) = 0$ or, alternately,

$$\frac{df(s)}{ds} = \frac{f(s)}{1-s}.$$

(b) Interpret the above condition in economic terms associated with the cost of producing a product, where producing more reduces the average cost of production but increases the marginal production cost (i.e., the instantaneous rate of change of costs go up).

14. Sociobiology

During the past decade or so, there has emerged an amalgamation of the fields of genetics, evolutionary biology, ethology and anthropology termed "sociobiology," which can be broadly construed as the study of the biological basis of behavior in humans and primates. Basically, what this amounts to is the investigation of the degree to which changes at the genetic level *determine* behavioral patterns at the level of the phenotype. In more prosaic terms, the central issue is whether human behavioral patterns are "preprogrammed" by the genes. Hard-core sociobiologists claim the answer is

yes, saying that cultural processes play only a secondary role in the determination of the individual propensity to behave in certain ways. As one
might suspect, such a claim strikes a raw nerve in those circles devoted to
the position that humans, by virtue of culture, are immune somehow from
the principles by which the rest of the animal world operates, and that by
denying this special role of humans, sociobiologists only provide ammunition
and the weight of science to those who wish to adopt political and sociological positions of genetic superiority in matters like intelligence, health, and
industrial productivity. Other critics have attacked sociobiology on more
scientific grounds, stating that there is no recognizable difference between
sociobiological theory and traditional Darwinian evolution. In short, sociobiology is nothing more than classical Darwinism dressed up in new clothes.
In our brief discussion of the matter here, we take no position on these controversial emotional and scientific debates, leaving the interested reader to
consult the items listed in the Notes and References for a full account of all
competing positions. Our goal is merely to give a short introduction to the
mathematical relationships involved in bridging the gap between the level
of genes and the level of behavioral phenotypes.

As a discipline, sociobiology is concerned with the implications of natural selection and evolutionary biology for the social behavior of organisms.
An animal's behavior can be examined for its effect on the animal's fitness
and for the resultant changes in the frequency of the genes associated with
or "causing" that behavior. From cost/benefit analyses of possible behaviors relative to their effects on fitness and gene frequencies, various models
of animal behavior are then derived. The basic behavioral modes that have
been examined in this regard tend to fall into just a few distinct categories:

• *Sex and Mating Systems*—Mating behavior relates to the strategies
males and females adopt in trying to maximize the likelihood of getting their
genes into the next generation. The majority of these studies focus upon
the issue of monogamous versus polygamous mating strategies.

• *Aggression*—Active aggression is one way in which an animal can gain
resources, thereby increasing its individual fitness. A cost/benefit analysis
determines the adaptive value of aggression in any given situation. Studies
in this area have been mostly devoted to an analysis of factors that tend
either to increase or decrease aggressive behavior, as in our simple Hawk-
Dove game discussed earlier in the chapter.

• *Dispersal*—The desire on the part of males to increase their gene
frequency in the next generation often leads to behavior patterns involving
migration, i.e., males leaving their social group and either joining another or
carrying on in a solitary fashion. Dispersal studies center upon conditions
under which such behavior is optimal, as well as on the form such migratory
patterns take for different species.

• *Altruism*—Since altruism is the central theoretical problem of socio-biology, let's take a closer look at what is involved in the analysis of this class of behaviors.

On classical Darwinian grounds, an altruistic act involving the sacrifice of an individual for the sake of another is difficult to comprehend. Darwinian theory claims that an individual acts so as to maximize the chance of its own survival. This dictate leads to behaviors that, by definition, can never be altruistic. Thus, the existence of altruistic behaviors stands as an observational counterexample to the strong Darwinian claim that behavioral adaptations take place so as to maximize individual fitness. The concept of kin selection and inclusive fitness provides a framework within which we can make sense of both altruistic acts and evolutionary change by gene frequency modification.

If one accepts the contention that the only important factor in evolution is a change in gene frequency independent of how that change comes about, then it can be argued that a definition of fitness limited to the number of offspring of an organism and its direct descendants is too narrow. This is because an organism could increase the frequency of its genes in the next generation by either increasing its own fitness or the fitness of its genetic relatives. This idea forms the basis for the theory of *kin selection* and leads to the concept of inclusive fitness as the sum of individual fitness and kin fitness.

As a simple example of how kin selection works, consider the actions of an animal A sighting an approaching predator. A can either give an alarm call (an altruistic act) and warn, say, three fellow group members. This action attracts attention to A, leading to predation. On the other hand, A can remain silent and hide (a selfish act), thereby leading to its survival but resulting in the loss of the others in the group. Question: Should A give the alarm call or not?

Let's assume that r represents the level of genetic relatedness between A and the other three group members. This number represents the proportion of genes that two individuals share and are common by descent. In sexually reproducing species (without considering inbreeding), if the group members are full siblings of A we have $r = \frac{1}{2}$. Thus, if A gives the alarm call, the genetic advantage to A is

$$(\tfrac{1}{2})A + (\tfrac{1}{2})A + (\tfrac{1}{2})A - A = +(\tfrac{1}{2})A.$$

But if A does the selfish thing and hides, the genetic calculus gives A's advantage as

$$A - (\tfrac{1}{2})A - (\tfrac{1}{2})A - (\tfrac{1}{2})A = -(\tfrac{1}{2})A.$$

Clearly, the *inclusive* fitness of A is greater if A gives the call. However, it's easy to see that if the group members were only *cousins* rather than full siblings, then $r = \frac{1}{8}$, and it would be best for A to remain silent. These simple arguments lead to the altruistic inequality $K > 1/r$, expressing the fact that K, the ratio of benefit to cost, must be greater than the reciprocal of the genetic relatedness between the recipient and the beneficiary of the act in order for an altruistic act to be evolutionarily advantageous. Despite the obvious artificiality of this example, there are numerous situations in nature where kin selection has been found to be a dominant behavioral mode. Now let's shift over to math mode and take another look at the issue of aggression, but this time by explicitly linking-up the genetic level with the phenotypic.

In the genetic model considered earlier in the chapter, we assumed that selection operated through viability differences between genotypes, with viability meaning the probability of survival from zygote to adult. Let's now expand upon this model by also taking into account fertility differences between mating pairs.

Let x_{ii} represent the frequency of the homozygote genotype A_iA_i, and let $2x_{ij}$ be that of the heterozygote $A_iA_j, i \neq j$, so that

$$\sum_{i,j} x_{ij} = 1.$$

Let $x_i = \sum_j x_{ij}$ be the frequency of allele A_i in the gene pool, and let $w(ij, st)$ denote the average fecundity (number of offspring) of a mating between an ij-male and an st-female. Assuming random mating, the frequency of the genotype A_iA_j in the next generation is

$$x'_{ij} = \frac{1}{2\phi} \sum_{s,t} [w(is, jt) + w(jt, is)] x_{is}x_{jt},$$

with ϕ a normalizing constant that corresponds to the mean fecundity of the population.

This relation is far too general. So let's further assume that the fertility of a couple can be decomposed into a male and a female contribution. Thus,

$$w(ij, st) = m(ij)f(st), \qquad 1 \leq i,j,s,t \leq k.$$

In this case, the average fecundity of gene A_i in the male gene pool is

$$M(i) = \sum_j m(ij)x_{ij},$$

whereas the average fecundity of gene A_j in the female pool is

$$F(j) = \sum_k f(jk)x_{jk}.$$

This leads to the differential equation for the genotype $A_i A_j$ as

$$\dot{x}_{ij} = \tfrac{1}{2}[M(i)F(j) + M(j)F(i)] - x_{ij}\phi,$$

with

$$\phi = \sum_i M(i) \sum_j F(j).$$

If we further assume equality of the sexes so that $m(ij) = f(ij)$, the above relation simplifies to

$$\dot{x}_{ij} = M(i)M(j) - x_{ij}\phi,$$

with

$$\phi = \left(\sum_i M(i) \right)^2.$$

Now we return to the sociobiological situation and consider how the above genetic changes affect behaviors. Assume the behavior is determined by a single gene locus with alleles $A_1, \ldots, A_k$. The payoff, which is dependent on the strategy employed by the individual and on that of its competitors, must somehow be related to reproductive success. There are many ways to express this, one of the simplest being to assume that the payoff is independent of sex and that the number of offspring of a given couple is proportional to the product of parental payoffs. Such a situation would correspond, for example, to fights that are not sex-specific like competition for food.

Under the foregoing assumption, to each genotype $A_i A_j$ there corresponds one of the strategies $E_1, \ldots, E_n$ or, more generally, a mixed strategy

$$P(ij) = (p_1(ij), \ldots, p_n(ij)) \in S^n.$$

The frequency of strategy E_i in the population is then given by

$$b_i = \sum_{st} p_i(st) x_{st},$$

where the average payoff for strategy E_k is

$$a_k = \sum_\ell a_{k\ell} b_\ell.$$

Here the numbers $\{a_{k\ell}\}$ are given by the payoff matrix. Consequently, the fitness for the genotype $A_i A_j$ is

$$m(ij) = \sum_k p_k(ij) a_k = \sum_{k\ell st} a_{k\ell} p_k(ij) p_\ell(st) x_{st}.$$

If we assume that the number of offspring of an $A_i A_j$ male/female is proportional to the fitness $m(ij)$, we are led to exactly the same dynamics for x_{ij} as above—with the significant difference that now the quantity $m(ij)$ is not a constant as before, but is linearly dependent upon x_{ij} as in the above expression. Hence, the relation for the gene frequency change dx_{ij}/dt is of fourth-degree in the genotype frequencies.

The above equations can be used to analyze the Hawk-Dove game considered earlier in the chapter. Such an analysis shows that the genetic constraints considered here may permit the establishment of an evolutionary stable equilibrium in the population, the corresponding mixture of strategies being exactly the same as that arising from the stable equilibrium of the above dynamical equations. However, it should be noted that there are cases in which the genetic structure *prevents* the establishment of an ESS equilibrium. But in most situations the genetic fine structure does not conflict with the purely game-theoretic analysis.

In passing, it's amusing to see how an act of altruism might come about as the rational, selfish choice in a purely economic context. Suppose there are two individuals, an egoist E and an altruist A. By the definition of altruism, A is willing to give some of his wealth to E. The question is: How much? The answer, of course, depends on A's degree of altruism, the relative wealth of E and A, the "cost" of giving and so forth.

In economics, it's often assumed that an individual's behavior ultimately comes down the maximization of his or her utility function. So to analyze this situation further, let's assume that A's utility function is given by $U_A = U_A(c_A, c_E)$, where c_A and c_E are the consumptions of A and E, respectively. We can also write A's budget constraint as

$$I_A = t_E + p c_A,$$

where I_A is A's total income, t_E is the amount A gives to E and p is a multiplier representing the fraction of A's overall income that is not spent or given to E. This relation yields the income of E as

$$p c_E = I_E + t_E.$$

Substituting this relation into that above for A, we find A's overall budget constraint

$$p c_A + p c_E = I_A + I_E \doteq S_A,$$

where S_A is called A's "social income."

If our altruist A wants to maximize his utility, subject to the social income constraint on his budget, then we must have

$$\frac{\partial U_A/\partial c_A}{\partial U_A/\partial c_E} = 1.$$

This means that A would transfer just enough money to E so that he would receive the same utility from increments to his own or to E's consumption. In other words, A would suffer the same loss in utility from a small change in his own or E's consumption.

It's interesting to note that this condition carries within it the possibility that A's own consumption might not be less than E's, because E would consider the effect of his own behavior on A's consumption. This indirect effect on the behavior of E might dominate the direct "disadvantages" of being altruistic. For example, assume that A transfers \$1,000 and that E could increase his own income by \$800 at the cost of harming A by \$5,000. Since E would not take these actions, A's altruism has increased his income by \$5,000, or by five times the amount given away to E. Now let's leave this small economic digression and get back to the main biological line of argument.

We have already noted that the key question in sociobiology is how cooperative behavior (altruism) can emerge in a group of individuals The preceding arguments suggest that the concepts of kin selection and inclusive fitness provide a basis for an explanation of such behavior at the genetic level. But it's also of considerable interest to consider the same question from the purely phenotypic level, looking at how cooperative behavior might emerge in situations where the *rational* strategy of an individual is to act selfishly.

A stylized version of this kind of situation is provided by the Prisoner's Dilemma game considered earlier, where we saw that the rational choice for each player was to defect, leading to a nonoptimal *joint* outcome for each. This noncooperative suboptimal result was obtained under the crucial assumption that the game is to be played only once. We now show how cooperation can emerge under the condition that the players will play the game many times, having the opportunity to learn from past interactions about the strategy employed by others in the population. As we will see, introducing repeated interactions dramatically changes the structure of the Prisoner's Dilemma situation, opening up the possibility for cooperative actions to emerge naturally from selfish motives.

Exercises

1. In the text we have emphasized kin selection as the process whereby altruistic behavior may become established in a population. But other possibilities have been proposed: (i) *reciprocity,* in which the altruistic party helps a non-relative in the hopes that some other non-relative will help it; (ii) *group selection,* in which the altruistic individual helps others within its group as a result of which the overall fitness of the group goes up; (iii) *parental manipulation,* whereby a parent forces one or more offspring to assist siblings, thereby costing the helpers but enhancing the parent's overall fitness. The

diagram below schematically depicts these four types of altruism-generating mechanisms. Can you match up the mechanisms with the corresponding diagram—part (a), (b), (c) or (d)—that represents it?

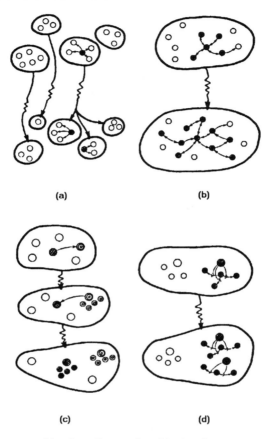

(a)   (b)

(c)   (d)

Altruism-Generating Mechanisms

2.   Calculate the coefficient of relatedness r for: (a) half brothers, (b) second cousins, (c) an uncle and nephew.

3. In the order *Hymenoptera,* which includes the ants, wasps and bees, the sex of offspring is determined in an unusual way. Specifically, females are diploid, developing from fertilized eggs. Thus females have both a mother and a father. Males, on the other hand, are haploid, developing from un- fertilized eggs. So males share genes only with their mother (the queen). (a) Suppose Ego is a female ant. What is the coefficient of relatedness r between Ego and one of her full sisters? (b) What is r for the relationship between Ego and one of her daughters? (c) On the basis of kin selection and inclusive fitness, what action would most benefit Ego in her goal of propa- gating as many of her genes into the next generation as possible? (d) Use

inclusive fitness to explain why we don't ever see any worker males. (e) What sex ratio would you expect to find in *Hymenoptera?*

4. The idea that genetic makeup is a strong determinant of social behavior is reasonably well-established in the animal world, at least for a large spectrum of behavioral acts. The sociobiology controversy revolves about the degree to which the same arguments apply to the species *Homo sapiens.* Consider arguments for and against the genetic determination of human behavior.

15. The Evolution of Cooperation

Consider again the Prisoner's Dilemma with the payoff matrix

Player I

		C	D
	C	(R,R)	(S,T)
Player II			
	D	(T,S)	(P,P)

where the choices available to each player are either Cooperation (C) or Defection (D). We assume the payoffs satisfy the ordering

$$T > R > P > S,$$

where T is the "temptation" payoff associated with defecting when the opponent cooperates, R is the "reward" obtained when both players cooperate, P is the "punishment" received by a pair of defecting players, and S is the "sucker's" payoff that goes to a player who cooperates when the other player defects. As we have already seen, when the game is played only once the rational strategy for each player is Defection, leading to each player receiving the suboptimal joint punishment payoff P. Now let's change the ground rules and assume the players are to engage in a sequence (finite or infinite) of plays, with each player's total payoff being the sum of his payoffs from each play. This is the so-called *iterated Prisoner's Dilemma.* Our interest will be in identifying circumstances under which joint cooperation may emerge as a rational strategy when the players know they will have to face each other many times. In such iterated situations, we make the additional assumption that

$$(T + S)/2 < R,$$

ensuring that if the two players somehow get locked into an "out of phase" alternation of cooperations and defections, each will do worse than if they had both cooperated on each play.

First of all, assume that the game will be played a *finite* number of times N, with N being known to both players *before* the first play of the game. It's easy to dispose of this case, since it's clear that the optimal choice at the last play N is for each player to defect; hence, we then have an $(N-1)$-stage process where again the optimal choice is defection at play $N-1$, and so on. Consequently, in an iterated Prisoner's Dilemma for which the number of plays is finite and known in advance, the optimal strategies are the same as in the "single-shot" game: always defect! So the only possibility for cooperation to emerge as the preferred course of action is for the number of interactions to be potentially infinite, or what amounts to the same thing, that there is a probability $w > 0$ that the game will continue for another round after any play of the game is completed.

In the biological context, the quantity w measures the likelihood of two individuals (or behavioral phenotypes from the same species) competing again after an initial interaction. Factors such as average life span, mobility and individual health can affect w, which we shall assume is held constant. The parameter w can also be interpreted as a *discount factor,* enabling us to compare the value of future payoffs with the return received from the initial round of play. As a simple example, suppose the players each employ the strategy of always defecting (ALL D). Then they each receive a payoff P per round, so that the discounted total payoff each player receives is

$$V(D \mid D) = P(1 + w + w^2 + w^3 + \ldots) = P\left(\frac{1}{1-w}\right).$$

In the iterated Prisoner's Dilemma, a *strategy* is any rule that tells a player what option to choose (C or D) in light of the entire past history of choices made by both sides. So, for instance, the selfish rule ALL D is a strategy; so is the slightly more complicated rule: Cooperate (C) on the first round, and thereafter do whatever your opponent did in the previous round. This is the well-known strategy TIT FOR TAT. As we shall see, this strategy plays a central role in analysis of the iterated Prisoner's Dilemma.

For any value of w, the strategy of unconditional defection (ALL D) is an ESS; however, other strategies may also be ESS. In fact, we can show that when w is sufficiently close to 1, there is no single best strategy. For example, suppose that the other player is using ALL D. Since your opponent will never cooperate, the best you can do yourself is also to adopt ALL D. Now suppose your opponent is using a strategy of "permanent retaliation," i.e., cooperating until you defect, and then always defecting after that. In this case, your best strategy is to never defect yourself— provided that the temptation to defect on the first round will eventually be more than compensated for by the long-term disadvantage of getting nothing but the punishment payoff P instead of the reward R on future plays. This will be true whenever the discount parameter w is close enough to 1. Consequently, whether or not you should cooperate—even on the first move—depends on the strategy employed by your opponent. So if w is sufficiently large, there is no one best strategy.

Now we turn to the evolution of cooperative strategies that are ESS. This problem can be decomposed into three separate questions:

• *Robustness*—What type of strategy can thrive in an environment composed of players who each use a variety of more or less sophisticated strategies?

• *Stability*—Under what conditions can such a strategy, once established, resist invasion by mutant strategies?

• *Initiation*—Even if a strategy is both robust and stable, how can it ever get started in an environment that is predominantly hostile, i.e., noncooperative?

To test the kind of strategies that can be robust, Robert Axelrod conducted a well-chronicled computer tournament in which a variety of strategies submitted by game theorists, evolutionary biologists, computer scientists, political scientists and others of this ilk were pitted against each other in a round-robin competition. The result of the tournament was that the highest average score was attained by TIT FOR TAT, the simplest of all the strategies submitted. An analysis of the results showed that the robustness of TIT FOR TAT was due to three factors: (1) it was never the first to defect, (2) it could be provoked into retaliation by a defection of the other party, and (3) it was forgiving after just a single act of retaliation. As an added measure of the robustness of TIT FOR TAT, as the less successful rules were weeded out of the tournament by better ones, it continued to do well against the other rules that initially scored at or near the top. And in the long run, TIT FOR TAT displaced all other rules. This experiment provides strong empirical evidence that TIT FOR TAT is a robust strategy, capable of thriving in an environment in which there are many players using different types of strategies.

Once a strategy has become established, the question arises of how well it can resist invasion by mutants. It's a rather simple matter to show that once TIT FOR TAT is established in a community of players, it's impossible for any mutant strategy to displace it—provided the probability of future interactions w is great enough. So if there is a high likelihood that two players will encounter each other again during the course of the game, and if the dominant strategy in the community is TIT FOR TAT, then any deviation from this strategy will result in a loss in fitness to the mutant. Thus, TIT FOR TAT is a "weak ESS" strategy. We term it weak because TIT FOR TAT is no better as a reply to ALL C than ALL C itself, only just as good. Therefore, the stability requirement for an ESS is only "weakly" satisfied. In what follows, we'll use ESS in this weak sense. In fact, we can give a precise value for how large w must be in order for TIT FOR TAT to be an ESS.

COOPERATION THEOREM. *In the iterated Prisoner's Dilemma, TIT FOR TAT is an ESS strategy if and only if*

$$w > \max \left\{ \frac{T-R}{T-P}, \frac{T-R}{R-S} \right\}.$$

We have already noted that TIT FOR TAT is not the only ESS for the iterated Prisoner's Dilemma; ALL D is another. So how could TIT FOR TAT ever get a foothold in a population initially consisting of all defectors? There appear to be at least two different mechanisms that provide a plausible path whereby TIT FOR TAT could get started, even in the face of a fundamentally hostile environment. The first such mechanism is *kin selection,* a concept we used earlier to help explain how altruism could emerge in Darwin's universe. Not defecting in a Prisoner's Dilemma game is altruism of a kind, since the altruistic individual is foregoing returns that might have been taken. So in this way cooperation can evolve if two players are sufficiently closely related. In effect, recalculation of the payoff matrix in such a way that an individual has a part interest in a partner's success (i.e., computing the payoffs in terms of inclusive fitness) can often eliminate the inequalities $T > R$ and $P > S$, in which case cooperation becomes unconditionally favored. Thus, the benefits of cooperation can be obtained by groups of sufficiently closely-related players. And once the genes for cooperation exist, selection will promote strategies relying on cooperative behavior.

Clustering is another mechanism by which cooperation can emerge in an essentially ALL D environment. Suppose that a small group of individuals is using a strategy like TIT FOR TAT, and that a certain proportion p of the interactions of members of this cluster is with others from the cluster. Then the average return to a member of the cluster if they are all playing TIT FOR TAT is

$$\frac{pR}{1-w} + (1-p)\left(S + \frac{wP}{1-w}\right).$$

If the members of the cluster constitute a negligible fraction of the interactions for the rest of the population, then the return obtained by those using ALL D is still $P/(1-w)$. If p and w are large enough, a cluster of TIT FOR TAT individuals can then become viable, even in an environment in which the overwhelming majority of the players are using ALL D. We note, in passing, that clustering is often associated with kinship, and the two mechanisms can often reinforce each other. But it's still possible for clustering to be effective even without kinship.

Example: Sea Bass Spawning

A case of cooperation that fits into the above framework involves the spawning behavior of sea bass. These fish have both male and female sexual organs, form pairs, and can be said to take turns at being the high investment (laying eggs) and the low investment (providing sperm to fertilize the eggs) partner. Up to ten spawnings occur in a single day, with only a few eggs provided each time. Pairs of fish tend to break up if sex roles are not divided evenly. It appears that such cooperative behavior got started at a time when the sea bass was scarce as a species, leading to a high level of inbreeding. Such inbreeding implies a relatedness in the pairs and this would have initially promoted cooperation without the need for further relatedness.

Both the experimental evidence, as well as the results of the computer tournament pitting strategies against each other, suggest that in order to do well in an iterated Prisoner's Dilemma, a strategy should possess several features that have been characterized in the following way by Axelrod:

- *Nice*—A strategy should never be the first to defect.

- *Retaliatory*—A successful strategy always punishes defections by the opponent.

- *Forgiving*—A good strategy is not vindictive: punishment is meted out to fit the crime, and the punishment ceases as soon as the other party begins to cooperate again.

- *Optimistic*—A winning strategy will be maximally discriminating in the sense that it will be willing to cooperate even if the other party has not yet cooperated.

The preceding results were obtained under a host of assumptions, each of which can be questioned on real-world grounds. Let's conclude our discussion of the evolution of cooperation by examining briefly work that has been done on relaxing one or another of these restrictions.

- *Interactions:* Axelrod's tournaments assumed that interactions are between pairs of players. But in some circumstances interactions could involve more than just two players. In this case we end up with an *n*-person Prisoner's Dilemma game, in which the players make a choice of C or D that they then play against all the other competitors.

It has been shown that increasing the number of players in the *n*-person game makes cooperation more difficult. In fact, for cooperation to be part of an equilibrium in an *n*-person Prisoner's Dilemma game when some players are playing ALL D, it's necessary for either the shadow the future to be very long or the number of cooperators to be very great.

- *Choices:* In most real-life situations the players have more than two choices of action at their disposal. In many situations there is the option to

leave the game (exit) or to force others to leave (ostracism). Studies have
shown that in the latter situation, for instance, the ostracism option acts to
induce cooperation on the part of the players.

In another direction, one might think of relaxing the condition that
the players have to make their choices simultaneously. It's been shown that
in the nonsimulataneous-choice situation, the stable equilibria are either
ALL D or a form of conditional cooperation. So removal of the simultaneity
condition acts to undermine the emergence of cooperation.

• *Payoffs:* Earlier we saw that there were certain structural similarities
between the Prisoner's Dilemma and Chicken, a game that models crisis
bargaining, despite the fact that their payoff structures differ. In a computer
tournament similar to the one organized by Axelrod, it was found that
playing PERMANENT RETALIATION, in which one cooperates until the
first defection by the opponent and thereafter always defects, is a strategy
that outdoes TIT FOR TAT in the iterated Chicken game. So the payoff is
an important determinant of whether or not cooperation of the TIT FOR
TAT-type can be expected to emerge.

• *Noisy Communication:* In real life it's not always easy to know how
to categorize the actions taken by others. If someone let's you cut in front
of him at a supermarket checkout counter, is this cooperation? Or is it
a means to get in back of you to relieve you of the wallet in your back
pocket? This is a rather fanciful illustration of one type of noise that can
enter into the Prisoner's Dilemma game—noise arising from a misperception
of your opponents action. Of course, if there is no way of distinguishing an
intentional from an unintentional defection or cooperation, then there is no
game-theoretic difference between the noise-free game and its counterpart
with misperception.

Another kind of noise arises when there is faulty transmission of the
strategies. For example, a player decides to cooperate but the decision that
is actually transmitted is defection. It can be shown that if there is any
probability at all of this kind of noisy transmission of choices, then two TIT
FOR TAT players will, in the long-run, average the payoffs of two interacting
RANDOM players, each of whom cooperates or defects with probability $\frac{1}{2}$.
Thus, noisy transmission of this sort severely undermines the effectiveness
of reciprocating strategies.

• *Importance of the Future:* The classic iterated Prisoner's Dilemma
assumes an indefinite ending to the game, the probability being $1 - w$ that
the game will end with the current move. The parameter w was termed the
"shadow" of the future. An important generalization of this idea is to drop
the assumption that w is fixed. Rather, we let the players choices affect the
likelihood of future interactions. There are many different ways that this
might be done, with most studies concluding that beliefs about the payoffs

and strategies, the computational resources available to the players, and the size of the overall population all enter into whether or not stable cooperative equilibria will emerge.

• *Population Dynamics:* In showing how cooperation can emerge in a population of egoists, Axelrod assumed that there was only a single mutant: TIT FOR TAT. It turns out that if one allows invasion by multiple mutants, there is no single strategy that is ESS in the Prisoner's Dilemma game. More specifically, if $w > 0$, then for each strategy X there is some set of strategies Z such that X is not an ESS against Z. Results of this sort emphasize the point that it's only by understanding the set of possible competitors that we have a chance of finding evolutionary paths leading to cooperation.

The ideas we have touched upon in this chapter only begin to scratch the surface of the manifold complexities and subtle complications involved in using evolutionary ideas in social, psychological and cultural environments only loosely related to biological settings. Since this is an area of considerable intellectual as well as practical importance, a good part of Chapter 7 will be devoted to the exploration of this theme. In addition, the reader is urged to consider carefully the Discussion Questions and Problems, where many additional aspects of evolutionary theory and cooperation are considered.

Exercises

1. In the text we saw that if w, the shadow of the future, is large enough, then no mutant strategy can invade a population of TIT FOR TAT players. This critical value of w depends on the payoff parameters P, R, S and T. Show that if w falls below this critical level, it then pays to defect on alternate moves.

2. Prove that any strategy that may be the first to cooperate can be an ESS only if w is large enough.

3. Use the payoff structure given in the text for one of the other mixed-motive games, like Chicken or Battle of the Sexes, and carry out a computer tournament of the type that Axelrod did for Prisoner's Dilemma. As a basis of comparison, include among your strategies some of the high-flyers in the Prisoner's Dilemma game, such as TIT FOR TAT, ALL D, and ALTERNATE.

Discussion Questions

1. The notion of adaptation to the environment plays a key role in Darwin's picture of evolution, in the sense that the selection criterion is based upon

the ability of a phenotype to survive in a given environment. If we *define* evolution to be a change in genetic frequency and employ the foregoing environmentally-based fitness criterion, are we justified in assuming that evolution will stop in an unchanging, i.e., constant environment?

2. In his book *The Battle for Human Nature*, Barry Schwartz draws striking parallels between the arguments of sociobiology, involving the "selfish gene" and its concern with its own proliferation, the so-called rational economic agent, who behaves so at to maximize utility, and the picture of pleasure maximization by reinforcement as advocated by behavioral psychologists. It's often argued that these three quite disjoint fields—evolutionary biology, economics, and behavioral psychology—provide *scientific* support for the claim that self-interest is a human trait originating in *natural* rather than *moral* law. Consider the pros and cons of this line of argument.

3. The *Red Queen Hypothesis* states that any evolutionary change in one species is experienced by coexisting species as a deterioration of their environment. Hence, a species must evolve as fast as it can in order to continue its existence. And if it doesn't evolve as fast as it can, a species will become extinct. (The terminology comes from the Red Queen in *Alice in Wonderland,* who remarked that "it takes all the running you can do, to keep in the same place.") Thus, the Red Queen Hypothesis argues that evolution is a smooth, on-going process with no jumps or discontinuities. Does this hypothesis seem plausible to you? Imagine that evolution is of the Red Queen type, continuing forever in a physically stable environment. Does it then follow that increasingly more complex forms (defined in some appropriate way) tend to appear?

4. The following payoff matrix represents a simplified version of the situation governing the Cuban Missile Crisis of 1962.

		USSR	
		Withdraw	Maintain
USA	Blockade	(3, 3) *(Compromise)*	(2, 4) *(USSR victory)*
	Air strike	(4, 2) *(US victory)*	(1, 1) *(Nuclear war)*

a) Does this represent a mixed-motive game? If so, what type is it (Chicken, Prisoner's Dilemma, etc.)?

b) What strategies were actually employed to settle the crisis?

c) Construct additional "games" of this sort to represent other types of political conflict. For example, the seemingly interminable Middle East situation, the conflict in Northern Ireland, the Turkish-Cypriot situation, or the Yugoslavian civil war.

d) Referring to the discussions of Chapter 2, try modeling these crises using catastrophe theory representations. Then compare your models and conclusions with those obtained using more conventional game-theoretic approaches.

5. Adam Smith claimed that the individual's pursuit of self-interest in economic transactions promotes the welfare of society as a whole more effectively than by any directed effort to do so. Smith's view was that the "invisible hand" of the economy would set things in order, and that no concerted collaborative effort should be needed. Consider this view in relation to the situation in collective bargaining in which it's in the interest of an individual union to achieve a wage settlement that is in excess of the inflation rate, *irrespective* of whether other unions exercises restraint in their own wage demands. However, if all unions adopt this selfish policy the prices of goods and services go up, with everyone being worse off than if they had all exercised restraint. Thus, the invisible hand becomes the invisible claw that can tear apart entire fabric of the economy. Try to formulate this situation as an *n-player Prisoner's Dilemma*. Can you think of other situations in which the *n*-person Prisoner's Dilemma accurately reflects social conflict?

6. Classical equilibrium-centered economics regards the operation of a firm as a sequence of choices taken to maximize some global objective function characterizing the firm's overall "profitability." This is again Adam Smith's world stepped up from the individual to the level of the firm. Recently, spearheaded by the work of Richard Nelson, Sidney Winter, and others, an alternative *evolutionary* view of the firm has been proposed. In this picture the firm's operation is seen as the execution of a set of "routines" forming the genotype of the firm. The theory assumes that firms make incremental changes (mutations) from time to time in their operating routines, the successful mutations "selected" in accordance with various criteria of profitability or enhanced ability of the firm to compete.

Compare the similarities and differences between evolution in a biological sense and evolution in the economic sense represented in this view of industrial operation. What are the relative strengths and weaknesses between firms operating according to Adam Smith's prescription and those carrying out their activities in an evolutionary mode? Are there any points of contact between the two theories? How would you go about formalizing such an evolutionary theory of economic processes using, say, replicator dynamics?

7. In Oswald Spengler's view of historical change, as well as the cyclic pattern of the rise and fall of civilizations described by Arnold Toynbee,

evolutionary concepts occupy a central position. Recently, Paul Colinvaux has outlined a similar view of historical change, emphasizing the biologically-based idea that the engine of change is the desire by humans to broaden their "eco-niche," defined as a set of capabilities for extracting resources, surviving hazard and competing with others in a given environment. The primary thrust of this view is that human demography can be explained by the process of regulating family size to accord with both perceived niche and available resources. A corollary of this theory is that society is then divided into "castes," in which those of high caste occupy a broader niche than those of lower castes. As population grows, there are only a few ways in which higher-quality niches can be found: technical innovation, trade, colonies, empire building and oppression. Thus, Colinvaux interprets the historical record as the result of group phenomena arising out of individuals acting independently under the dictates of natural selection.

a) Compare this theory of historical change with the ideas underlying the sociobiological view of human behavior.

b) The ecological view of history seems to downplay "great man" theories, which assert that the moving force of historical change is attributable to the doings of a few individuals like Alexander the Great, Napoleon, or Hitler. What role could such great men play in an evolutionary view of history? How would you formalize the role of such dominant figures so as to incorporate their "great acts" into a mathematical model?

c) Is such an ecological view of history *falsifiable;* i.e., can you think of any experiments that could be performed using the historical record that would confirm or deny the evolutionary picture of historical change? (Remark: It was the lack of such experimentally testable hypotheses that led the philosopher of science Sir Karl Popper to condemn the historical views of Marx, Spengler, and others as being nonscientific. See Chapter 10 for a fuller account of these matters.)

8. The term "culture" is generally assumed to include those aspects of thought, speech, behavior and artifacts that can be learned and transmitted. Does the Darwinian equation

$$\text{heredity} + \text{variation} + \text{selection} = \text{adaptation},$$

appear to be relevant to the problem of cultural change? How would you define the analogues of "adaptation," "variation" and "selection" in the cultural context? Do you see any significant differences between the way phenotypes change in a biological sense, and the way that cultural patterns in a society shift over the course of time? Consider changes in language as a specific example, where new words emerge and enter the language of a particular social group while old words become archaic and eventually disappear.

9. The British biologist Richard Dawkins has argued eloquently the position that the basic unit upon which natural selection operates is not the phenotype, but rather the gene itself. This "selfish gene" theory assumes that the body is just a vehicle that the genes commandeer to use as survival machines. Thus, the conventional interpretation of phenotypic fitness as the yardstick of adaptability is only a means to the greater end of the genes propagating themselves into the next generation.

a) Consider the merits of the selfish gene theory as compared with the traditional Darwinian view of the entire organism as the basic unit of selection.

b) How can you explain the emergence of altruism in the selfish gene context?

c) Dawkins also introduced the notion of a "meme" as a unit of cultural transmission that would propagate itself through a "meme pool" in much the same ways that genes propagate themselves through a gene pool, spreading ideas, catchwords, cooking recipes, techniques for making houses, and the like. Thus, the memes act as agents of cultural evolution. Compare this idea of a cultural meme with that of a biological gene, both in structure and in mode of transmission. How would you go about trying to model such a cultural transmission process using replicator dynamics?

10. An idea closely related to Dawkins' memes is the concept of the *sociogene* postulated by Carl Swanson. He regards both a biological gene and a sociogene as carriers of information.

a) If we think of the information content of a gene as being coded into the DNA, how could the cultural information of a sociogene be carried?

b) Information has many attributes: sources, replication, transmission, compression, mutation, and so on. Compare these attributes for both "biogenes" and sociogenes. Are there any significant differences?

c) In a viral infection, the invading virus parasitizes the genetic machinery of the host cell, inducing the cell to make copies of the virus rather than copies of the cell itself. Does the spread of a "fad" like the hula hoop, designer jeans or Madonna videos correspond to a kind of cultural infection of this sort? If so, how does the virus spread? And what would it take to get rid of the "bug"?

11. In earlier chapters we've considered a number of alternative formulations for measuring the *complexity* of a dynamical process. It's often claimed that the very nature of evolution requires that replicators evolve from simpler to more complex forms or face extinction. How could you formulate a research program for testing this hypothesis using the replicator equations of the text? Do you think it makes any difference which definition of complexity you use? Does the answer to the question depend upon

whether you're characterizing evolutionary processes at the biological, cultural or historical level? Do you think any formal theory will discriminate between the complexity of a human and an ape? Or a whale? Or a bird?

12. Evolution is an historical process, a unique sequence of events. Theories about evolution come in two types:

A) General theories that say something about the mechanisms that underlie the whole process.

B) Specific theories that account for particular events.

It's often held that theories of Type B are untestable, hence unscientific, since it is impossible to run the historical process again with only a single factor changed to see whether the end result is different. Such theories are unfalsifiable in a Popperian sense. Do you think a strict adherence to the criterion of falsifiability is a valid way to assess the merits of a specific evolutionary theory (or model)? Can you think of any simple models of evolutionary behavior that are heuristically valuable, yet untestable? Is testability only relevant when we want to use the models to say something about a real-world situation?

13. Consider the following statements about altruism:

I. Within each group of individuals, altruists are at a reproductive disadvantage compared with nonaltruists.

II. In the ensemble of groups (within which a group selection process acts), whether altruists have a higher average fitness than nonaltruists is a contingent empirical matter not settled by statement A.

These two statements appear to be contradictory. How can you reconcile them? (Hint: Consider the distinction between two events being *causally related* as opposed to their being *correlated*.)

14. Darwinian evolution predicts a continuous change of phenotypic form as a result of genetic mutations and natural selection; the fossil record, on the other hand, shows strong evidence of discontinuities in the pathway from ancient forms to those seen today. The "punctuated equilibrium" theory of evolutionary change postulates that evolution moves in fits and spurts, a far cry from the steady, if boring, pace of the Darwinian picture. If the "punctuationist" vision is correct, what factors can you identify that would account for the long periods of stasis, followed by the rapid appearance of new species? Do you think that mass extinctions, like the one that presumably killed off the dinosaurs 65 million years ago, form an integral part of the punctuationist view? Do you think that catastrophe theory could be used to formalize the main features of the punctuated equilibrium view of evolutionary change?

15. In most evolutionary models, the selection mechanisms are based upon optimization principles of one sort or another. Mathematically, all optimization problems involve the same set of basic components:

A) A state set consisting of the variables of the problem.

B) A strategy set describing the admissible courses of action that can be taken.

C) A constraint set specifying various physical and mathematical limitations on the strategies that can be pursued.

D) A performance function by which to measure the "goodness" (or fitness) of any particular strategy.

In evolutionary models there are serious difficulties in specifying completely any of these components. Discuss these difficulties within the specific context of some of the examples given in the text. Under what circumstances do you think optimization models would be useful in analyzing evolutionary change?

16. Conventional Darwinian evolution is a fundamentally *reductionistic* theory, the principal thesis being that atomistic changes at the genetic level somehow percolate upward to the phenotypic form, at which point the inexorable pressures of natural selection can work to separate the winners from the losers in the gene pool. Recently, Rupert Sheldrake has postulated a more *holistic* process for the emergence of physical form via the medium of what the geneticist Conrad Waddington has termed a *morphogenetic field.*

Basically, the idea is that there exist fields of potential for biological form, just as there are fields for other physical processes like gravity and electromagnetism. Sheldrake claims that these fields shape the forms of both living and nonliving things, the initial form emerging more or less by chance (Sheldrake's theory is a little vague on this point). But once a particular form has gained a foothold, those forms that have been successful in the past have a greater chance of being chosen in the future for organisms of the same species. Such primal forms are called "morphogenetic fields," and bear the same role to the physical form of biological organisms that, say, the electromagnetic field bears to the physical structure of an electron.

a) Can you think of any experimental tests that might be performed to test Sheldrake's theory?

b) In physics we have equations like Maxwell's field equations, the Schrödinger equation and Einstein's equations to describe various fields of importance. Each of these equations implicitly assumes that the fields they describe arise out of material entities like electrons and other sorts of elementary particles. Assuming Sheldrake's theory is correct, it would follow that there should exist analogous field equations describing how biological matter would physically organize itself into various shapes. What do you think the mathematical structure of such equations might be?

c) If morphogenetic fields do indeed exist, how would one go about measuring them?

d) Do you think there could be morphogenetic fields for such nonphysical structures as the information patterns seen in languages, fashions, and tunes?

17. The French ethnologist Claude Lévi-Strauss has distinguished between two basic forms of human societies, the "clockworks" and the "steam engines." The clockwork societies live practically historyless in a sociocultural equilibrium without the evolution of new structures. The steam-engine societies, by way of contrast, undergo vivid evolution of the sort seen in most modern industrialized countries. According to Lévi-Strauss, the distinction between them is due to writing. Explain how you think writing could have come about as an evolutionarily adaptive trait. Within the framework of your explanation, discuss why all societies haven't adopted written language. Does conventional evolutionary theory give a credible account of this phenomena?

18. Many species have approximately equal numbers of males and females, although they could produce just as many offspring per generation if there were fewer males, who each fertilized several females. Why are there so many males?

19. In his fascinating book *The Evolution of Cooperation*, Robert Axelrod gives the following precepts for improving cooperation in iterated Prisoner's Dilemma situations:

• *Enlarge the Shadow of the Future*—Make sure the future is sufficiently important relative to the present.

• *Change the Payoffs*—Rearrange payoffs so as to decrease the incentives for double-crossing your partner.

• *Teach People to Care about Each Other*—Cooperation can begin, even in a hostile environment, if a sufficient degree of altruism is present.

• *Teach Reciprocity*—Insist that defectors be punished and cooperators rewarded.

• *Improve Recognition Abilities*—Teach techniques for recognizing players that you have interacted with in the past, so that it will be possible to employ various types of reciprocity.

a) Discuss how you would apply these principles to the Arms Race situation involving the United States and the USSR. Do you think this sort of analysis would be appropriate for the Cuban Missile Crisis discussed earlier in Discussion Question 4?

b) In America's early days the members of the Congress were known to be quite unscrupulous and deceitful. Yet over the years cooperation developed and proved stable. Explain how this could have come about using the principles of "cooperation" theory.

20. The action of natural selection has sometimes been compared to that of an engineer, in the following senses: (1) it works according to a preconceived plan; (2) nature has at its disposal both material and machinery specially prepared for production of the desired end product, and (3) the objects produced approach the level of perfection made possible by the technology of the time. By way of contrast, others, including Darwin, have argued that evolution is far from perfection. Rather, they say, natural selection acts more like a backyard tinkerer, managing to produce something usable in a number of different ways from odds and ends that is related to no special product and results from a series of contingent events. Consider the pros and cons of these two views of evolution.

21. The optimization approach toward analyzing evolutionary adaptation has been criticized by people like Stephen Jay Gould and Richard Lewontin, who argue that much of molecular variation is selectively neutral and that unpredictable events have had a major effect on evolution. These critiques are based on the assumption that optimization advocates believe that animals and plants are optimally adapted, or at least that the goal of evolution is to make them so. Does this argument against models based on optimization-theoretic ideas seem plausible to you? In particular, if selection can produce optimally-adapted organisms, why do we see so much natural variability within a given population?

22. Economist Jack Hirschleifer has observed that Shakespeare's play *King Lear* offers a good lesson in the potential limitation of the model of economic altruism discussed in the text. In that model, the paternal altruist (Big Daddy) and the selfish offspring (Rotten Kid) can affect the joint production opportunities determining the level of family income, termed S_A in the text. After the completion of all interactions, Big Daddy will be making unilateral altruistic transfers t_A to Rotten Kid. So, however selfish Rotten Kid really is, it's in his interest to help Big Daddy maximize the family income. By appeal to Shakespeare or otherwise, discuss the potential flaw in this line of reasoning. (Hint: Consider the case in which Big Daddy doesn't have the last word, i.e., does not control the last action taken in the temporal sequence of interactions with Rotten Kid.)

Problems

1. Consider the two-person, zero-sum game with payoff matrix $A = (a_{ij})$, i.e., if Player A uses strategy i and B uses strategy j, the payoff to A is a_{ij}, while the payoff to B is $-a_{ij}$, $i = 1, 2, \ldots, n; j = 1, 2, \ldots, m$.

a) Let A's mixed strategy be given by $x = (x_1, x_2, \ldots, x_n)$, i.e., x_i is the probability that A plays pure strategy i, $\sum_i x_i = 1$. Show that if all $a_{ij} \geq 0$, x can be determined as the solution to the *linear programming (LP)* problem

$$\min \sum_{i=1}^{n} x_i,$$

subject to the constraints

$$\sum_{i=1}^{n} a_{ij} x_i \geq 1, \qquad x_j \geq 0, \qquad j = 1, 2, \ldots, m.$$

b) If some $a_{ij} < 0$, show how to modify the payoff matrix to obtain the same reduction to an LP problem.

c) What LP formulation could you use to determine B's optimal mixed strategy?

d) Can this solution procedure work for determining the ESS for mixed-motive games? That is, can we also reduce the determination of the solution to a nonzero-sum game to a special linear programming problem?

2. In the text we have considered only continuous-time replicator dynamics.

a) Show that if replicators change in *discrete-time,* the corresponding dynamical equations are

$$x_i' = x_i \left[\frac{f_i(x)}{\Phi} \right],$$

where the normalization term Φ is given by

$$\Phi = \sum_{i=1}^{n} x_i f_i(x),$$

the same as in the continuous-time case.

b) In discrete-time, the genetic model discussed in the text is

$$x_{ij}' = \frac{1}{2\phi} \sum_{s,t} [w(is, jt) + w(jt, is)]\, x_{is} x_{jt},$$

for the frequency of genotype $A_i A_j$ in the next generation. Assuming multiplicative fecundity and symmetry between the sexes, we have

$$w(ij, st) = m(ij)m(st).$$

Letting

$$M(i) = \sum_j m(ij)x_{ij},$$

we obtain the dynamics

$$x'_{ij} = \frac{1}{\phi} M(i)M(j),$$

with

$$\phi = \left(\sum_i M(i)\right)^2.$$

If $x'_i \doteq \sum_j x'_{ij}$, show that $x'_{ij} = x'_i x'_j$. Hence, show that after one generation the mean fecundity ϕ increases. Is this result also true for the case of continuous-time dynamics?

c) How would you use the above result to quantitatively investigate the Red Queen Hypothesis considered in Discussion Question 3?

3. The classical Lotka-Volterra equation describing the predator-prey interactions among n species is given by

$$\dot{y}_i = y_i \left(c_i + \sum_j b_{ij}y_j \right), \qquad i = 1, 2, \ldots, n.$$

Such equations "live" on R^n_+ and generally do not satisfy replicator equations. However, show that by setting $y_n \equiv 1$, the *relative* densities, defined by the barycentric transformation

$$x_i = \frac{y_i}{\sum_{j=1}^n y_j}, \qquad i = 1, 2, \ldots, n-1,$$

do satisfy the replicator dynamics

$$\dot{x}_i = x_i[(Ax)_i - \Phi], \qquad x \in S^n - \{x : x_{n-1} = 0\}, \qquad i = 1, 2, \ldots, n-1,$$

where the elements of the matrix A are $a_{ij} = b_{ij} - c_i$.

4. Let S^n be the unit simplex in R^n and let $p \in$ int S^n. The *Shahshahani metric* on S^n is defined via the relation

$$< x, y >_p = \sum_i \frac{1}{p_i} x_i y_i.$$

a) Show that the product defined above satisfies the conditions for an inner product, i.e., $< x, x >_p \geq 0$, with equality only for $x = 0$, $< x, y >_p = < y, x >_p$ and $< x, y >_p \leq < x, z >_p + < z, y >_p$ (the triangle inequality).

b) Let $V: R^n \to R$ be a smooth function. Define the *Shahshahani gradient* of V to be

$$< \text{grad } V(p), y >_p \doteq DV(p)y,$$

for all $y \in T_p S^n$, the tangent space to S^n at p. Here $DV(p)$ is the derivative of V at p. Using the fact that $y \in T_p S^n$ if and only if $y \in R^n$ satisfies $\sum_i \dot{y}_i = 0$, show that the replicator equation

$$\dot{x}_i = x_i[f_i(x) - \Phi], \qquad i = 1, 2, \ldots, n,$$

is a Shahshahani gradient of V if and only if f is equivalent to grad V, in the sense that there exists a function $c: S^n \to R$ such that

$$f_i(x) - (\text{grad } V)_i(x) = c(x)$$

for all $x \in S^n$ and all $1 \leq i \leq n$.

c) Show that when $V = \frac{1}{2}\sum_{i,j} a_{ij} x_i x_j = \frac{1}{2}(x, Ax)$, with A symmetric, the Shahshahani gradient leads to the replicator equations

$$\dot{x}_i = x_i[(Ax)_i - \Phi], \qquad i = 1, 2, \ldots, n.$$

On the other hand, prove that if $V = \sum a_i x_i$, the Shahshahani gradient leads to the equations for haploid organisms

$$\dot{x}_i = x_i(a_i - \Phi), \qquad i = 1, 2, \ldots, n,$$

where x_i is the frequency of chromosome G_i and a_i is its fitness. In other words, the above replicator systems are *gradient* dynamical systems with respect to the Shahshahani metric.

d) Can the bifurcation behavior of the equilibria of the above gradient systems be analyzed using the tools of elementary catastrophe theory developed in Chapter 2?

e) Consider the case when V is a homogeneous function of degree s, i.e., $V(sx) = sV(x)$. Show that in this case $\Phi(x) = sV(x)$ and the average fitness Φ grows at the fastest possible rate, the orbits of the dynamics being orthogonal (in the sense of the Shahshahani inner product) to the constant level sets of Φ.

f) Prove that the first-order replicator equations

$$\dot{x}_i = x_i[(Ax)_i - \Phi], \qquad i = 1, 2, \ldots, n,$$

are obtainable from the Shahshahani gradient of a scalar function V if and only if

$$a_{ij} + a_{jk} + a_{ki} = a_{ji} + a_{ik} + a_{kj},$$

for all indices i, j and k. Show that this is equivalent to the requirement that there exist constants c_i such that $a_{ij} = a_{ji} + c_i - c_j$, $i, j = 1, 2, \ldots, n$.

5. Consider a replicator system of the form

$$\dot{x}_i = x_i[g_i(x_i) - \Phi], \qquad i = 1, 2, \ldots, n,$$

where, without loss of generality, we assume that

$$g_1(0) \geq g_2(0) \geq \cdots \geq g_n(0) > 0.$$

a) Show that the above system is a Shahshahani gradient system.

b) Prove that if the $g_i(\cdot)$ are monotonically decreasing, then there exists a number $K > 0$ and a point $p \in S^n$ such that

$$g_1(p_1) = \cdots = g_m(p_m) = K,$$
$$p_1 > 0, \ldots, p_m > 0, \, p_r = 0, \qquad r > m,$$

where m is the largest integer such that $g_m(0) > K$. Show that this implies the system has a unique, global attractor.

c) Prove that the coordinates of the point $p = (p_1, p_2, \ldots, p_n)$ satisfy the relations

$$\lim_{t \to \infty} x_i(t) = p_i, \qquad i = 1, 2, \ldots, n.$$

6. Consider a collection of N species forming a closed ecosystem. Let $\widehat{W}_i$ be the maximal possible fitness of species i in the current environment if the species had all possible favorable genetic alleles, and let $\overline{W}_i$ be the current average fitness of species i. Define the *evolutionary lag load* of species i as

$$L_i = \frac{\widehat{W}_i - \overline{W}_i}{\widehat{W}_i}, \qquad i = 1, 2, \ldots, N.$$

Let β_{ij} represent the increase in the lag load of species i due to a unit change in the lag load of species j.

a) If the average lag load for the ecosystem is given by $\overline{L} = (1/N) \sum_i L_i$, show that

$$\frac{d\overline{L}}{dt} = \frac{1}{N} \left[\sum_j (L_j \sum_i \beta_{ij}) - \sum_j L_j \right].$$

Show that this equation has a stationary equilibrium point only if $\sum_i \beta_{ij} \equiv 1$ for all j; otherwise $\overline{L}$ will either decrease (*convergent* evolution) or increase (*divergent* evolution), depending upon whether $\sum_i \beta_{ij} < 1$ (resp. > 1) for most j. (Remark: This result seems to cast doubt upon the plausibility of the Red Queen Hypothesis, since it implies that there is only a special set of values β_{ij} for which the average lag load remains constant, i.e., for which there is no overall evolutionary change.)

b) Show that the above conclusion is based upon the dubious assumption that the coefficients β_{ij} are constants, *independent* of the number of species N.

c) Consider the number of species N to be a variable, depending upon immigration, speciation, extinction, and so forth. Let the following model describe the change in average lag load and in number of species:

$$\frac{d\overline{L}}{dt} = (a + b\overline{L} + cN)\overline{L},$$

$$\frac{dN}{dt} = h + (d - e)\overline{L} + (f - g)N,$$

where we assume on physical grounds that $b < 0$, $h > 0$ and $(f - g) < 0$ (making no assumptions about the signs of a, c and $(d - e)$). Show that by suitable selection of the coefficients there can be either a steady Red Queen type of continued evolution, or a stationary state without any evolutionary change. The first corresponds to a gradualistic pattern of Darwinian type, whereas the second represents a pattern of punctuated evolution.

7. Consider the childhood game of "Rock–Scissors–Paper" (R-S-P) with the payoff matrix

$$
\begin{array}{c}
\begin{array}{ccc} R & S & P \end{array} \\
\begin{array}{c} R \\ S \\ P \end{array}
\left(
\begin{array}{ccc}
-\epsilon & 1 & -1 \\
-1 & -\epsilon & 1 \\
1 & -1 & -\epsilon
\end{array}
\right).
\end{array}
$$

Show that if $\epsilon > 0$ (i.e., there is a small positive payment to the bank for a draw), the mixed strategy $(\frac{1}{3}, \frac{1}{3}, \frac{1}{3})$ is ESS, but if $\epsilon < 0$ there is no ESS. What about the usual case when $\epsilon = 0$?

8. We can "soup up" the Hawk-Dove game of the text by adding a third strategy RETALIATOR, which behaves like a DOVE against DOVE. But if

its opponent escalates, RETALIATOR then turns into a HAWK. Suppose the payoff matrix for such a Hawk-Dove-Retaliator game is

$$
\begin{array}{c c}
 & \begin{array}{c c c} H & \quad D & \quad R \end{array} \\
\begin{array}{c} H \\ D \\ R \end{array} &
\left(\begin{array}{c c c}
-1 & 2 & -1 \\
0 & 1 & 0.9 \\
-1 & 1.1 & 1
\end{array} \right)
\end{array}
$$

a) Show that $(0,0,1)$, i.e., always play RETALIATOR, is an ESS.

b) If RETALIATOR is a strategy that is a "compromise" between HAWK and DOVE, is the strategy $(\frac{1}{2}, \frac{1}{2}, 0)$, i.e., "HALF HAWK, HALF DOVE" also an ESS?

c) Show that, in general, if any entry on the diagonal of a payoff matrix is greater than any entry in the same column, the corresponding pure strategy is ESS (as it is for RETALIATOR in this problem).

9. In *asymmetric* contests, the two participants have different roles to play that may affect both their actions and their payoffs. For example, in the Hawk-Dove game each contest may be between the owner of a property and an intruder, with each participant knowing beforehand the role he or she will play. In such situations, Reinhard Selten has proved that there can be no mixed strategy that is ESS.

a) Can you prove Selten's result? (Remark: It *is* possible for a mixed strategy I to be *neutrally* stable, in the sense that I is as good as a pure strategy.)

b) Show that Selten's Theorem does not hold if the roles of the two participants are not known in advance.

10. Let I be a mixed ESS with *support* $A, B, C, \ldots$. Here the support is simply the strategies that are played with nonzero probabilities in I. Let $E(X, I)$ be the expected return from playing pure strategy X against the mixed ESS strategy I.

a) Prove the Bishop-Canning Theorem that

$$
E(A, I) = E(B, I) = \cdots = E(I, I).
$$

b) Use the Bishop-Canning Theorem to show that the payoff matrix

$$
\begin{array}{c c}
 & \begin{array}{c c} P & \quad Q \end{array} \\
\begin{array}{c} P \\ Q \end{array} &
\left(\begin{array}{c c}
a & b \\
c & d
\end{array} \right)
\end{array}
$$

admits an ESS if $a < c$ and $d < b$. In such a case, prove that the ESS says to play strategy P with probability

$$
p = \frac{b - d}{(b + c - a - d)}.
$$

c) Apply this result to calculate the ESS for the Hawk-Dove game of the text.

11. Consider a situation in which food or some other resource is patchily distributed, with exploitation of a patch yielding diminishing returns. Assume that when one patch is abandoned, appreciable time is needed to find or travel to the next one. Further, assume that the foragers search the various patches randomly. Prove the Marginal Value Theorem stating that the overall benefit is maximized by exploiting each patch until the rate of benefit falls to the maximum *mean* rate that can be sustained over a long period.

As a concrete illustration of the Marginal Value Theorem, consider the feeding pattern of ladybird larvae. Such larvae feed on aphids, eating the soft tissue and leaving behind the exoskeleton. Assume that a larvae spends an amount of time t feeding on each aphid, and that the mass of food extracted in this time is $m(t)$. Let T be the average time required to find a new aphid after leaving a partially eaten one. The mean rate of food intake is then

$$Q = \frac{m(t)}{T + t}.$$

The Marginal Value Theorem now tells us that the optimal value of t is that which maximizes Q. Illustrate this result graphically.

12. In the text we've considered only *symmetric* contests, in which each participant was assumed to have the same strength and the same strategies and payoffs. In actual contests this condition is seldom satisfied, and one has to deal with *asymmetric* conflicts—owner and intruder, male and female, and so forth.

Let $E_1, E_2, \ldots, E_n$ be the strategies available to Player I and $F_1, \ldots, F_m$ those available to Player II. If an E_i-strategist meets an F_j-strategist, the payoff to the E_i player is a_{ij}, whereas the opponent then receives the amount b_{ji}, $1 \le i \le n$, $1 \le j \le m$.

a) Show that the replicator dynamics for such asymmetric contests are given by

$$\dot{x}_i = x_i[(Ay)_i - \phi], \qquad i = 1, 2, \ldots, n,$$
$$\dot{y}_j = y_j[(Bx)_j - \psi], \qquad j = 1, 2, \ldots, m.$$

On what space does this dynamic unfold?

b) An evolutionarily stable equilibrium is now given by a pair of points (p, q), $p \in S^n$, $q \in S^m$. What are the conditions for (p, q) to be evolutionary stable?

c) Prove that a mixed equilibrium (p, q) can never be a sink; i.e., the characteristic values of the linearized system at (p, q) cannot all have negative real parts. Thus, show that there can never exist a mixed ESS. (Remark: This result proves Selten's Theorem stated earlier in Problem 9.)

d) What are the corresponding discrete-time replicator dynamics for asymmetric contests?

13. In developing the ESS, we assumed that each individual in a population plays a *pure* strategy. Under this assumption, the ESS determines only the steady-state fraction of the population that adopts one or another of the pure strategies. Now consider the more general case in which *individuals* can play mixed strategies. Suppose that an individual plays strategy i with probability $p_i, i = 1, 2, \ldots, n$. Thus, an individual is represented by a point $p \in S^n$, and the population is represented by a *distribution function* f on S^n such that

$$\int df = \int_{S^n} f(p)\, dp = 1.$$

a) Show that the *mean* of f (the probable strategy of an opponent) is given by

$$x = \int pf(p)\, dp,$$

and that the *covariance matrix* of the population f is given as

$$F = \int (p - x)(p - x)'\, df(p).$$

b) Assume, as before, that the growth rate of a strategy equals its differential advantage. Show that this means

$$\frac{df(p)}{dt} = f(p)[(p, Ax) - (x, Ax)],$$

where x is the mean defined above and A is the payoff matrix of the game.

c) Prove that the mean changes according to the dynamical equation

$$\dot{x} = FAx.$$

d) Use the above results to prove that if a game has an ESS at a vertex or an edge of the simplex S^n, then it is an attractor both for populations playing pure strategies and for the means of populations playing mixed strategies.

14. **War of Attrition** games are contests between pairs of animals competing for a prize (a mate, territory, or whatever), in which each competitor engages in display behavior as a way of inhibiting attempts by its opponent to take possession of the prize. In such games one of the combatants eventually departs the playing field, leaving the prize to the one who held out the longest.

Assume the cost of display is proportional to its duration and that the value of the prize is $V > 0$. Further, assume the competitors choose strategies x and y, respectively, where x denotes "wait an amount of time x," with y defined similarly. Let the payoffs of the game be given by

$$E[x, y] = \begin{cases} V - y, & \text{if } x > y, \\ \frac{1}{2}V - x, & \text{if } x = y, \\ -x, & \text{if } x < y. \end{cases}$$

a) Show that neither player has an optimal pure strategy by showing that no strategy x is best against itself.

b) Define a mixed strategy via a frequency density function $p(x)$ of a random variable X. That is, the mixed strategy is to wait an amount of time x, where x is determined as the realization of a random variable X having probability density function $p(x)$. Prove that the exponential density function

$$p(x) = \frac{\exp(-x/V)}{V}, \qquad x \geq 0,$$

is an ESS against any pure strategy for the War of Attrition by establishing the following relations:

$$E[x, p(y)] = 0 \text{ for all strategies } x,$$

and

$$E[p(x), y] > E[y, y] \text{ for all } y.$$

The first condition means that all strategies x (and their mixtures) are equally good against $p(y)$, while the second says that $p(x)$ is a better reply against y than y itself, for all strategies y.

c) The results of part (b) show that $p(x)$ is ESS against any *pure* strategy y. Prove that the same result also holds for any *mixed* strategy y.

15. Consider the following model for a mutation and selection process. Let there be one gene locus with alleles $A_1, \ldots, A_n$ and let $x_1, \ldots, x_n$ be their relative frequencies in the gene pool of the population. Due to natural selection, only a fraction $w_{ij} x_i x_j$ of the gametes (genotype $A_i A_j$) will survive to procreative age, where $w_{ij} = w_{ji} \geq 0$ are the *fitness parameters* of the population. Let ϵ_{ij} be the *mutation rate* from A_j to A_i, $(i \neq j)$, where $\epsilon_{ij} \geq 0$, $\sum_{i=1}^{n} \epsilon_{ij} = 1$, for all $j = 1, 2, \ldots, n$.

a) Prove that the frequency x_i' of genotype A_i in the next generation is given by

$$x_i' = \sum_{j=1}^{n} \frac{\epsilon_{ij} x_j (Wx)_j}{W(x)},$$

where $W(x) = (x, Wx)$.

b) Show that the continuous-time version of the above result is

$$\dot{x}_i = W(x)^{-1} \sum_{j,k} \epsilon_{ij} w_{jk} x_j x_k - x_i, \qquad i = 1, 2, \ldots, n.$$

c) Suppose the model without mutation admits a stable interior equilibrium. Under the conditions of parts (a) and (b), prove that for mutation rates of the form $\epsilon_{ij} = \epsilon_i$, $i \neq j$, the differential system has a globally stable interior equilibrium, while the difference equation has an equilibrium that is at least locally stable.

d) Show that the differential system of part (b) forms a gradient system using the Shahshahani gradient for the function

$$V(x) = \frac{1-\epsilon}{2} \log W(x) + \sum_{i=1}^{n} \epsilon_i \log x_i,$$

where $\epsilon = \sum_{j=1}^{n} \epsilon_j$. Thus, conclude that the change in gene frequency occurs so as to maximize the increase in mean fitness.

16. By a *learning rule,* we mean a rule specifying which of a set of possible actions A, B, ... an animal will perform on any occasion given its previous experiences. A "rule for ESS's " is a rule such that if all the members of a population adopt it, they will in time come to adopt the ESS for the game in question.

a) Show that the rule

$$\text{probability of A} = \frac{\text{total payoff received so far for doing A}}{\text{total payoff received so far for all actions}},$$

is an ESS learning rule.

b) Suppose different individuals in a population adopt different learning rules and reproduce their kind. Further, suppose the number of offspring are proportional to the payoff an animal accumulates by playing games against others in the population. If after a time a given learning rule evolves for the population, we call it an "evolutionary stable learning rule," or ES. Note that an ES and a rule for ESS's are two quite different concepts: one is the learning rule we expect to evolve; the other is a rule that takes the population to the ESS of a game. Nevertheless, show that an ES rule is necessarily a rule for ESS's.

17. In the iterated Prisoner's Dilemma, we say a strategy A *invades* a strategy B if A receives a higher score against B than B gets against itself. The strategy B is called *collectively stable* if B cannot be invaded.

a) Prove the following theorem:

CHARACTERIZATION THEOREM. *A strategy B is collectively stable if and only if B defects on move n whenever the opponent's score so far is sufficiently large.*

b) Show that for a nice strategy to be collectively stable, it must defect upon the very first defection of the other player.

c) We say that a strategy A *territorially invades* B if every location in the territory will eventually convert to A. B is called *territorially stable* if no strategy can territorially invade it. Prove that if a strategy is collectively stable, then it is also territorially stable. As a result, protection from invasion is at least as easy in a territorial system as in a freely mixing system. Hence, mutual cooperation can be sustained in a territorial system at least as easily as in a freely mixing system.

18. Consider a population playing the iterated Prisoner's Dilemma with the strategy TIT FOR TAT.

a) Prove that if the mutant strategies ALL D and ALTERNATE (i.e., play DCDCDC...) cannot invade (displace) TIT FOR TAT, then *no* mutant strategy can invade the population.

b) Prove that neither ALL D nor ALTERNATE can invade if the probability of further interaction w is large enough.

c) Show that if a nice strategy cannot be invaded by a single individual, then it cannot be invaded by any cluster of individuals either.

19. Let G be a non-oriented graph on the vertices $I = \{1, 2, \ldots, n\}$. Form the interaction matrix A of a replicator system by letting its entries be given by

$$A_{ij} = \begin{cases} 1, & \text{if there is an arc connecting vertices } i \text{ and } j \text{ in } G, i \neq j, \\ 0, & \text{if vertices } i \text{ and } j \text{ are not connected}, \\ \frac{1}{2}, & \text{if } i = j. \end{cases}$$

Such a matrix A is usually termed an *incidence matrix*. Define a *clique* in G to be a maximal subset of I such that any two elements in I are connected by an arc.

a) Show that the cliques of G correspond to the asymptotically stable equilibria of the replicator system having matrix A.

b) Suppose that for an incidence matrix A there exists an asymptotically stable equilibrium p involving $n - i$ alleles (i.e., the corresponding clique contains $n - i$ elements). Prove that there can then be at most 2^i asymptotically stable equilibria.

20. Consider the three-dimensional Lotka-Volterra system

$$\dot{x}_i = x_i \left(c_i + \sum_{j=1}^{3} a_{ij} x_j \right), \qquad i = 1, 2, 3,$$

with intraspecies competition $(a_{ii} < 0)$. Prove that this system is perseverant if and only if the following conditions hold:

i. There exists an interior equilibrium point x^*.

ii. $\det(-A) < 0$.

iii. If the system without species k admits a unique interior equilibrium, the corresponding principal minor of $-A$ is positive (i.e., the minor formed from $-A$ by deleting row and column k).

21. Prove that for the replicator dynamics, there exists an interaction matrix A for which the system admits a Hopf bifurcation if $n = 4$. (Hint: Consider the matrix

$$A = \begin{pmatrix} 0 & 0 & -\alpha & 1 \\ 1 & 0 & 0 & -\alpha \\ -\alpha & 1 & 0 & 0 \\ 0 & -\alpha & 1 & 0 \end{pmatrix},$$

where α is a real parameter.)

22. Consider the following assumptions about a group of n teams engaged in a sports draft of the type practiced by professional football, basketball and baseball teams in the United States:

- *Strict preferences and partial ordering*—each team has a strict preference ordering on the players, which induces a partial ordering on any subset of k players it might receive in the draft. This partial ordering is by pairwise comparison. So, for instance, if team X has a preference ordering 123456 on six players, then the subset $\{1, 4, 5\}$ will be preferred to $\{2, 4, 6\}$. But the subsets $\{1, 2, 6\}$ and $\{2, 3, 4\}$ will be incomparable in X's judgment.

- *Self-interest*—Each team's goal is to benefit itself, not to hurt other teams. Thus, no team will select a player solely on the grounds of denying that player to another team.

- *Independence*—Each team acts independently of the others. In other words, there are no coalitions or hidden agreements.

- *Complete information*—Each team knows the preference orderings of the other teams.

Define a team's choice of a player to be *sincere* if among the players remaining to be selected, that player is the highest one on the team's preference list.

a) Assume there are only two teams. Consider the following algorithm due to Kohler and Chandrasekaran for allocating the players between Teams A and B:

Kohler-Chandrasekaran Algorithm

1. Team B's last choice will be the player ranked last in Team A's preference ordering.

2. Cross that player off both preference lists.

3. Team A's last choice is now the player ranked last on Team B's reduced list.

4. Cross that player off both reduced lists.

5. Continue in this fashion until all players have been chosen.

Prove that the Kohler-Chandrasekaran algorithm leads to a Pareto optimal allocation A with respect to pairwise comparison. That is, there is no different allocation A' of players such that each team either gets the same players under both A and A', or prefers the A' to A under pairwise comparison. (Remark: This result shows that no Prisoner's Dilemma situation can arise whereby both teams suffer if they play optimally.)

b) Prove that for any number of teams, the sincere outcome is always Pareto optimal with regard to pairwise comparison. That is, if each team always chooses what they regard as the highest-ranking player left in the pool, the final outcome will be Pareto optimal.

c) Prove that if there are three or more teams, optimal "sophisticated" play, i.e., not necessarily sincere, may lead to an outcome that is not Pareto optimal by pairwise comparison. Even more strongly, show by example that the outcome may be strictly worse than the outcome each team would receive if they chose sincerely. In short, show that if each team follows an individually optimal choice strategy, the final result can be a Prisoner's Dilemma situation in which **each** team is worse off than if they had all chosen sincerely. (Remark: This shows that a team may do better for itself by occupying a later position in the draft.)

23. Consider the three-player Hawk-Dove-Bully (H–D–B) game, whose payoff matrix is given by

$$A = \begin{array}{c} \\ H \\ D \\ B \end{array} \begin{pmatrix} \begin{array}{ccc} H & D & B \end{array} \\ \begin{array}{ccc} 0 & 4 & 3 \\ 2 & 0 & -3 \\ 2 & 4 & 0 \end{array} \end{pmatrix}.$$

(a) Demonstrate that the corresponding replicator system has the flow pattern (on the H–D–B 2-simplex) displayed below. (b) Show that the point Q has the coordinates $Q = (\frac{3}{5}, \frac{2}{5})$.

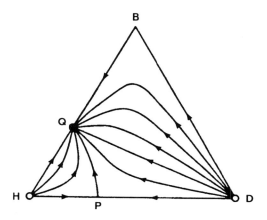

Flow Pattern for the Hawk-Dove-Bully Game

Notes and References

§1. The philosophical underpinnings of evolutionary epistemology have far broader currency than just to support the way biological organisms change in response to their environment. For introductory accounts of evolution as a conceptual theme in the philosophy of science, see the volumes

Wuketits, F., ed., *Concepts and Approaches in Evolutionary Epistemology,* Reidel, Dordrecht, 1984,

Popper, K. R., *Objective Knowledge: An Evolutionary Approach,* Clarendon Press, Oxford, 1972,

Jantsch, E., *Design for Evolution: Self-Organization and Planning in the Life of Human Systems,* Braziller, New York, 1975,

Hull, D., *Science as a Process,* University of Chicago Press, Chicago, 1988.

For those interested in the long-running, always entertaining battle being waged between the lab and the pulpit over the account of life given in the book of Genesis, we recommend

Science and Creationism, A. Montagu, ed., Oxford University Press, Oxford, 1984,

Scientists Confront Creationism, L. Godfrey, ed., Norton, New York, 1983,

Ruse, M., *Sociobiology: Sense or Nonsense?,* 2nd Ed., Reidel, Dordrecht, 1985,

The Sociobiology Debate, A. Caplan, ed., Harper & Row, New York, 1978.

§2. For an easily digestible, nonmathematical overview of both the Darwinian and neo-Darwinian theories of evolution, see

Arthur, W., *Theories of Life,* Penguin, London, 1987,

Dawkins, R., *The Blind Watchmaker,* Longman, London, 1986.

More technical expositions are available in

Mayr, E., *The Growth of Biological Thought,* Harvard University Press, Cambridge, MA, 1982,

Smith, J. Maynard, *The Theory of Evolution,* Penguin, London, 1975.

A critical account of the entire issue of fitness and selection is found in

Sober, E., *The Nature of Selection: Evolutionary Theory in Philosophical Focus,* MIT Press, Cambridge, MA, 1984.

For an account of Weissman's theory of the "germ plasm," see

Webster, G., and B. C. Goodwin, "The Origin of Species: A Structuralist Approach," *J. Soc. Biol. Struct.,* 5 (1982), 15–47,

Stenseth, N., "Darwinian Evolution in Ecosystems: A Survey of Some Ideas and Difficulties Together with Some Possible Solutions," in *Complexity, Language and Life: Mathematical Approaches,* J. Casti and A. Karlqvist, eds., Springer, Heidelberg, 1986, pp. 105–145.

§3. The theory of genetics was put on a solid mathematical footing by the work of the British researchers Ronald A. Fisher and J. B. S. Haldane. Their classic works in the area are

Fisher, R. A., *The Genetical Theory of Natural Selection,* Clarendon Press, Oxford, 1930,

Haldane, J. B. S., *The Causes of Evolution,* Longman, London, 1932 (reprinted by Princeton University Press, Princeton, 1990).

More recent accounts are

Futuyama, D., *Evolutionary Biology,* Sinauer, Sunderland, MA, 1979,

Roughgarden, J., *Theory of Population Genetics and Evolutionary Ecology: An Introduction,* Macmillan, New York, 1979,

Birkett, C., *Heredity, Development and Evolution,* Macmillan, London, 1979.

The classic paper sparking off the explosion of interest in game-theoretic ideas in the cause of evolutionary ecology is

Smith, J. Maynard, and G. Price, "The Logic of Animal Conflict," *Nature,* 246 (1973), 15–18.

A much more complete account, together with peer commentary, is given in the review articles

Smith, J. Maynard, "Game Theory and the Evolution of Behavior," *Behav. & Brain Sci.,* 7 (1984), 94–101,

Parker, G. and J. Maynard Smith, "Optimality Theory in Evolutionary Biology," *Nature,* 348 (1990), 27-33,

and in the book

Smith, J. Maynard, *Evolution and the Theory of Games,* Cambridge University Press, Cambridge, 1982.

§4. Seldom can the origin of a field of intellectual activity be traced specifically to a single book or paper. Game theory is one of the few exceptions. The classic work from which the subject sprung forth almost full-grown is

von Neumann, J., and O. Morgenstern, *The Theory of Games and Economic Behavior,* Princeton University Press, Princeton, 1944.

Since the appearance of the above "encyclopedia" of games of strategy, numerous introductory accounts have appeared. Some of the best are

Williams, J., *The Compleat Strategist,* McGraw-Hill, New York, 1954,

Dresher, M., *The Mathematics of Games of Strategy,* Prentice-Hall, Englewood Cliffs, NJ, 1961,

Hamburger, H., *Games as Models of Social Phenomena,* Freeman, San Francisco, 1979,

Jones, A. J., *Game Theory: Mathematical Models of Conflict,* Ellis Horwood, Chichester, UK, 1980.

§5. The material of this section follows the treatment given in the following book, which is notable for its detailed account of experimental research on strategic interaction, together with a wealth of examples taken from political science, philosophy, biology and economics:

Colman, A., *Game Theory and Experimental Games,* Pergamon, London, 1982.

The Prisoner's Dilemma seems to have been explicitly formulated originally by Merrill Flood at the RAND Corporation and named by Albert Tucker in the early 1950s. It has been the subject of an extensive literature, which by now numbers well over one thousand papers and books. Good introductory accounts of the intricacies of the game are found in

Rapoport, A., and A. Chammah, *Prisoner's Dilemma: A Study in Conflict and Cooperation,* University of Michigan Press, Ann Arbor, MI 1965,

Rapoport, A., *Mathematical Models in the Social and Behavioral Sciences,* Wiley, New York, 1983,

Brams, S., *Game Theory and Politics,* Free Press, New York, 1975.

§6–7. The best compact reference for the mathematical details of replicator systems is the overview given in

Sigmund, K., "A Survey of Replicator Equations," in *Complexity, Language and Life: Mathematical Approaches,* J. Casti and A. Karlqvist, eds., Springer, Heidelberg, 1986, pp. 88–104,

and the book

Hofbauer, J. and K. Sigmund, *The Theory of Evolution and Dynamical Systems,* Cambridge University Press, Cambridge, 1988.

The example of the hypercycle is treated in great mathematical and chemical detail in

Eigen, M., and P. Schuster, *The Hypercycle: A Principle of Natural Self-Organization,* Springer, Berlin, 1979,

Eigen, M., J. McCaskill and P. Schuster, "Molecular Quasi-Species," *J. Phys. Chem.,* 92 (1988), 6881-6891,

Fontana, W., W. Schnabl and P. Schuster, "Physical Aspects of Evolutionary Optimization and Adaptation," *Physical Rev. A,* 40 (1989), 3301-3321,

§8. In the ecological community, the phenomenon we have termed perseverance is often called "persistence." Unfortunately, this latter term is also used to mean something quite different in circles outside population ecology. Consequently, we prefer to employ the related word perseverance, which captures equally well the concept as used in this section. Important references on this topic are

Hutson, V., and C. Vickers, "A Criterion for Permanent Coexistence of Species with an Application to a Two-Prey/One-Predator System," *Math. Biosci.*, 63 (1983), 253–269,

Amann, E., *Permanence for Catalytic Networks,* Dissertation, Dept. of Mathematics, University of Vienna, Vienna, 1984,

Hofbauer, J., "A Difference Equation Model for the Hypercycle," *SIAM J. Appl. Math.*, 44 (1984), 762–772.

§9. The mathematical interconnections between two-player games and replicator dynamics, including a very detailed consideration of the classification scheme considered here, may be found in

Zeeman, E. C., "Population Dynamics from Game Theory," in *Global Theory of Dynamical Systems, Proceedings,* Springer Lecture Notes in Mathematics, Vol. 819, New York, 1980, pp. 471–497.

Another paper containing further developments along the same lines is

Zeeman, E. C., "Dynamics of the Evolution of Animal Conflicts," *J. Theor. Biol.*, 89 (1981), 249–270.

§10. For further details on the role of ESS in replicator dynamics, see

Hofbauer, J., P. Schuster and K. Sigmund, "A Note on Evolutionary Stable Strategies and Game Dynamics," *J. Theor. Biol.*, 81 (1979), 609–612,

Schuster, P., and K. Sigmund, "Towards a Dynamics of Social Behavior: Strategic and Genetic Models for the Evolution of Animal Conflicts," *J. Soc. & Biol. Struct.*, 8 (1985), 255–277.

§11–12. The desert plant example, together with its extension to the case of many players as well as a consideration of the optimality versus stability issue, is given in

Riechert, S., and P. Hammerstein, "Game Theory in the Ecological Context," *Ann. Rev. Ecol. Syst.*, 14 (1983), 377–409.

For an account of field experiments supporting the theoretical properties of ESS in an ecological setting, see

Riechert, S., "Spider Fights as a Test of Evolutionary Game Theory," *American Scientist,* 74 (1986), 604–610.

§13. For additional information on the matter of plants and their game-theoretic flowering strategies, see

Vincent, T., and J. Brown, "Stability in an Evolutionary Game," *Theor. Pop. Biol.,* 26 (1984), 408–427.

§14. The classic work that started all the ruckus over the degree to which social behavioral patterns are biologically determined is

Wilson, E., *Sociobiology,* Harvard University Press, Cambridge, MA, 1975.

For an introductory layman's account, see Chapter Three of

Casti, J., *Paradigms Lost: Images of Man in the Mirror of Science,* Morrow, New York, 1989 (paperback edition: Avon, New York, 1990).

Other introductory accounts of the basic ideas forming the sociobiologist's creed are

Barash, D., *Sociobiology and Human Behavior,* Elsevier, New York, 1977,

Alexander, R., *Darwinism and Human Affairs,* University of Washington Press, Seattle, 1979,

Baer, D., and D. McEachron, "A Review of Selected Sociobiological Principles: Applications to Hominid Evolution," *J. Soc. & Biol. Struct.,* 5 (1982), 69–90.

The very idea that human behavior might be determined by biological "programming" seems to strike a raw nerve with those devoted to the bizarre notion that we humans are somehow masters of our own destiny. For blow-by-blow accounts of this highly emotionally-charged debate, see the Ruse and Caplan volumes cited under §1, as well as

Kitcher, P., *Vaulting Ambition,* MIT Press, Cambridge, MA, 1985,

Lewontin, R., S. Rose and L. Kamin, *Not in Our Genes,* Pantheon, New York, 1984,

Kaye, H., *The Social Meaning of Modern Biology,* Yale University Press, New Haven, CT, 1986.

For a fuller account of the genotype-to-phenotype model of animal behavior, see the Schuster and Sigmund article cited under §10 above, as well as the papers

Bomze, I., P. Schuster and K. Sigmund, "The Role of Mendelian Genetics in Strategic Models of Animal Behavior," *J. Theor. Biol.*, 101 (1983), 19–38,

Treisman, M., "Evolutionary Limits to the Frequency of Aggression between Related or Unrelated Conspecifics in Diploid Species with Simple Mendelian Inheritance," *J. Math. Biol.*, 93 (1981), 97–124.

The example of altruistic behavior emerging out of pure utility maximization is first presented in

Becker, G., "Altruism, Egoism, and Genetic Fitness: Economics and Sociobiology," *J. Econ. Lit.*, 14 (1976), 817-826.

§15. The complete story of the computer tournament in which various strategies for the Prisoner's Dilemma game were pitted against each other in a round-robin competition is given in the fascinating book

Axelrod, R., *The Evolution of Cooperation*, Basic Books, New York, 1984.

Extensions and generalizations are taken up in the review article

Axelrod, R. and D. Dion, "The Further Evolution of Cooperation," *Science*, 242 (1988), 1385–1390.

Further discussion of the computer tournament, along with extensive commentary on the possible social implications of the results, can be found in

Hofstadter, D., "The Prisoner's Dilemma Computer Tournament and the Evolution of Cooperation," in *Metamagical Themas*, Basic Books, New York, 1985, pp. 715–734.

Some of the implications of the Prisoner's Dilemma for biological behavior are outlined in

Axelrod, R. and W. Hamilton, "The Evolution of Cooperation," *Science*, 211 (1981), 1390–1396.

DQ #2. For detailed arguments showing the parallels between economics, evolutionary biology and behavioral psychology, see

Schwartz, B., *The Battle for Human Nature*, Norton, New York, 1986,

von Schilcher, F., and N. Tennant, *Philosophy, Evolution and Human Nature*, Routledge and Kegan Paul, London, 1984.

DQ #3. The Red Queen Hypothesis was introduced into the ecological literature by Leigh Van Valen in

Van Valen, L., "A New Evolutionary Law," *Evol. Theory,* 1 (1973), 1–30.

For a review of this somewhat controversial hypothesis, see

Stenseth, N., "Darwinian Evolution in Ecosystems—The Red Queen View," in *Evolution,* P. Greenwood, *et al.,* eds., Cambridge University Press, Cambridge, 1985.

DQ #4. A detailed consideration of the Cuban Missile Crisis from a modeling and systems-analytic perspective is provided by the classic work

Allison, G., *Essence of Decision: Explaining the Cuban Missile Crisis,* Little Brown, Boston, 1971.

DQ #5. For a full discussion of the *n*-Player Prisoner's Dilemma, see the book by Colman cited under §5 above.

DQ #6. The idea of harnessing evolutionary principles to economic phenomena is not a new one, as is recounted in the Schwartz book cited under DQ #2. However, recently there has been considerable interest in utilizing the same approach to study economics at the level of the firm rather than at the level of the individual. A pioneering work in this area that builds upon the earlier work of Joseph Schumpeter is

Nelson, R. R., and S. G. Winter, *An Evolutionary Theory of Economic Change,* Harvard University Press, Cambridge, MA, 1982.

Two other volumes presenting work along the same lines are

Foster, J., *Evolutionary Macroeconomics,* Allen and Unwin, London, 1987,

The Economy as an Evolving Complex System, P. Anderson, K. Arrow and D. Pines, eds., Addison-Wesley, Redwood City, CA, 1988.

In a somewhat different direction, see also

Boulding, K., *Evolutionary Economics,* Sage, Beverly Hills, CA, 1981.

DQ #7. The classic works of Spengler and Toynbee on the dynamics of historical change are

Spengler, O., *The Decline of the West,* Knopf, New York, 1926,

Toynbee, A., *A Study of History,* Oxford University Press, Oxford, 1972.

An interesting use of linear system theory to support some of Toynbee's ideas is

Lepschy, A., and S. Milo, "Historical Events Dynamics and 'A Study of History' by Arnold Toynbee," *Scientia*, 11 (1976), 39–50.

The work of Colinvaux emphasizes the role of an eco-niche in determining the flow of historical events on a broad temporal and spatial scale. A detailed account is given in

Colinvaux, P., *The Fates of Nations: A Biological Theory of History*, Simon & Schuster, New York, 1980.

DQ #9. Dawkin's theory of the selfish gene is entertainingly presented in his well-known books

Dawkins, R., *The Selfish Gene,* Oxford University Press, Oxford, 1976,

Dawkins, R., *The Extended Phenotype,* Freeman, San Francisco, 1982.

For further discussion of the idea of a cultural meme, see the article

Hofstadter, D., "On Viral Sentences and Self-Replicating Structures," *Scientific American,* January 1983 (reprinted in *Metamagical Themas,* Basic Books, New York, 1985).

DQ #10. Swanson's ideas are given in detail in

Swanson, C., *Ever-Expanding Horizons,* University of Massachusetts Press, Amherst, MA, 1983.

DQ #14. The most prominent proponents of the punctuated equilibrium view of evolutionary change are the well-known paleontologists Stephen J. Gould and Nils Eldredge. Their views were first presented in

Eldredge, N., and S. Gould, "Punctuated Equilibria: An Alternative to Phyletic Gradualism," in *Models in Paleobiology,* T. Schopf, ed., Freeman, San Francisco, 1972.

A rather more introductory account is

Eldredge, N. and I. Tattersall, *The Myths of Human Evolution,* Columbia University Press, New York, 1982.

DQ #16. The Sheldrake theory of morphogenetic fields is viewed by mainline biologists with the same degree of enthusiasm that Stalin viewed Trotsky. Nevertheless, a number of predictions of the theory have been borne out by experiments carried out in recent years, and the idea now seems more alive than ever before. For an account of the underlying ideas, together with reprints of reviews of the first edition of the pioneering book and a description of some of the experimental evidence, see

Sheldrake, R., *A New Science of Life, A New Edition,* Anthony Blond, London, 1985,

Sheldrake, R., *The Presence of the Past,* Times Books, New York, 1988.

An introductory account of Sheldrake's ideas for the layman is found in Chapter Three of

Casti, J., *Searching for Certainty: What Scientists Can Know About the Future,* Morrow, New York, 1991.

DQ #19. See the Axelrod book cited under §15 for a detailed account of how the results of the Prisoner's Dilemma tournament suggest methods for improving cooperation among individually selfish competitors.

DQ #20-21. Whether or not evolution acts so as to adapt organisms optimally to their environment has long been a source of controversy among biologists and ecologists. The competing positions are well set out in

Jacob, F., "Evolution and Tinkering," *Science,* 196 (1977), 1161–1166,

Cody, M., "Optimization in Ecology," *Science,* 183 (1974), 1156–1164.

DQ #22. For more of Hirschleifer's ideas in this regard, see the commentary following the Becker article cited under §14.

PR #7–10. For these examples, as well as many, many more, including a discussion of the Bishop-Canning Theorem, see

Smith, J. Maynard, *Evolution and the Theory of Games,* Cambridge University Press, Cambridge, 1982.

PR #15. A fuller account of this problem can be found in

Hofbauer, J., "The Selection-Mutation Equation," *J. Math. Biol.,* 23 (1985), 41–53.

PR #17-18. These results are considered in greater detail in the Axelrod book cited under §15.

PR #22. Further results showing some of the other conundrums one can get into with professional sports drafts are given in the article

Brams, S. and P. Straffin, Jr., "Prisoner's Dilemma and Professional Sports Drafts," *American Math. Monthly,* 86 (1979), 80–88.

CHAPTER SIX

The Analytical Engine: A System-Theoretic View of Brains, Minds and Mechanisms

1. Brains and Minds

One of the standard conundrums of philosophy is the teaser, "What is mind?" Descartes' view that the brain and the mind are totally separate entities is aptly summed up in the epigram from Philosophy 101: "What is mind? No matter. What is matter? Never mind." This *dualist* interpretation of the mind-body problem stood more-or-less unchallenged until rather recently when neurophysiological advances, coupled with the growth of computer technology, brought forth a host of new insights, fresh ideas and competing views.

The brain-as-computer metaphor leads to the intuitively appealing picture of the brain as a piece of hardware serving as an information-processing device, while the mind is seen as the software. In other words, all the functions and features we usually associate with mind, like creative thought, emotions, pain and so forth, are simply programs run on the neurophysiological hardware of the brain. Roughly speaking, this is a weak form of what philosophers of mind call the *central-state identity* hypothesis, i.e., the claim that all mental events are identical with neurophysiological events in a material brain.

In this chapter we take no position on the brain-mind problem; rather, our goal is to indicate how a mathematical machine can serve to illuminate some aspects of how **any** type of information processing object like a brain might carry out functional activities such as remembering external stimuli, engaging in cognitive introspection, and executing instructions for activities like walking, seeing and speaking. We will also touch upon one of the most heated psychological debates of this century involving the competing claims of the behavioral and cognitive psychologists. This debate has by now been pretty much played out. But we shall see that rather simple system-theoretic arguments provide additional ammunition to sink the already floundering ship of behaviorism. These matters will be taken up in somewhat more detail later on. For now, let's take a longer look at the kind of mathematical machine that we'll use as a model for linking together the brain and the mind.

Exercises

1. Discuss the possible gaps in the analogy associating the brain with the hardware of a digital computer, while thinking of the mind as the software. In particular, consider how "programs" get written and whether or not there is any room in this analogy for computer concepts like various types of memory, e.g., read-only (ROM), write-once, read-many (WORM)) or something like "firmware" (software that's semi-permanently "burned-in" to, say, an EPROM chip.)

2. In the brain, the switches are mainly chemical; in a computer they are electronic. Besides the obvious difference in switching speeds that this fact gives rise to, are there any other important differences you can think of between these two kinds of switching circuits? Do you think it would be possible to build a "chemical computer" modeled along the lines of the switching circuits in a brain?

2. Input/Output Relations and Control Systems

In Chapter 3 we introduced the idea of a mechanism, or a machine, as a formal device to represent a causal, dynamic relationship between observables. The basic ingredients needed are sets of inputs, outputs and states, together with two rules expressing how inputs act to change states and how states give rise to outputs. In the automata considered in Chapter 3, the characterizing feature of the machines was that the cardinalities of all three of these sets were assumed to be at most *countable* (and very often just finite). But no assumptions were made about the mathematical structure of the state-transition and output maps other than that they be well-defined.

In this chapter we drop the countability assumption on the inputs, outputs and states in favor of another hypothesis about the size of these sets. We assume here that the input and output sets are finite-dimensional vector spaces. But we make no *a priori* assumptions about the state space. Instead we take the view that the states are objects to be constructed in a "natural" way directly from the system's input/output description.

So let's assume we're given the input/output relation $f : \Omega \to \Gamma$, where Ω and Γ are finite-dimensional vector spaces. Our task will be to construct a state-space X, an initial state x_0, and maps $g \colon \Omega \to X$, and $h \colon X \to \Gamma$ so that the diagram

$$\Omega \xrightarrow{\;f\;} \Gamma$$
$$g \searrow \quad \nearrow h$$
$$X$$

commutes. Furthermore, for technical reasons that will become apparent later, we require that the map g be onto X, while demanding that h be one-to-one. The commutativity of the diagram, together with the conditions on

g and h, constitute the requirements for $\Sigma = (X, g, h, x_0)$ to be what we call a *canonical model,* or *realization,* of the input/output relation $\mathcal{I} = (\Omega, \Gamma, f)$.

As we will see below, the abstract model Σ is equivalent to the dynamical system

$$x_{t+1} = \phi(x_t, u_t), \quad x\big|_{t=0} = x_0,$$
$$y_t = h(x_t),$$

which formally has the same appearance as the automata studied in Chapter 3. The only thing that's changed is the interpretations we attach to the quantities x, u and y. Part of our story in this chapter will be to see how this change in setting from one formal system (finite-state automata) to another (dynamical control systems) affects the nature of the questions that can be asked (and the answers that can be obtained) about real-world processes modeled by the input/output relation $\mathcal{I} = (\Omega, \Gamma, f)$. To make our treatment as painless as possible (mathematically, anyway), we'll restrict our attention to finite-dimensional, linear control systems, i.e., those for which the maps f, g and h are linear, while the spaces Ω, Γ and X are finite-dimensional vector spaces. As will be seen, even this simple setting contains plenty of surprises, and is rich enough to supply us with a considerable wealth of ideas about matters of current concern in neurophysiology and psychology, not to mention many other areas such as economics, electrical engineering and physics. Indications of how the basic ideas of the linear theory extend to nonlinear processes will then be taken up in the final two sections of the chapter. But before embarking upon the analytical pyrotechnics, let us pause for a moment to review some matters in psychology that serve to motivate some of our later discussions.

Exercises

1. Identify the spaces Ω, Γ and X, along with the maps f, g and h, for the cellular automata discussed in Chapter 3.

2. The commutative diagram above shows that the input/output map f factors through the states X into the maps g and h. Relate this factorization to the situation in set theory, where if we have a set S and an equivalence relation r defined on S, it's possible to factor the map r through the set of equivalence classes $\{[s]_r\}$.

3. Behavioral vs. Cognitive Psychology

In the early 1920s, John Watson made the radical suggestion that behavior does not have mental causes. Stimulated by the general idea of logical positivism, this thesis was further developed and modified by Clark Hull,

B. F. Skinner and others, and has come to be termed *psychological behaviorism*. A principal reason for adoption of the behaviorist view was to rid psychology of the dualist attitude that mind is a nonphysical entity, disjoint somehow from the physical brain. The behaviorist solution is to eliminate all notion of mind, mental states and mental representation from psychological investigation, concentrating solely upon externally observed *stimulus-response* behavior patterns.

By the early 1960s, it was recognized that both Descartes' dualist position on the brain and the mind and the behaviorist approach to human behavior were unattractive, and effort was focused upon developing a materialist theory of mind that did allow for unobservable mental causes. One such theory, termed *logical behaviorism,* was quite similar to classical behaviorism and is really just classical behaviorism in a semantic form. Another approach, the *central-state identity theory,* postulates that mental events, states and processes are identical with neurophysiological events in the brain. Thus, under the central-state identity hypothesis, a behavioral effect is the result of a causal pattern of physical events in the brain. The problem with this idea is that in either its weak or strong form—*token* and *type physicalism,* respectively—it asserts that all mental particulars that exist or could ever exist are neurophysiological. Thus, the logical possibility of machines or various types of disembodied spirits having mental properties is ruled out because they are not composed of neurons.

During the last decade or so, a way out of these dilemmas has been provided by the theory of *functionalism,* an outgrowth of that amalgam of philosophy, linguistics, neurophysiology, computer science and psychology loosely labeled the *cognitive sciences.* Functionalism is based upon the idea that a mental state can be defined by its causal relations to other mental states, and that such mental states can be realized by many systems. In essence, behavior is driven by software, not hardware. An account of these various theories is given in the popular articles and books cited in the Notes and References. Since it will not be necessary for us to distinguish between the central-state identity theory and functionalism, here we adopt the generic term *structuralism* to represent any theory of the mind that involves physical mental states, be they manifested in a human brain, a disembodied plasma cloud from space or a collection of silicon wafers in a machine.

One of the main goals of this chapter is to display a precise, system-theoretic argument for asserting the *abstract* equivalence of behaviorism and cognitivism, while showing that operationally only the cognitive view offers any hope for a predictive, causal theory of human behavior. This conclusion is a natural consequence of the so-called Realization Theorem of mathematical system theory. Following the path laid out by the cognitive framework, we then provide a fairly detailed mathematical description of the way in which a "brain" might process and store external stimuli in order to generate observed behavioral responses. At the end of the chapter we offer some speculations based upon the theory of system invariants for how thoughts could be generated almost as a by-product of the internal system dynamics.

Exercises

1. Explain in everyday terms why behaviorism cannot possibly provide a predictive, causal theory of behavior?

2. What kind of unobservable mental states are we speaking of in: (a) the central-state identity theory, (b) functionalism, (c) structuralism? Are these mental states in the brain or in the mind?

4. Stimulus-Response Patterns and External System Models

Suppose we have an information-processing object $\mathcal{I}$ (human being, machine, cloud, or whatever) consisting of the proverbial "black box" connected to its environment by certain input and output channels (Fig. 6.1). Assume that at any given moment t, the stimulus u_t is selected from some set of symbols U (not necessarily numbers), while the observed response belongs to another set of symbols Y. To simplify the exposition, assume t takes on only the discrete values $t = 0, 1, 2, \ldots$. Then a stimulus-response pattern of $\mathcal{I}$ is represented by the sequence $\mathcal{B}_{\mathcal{I}} = \{(u_t, y_{t+1})\}$, $t = 0, 1, 2, \ldots$. (Note: To respect causality, the first output appears one time unit after application of the first input.)

$$u \longrightarrow \boxed{\mathcal{I}} \longrightarrow y$$

Figure 6.1. An Information-Processing Object

If we let Ω denote the set of all admissible stimuli sequences, with Γ representing the set of all responses. Then the overall *external behavior* of the object $\mathcal{I}$ can be denoted by a stimulus-response map

$$f : \Omega \to \Gamma,$$
$$\omega \mapsto \gamma,$$

where the input sequence is

$$\omega = \{u_0, u_1, u_2, \ldots\}, \qquad \omega \in \Omega, \qquad u_i \in U,$$

while the output sequence is given by

$$\gamma = \{y_1, y_2, \ldots\}, \qquad \gamma \in \Gamma, \qquad y_i \in Y.$$

In the above expressions, the input and output spaces Ω and Γ are finite-dimensional vector spaces whose elements are sequences of vectors.

According to the behaviorists, all that can ever be known about $\mathcal{I}$ are the map f, together with the sets Ω and Γ. Or, put another way, a behaviorist would claim that to be given f would be to be given everything that could be known about the disposition of the object $\mathcal{I}$ to behave in a certain way. Moreover, it would be *unscientific* to assert the existence of any unobservable internal states or mechanisms generating the object's behavior. Mathematical system theory provides an honest, true, clear and direct refutation of this claim.

Exercise

1. The Russian psychologist Ivan Pavlov was the prototypical behaviorist, a man whose experiments with ringing bells and drooling dogs are by now part of the folklore of the world of psychology. Describe the sets Ω and Γ, as well as the map f, characterizing Pavlov's experiments.

5. *Cognitive States and Internal Models*

An internal model Σ of the behavior pattern f (or, equivalently, the behavior sequence $\mathcal{B}_\mathcal{I}$), involves postulating the existence of a set X of internal *state variables,* an initial state $x_0 \in X$, a map g linking the stimuli u and the states and a rule h specifying how internal states combine to generate the observed response y. More compactly, we have

$$x_{t+1} = \phi(x_t, u_t), \quad x\big|_{t=0} = x_0$$

$$\quad (\Sigma)$$

$$y_t = h(x_t),$$

$x_t \in X$, $u_t \in U$, $y_t \in Y$. We then say that Σ is an internal model of the observed behavior f if the input/output behavior of Σ agrees with that of $\mathcal{I}$. Note that in order for this to happen, we must be able to construct a set X, as well as maps

$$\phi : X \times U \to X,$$
$$h : X \to Y.$$

From an abstract point of view, the first step in the structuralist program is to ensure that for any given external model $\mathcal{I} = (\Omega, \Gamma, f)$, a corresponding internal model $\Sigma = (X, \phi, h, x_0)$ exists. If this is the case, it would be natural to associate the abstract states X with the postulated physical states of the brain in some fashion, while interpreting the maps ϕ and h

as means for encoding and decoding external stimuli and mental states, respectively. It's one of the great triumphs of mathematical system theory to have been able to provide a definite procedure for constructing these sets and maps. This chapter gives a detailed report of this solution, as well as a full account of how to explicitly construct the encoding/decoding operations from the given stimulus-response data $\mathcal{B}_\mathcal{I}$.

Exercises

1. The Swiss psychologist Jean Piaget was at the forefront of the movement away from a behavioral toward a cognitive view of human behavior. In particular, he argued that the mental development of a child could be divided into four stages—the sensorimotor, preoperational, operational and formal—each of which could be explained by appeal to the development of different, unobservable mental representations arising in the child's brain. Identify plausible sets Ω, Γ and X, as well as maps ϕ and h, that would characterize Piaget's theory.

2. Where does memory fit in to the cognitive scheme of things, i.e., in the sets and maps describing the internal model of a brain, where would we place memory?

6. Realizations and Canonical Models

Loosely speaking, we can phrase the behavioral-structuralist problem as follows:

> Given a stimulus-response pattern $\mathcal{B}_\mathcal{I}$ associated with an external description of $\mathcal{I}$, find an internal model Σ such that $\mathcal{B}_\mathcal{I} = \mathcal{B}_\Sigma$, where $\mathcal{B}_\Sigma$ is the stimulus-response pattern of Σ.

It turns out that the solution to this problem is trivially easy—there are an infinite number of models $\Sigma = (X, \phi, h, x_0)$ such that $\mathcal{B}_\mathcal{I} = \mathcal{B}_\Sigma$. So we refine the problem by asking: How can we single out a *good* model from this infinitude of candidates? The answer hinges upon invoking a system-theoretic translation of Occam's Razor for pinning down what we mean by a "good" model. Intuition and Occam suggest that a good model will be one that is "compact," or "minimal," in some well-defined sense. Let's see how to make this idea more precise.

Assume we are given *any* model $\Sigma = (X, \phi, h, x_0)$. Then we say that Σ is *completely reachable* if for any state $x^* \in X$, there exists an input sequence $\omega \in \Omega$ and a time T such that $x_T = x^*$. That is, the input ω transfers the system state from x_0 at time $t = 0$ to x^* at time T. Notice that the property of complete reachability depends upon Ω, T, and ϕ, but is independent of the output function h.

Example: An Unreachable System

Assume we have the scalar system given by the dynamics

$$x_{t+1} = x_t^2 + u_t, \qquad x_0 = 0, \qquad x_t \in X = R^1,$$

where the inputs $u_t \geq 0$, i.e., the inputs consist of the set

$$\Omega = \{(u_0, u_1, u_2, \ldots,) : \ u_i \geq 0\}.$$

In this case, it's clear that $x_t \geq 0$ regardless of what input sequence $\omega \in \Omega$ the system receives. So no states $x_T < 0$ can ever be reached, i.e., the system is **not** completely reachable. On the other hand, it's intuitively clear that if we drop the non-negativity constraint on u_t, then any state x_T can be reached from the origin by application of a suitable input sequence.

Now let's focus upon the output of Σ. We call Σ *completely observable* if any initial state x_0 can be pinned down uniquely from knowledge of the system input sequence, together with observation of the system output y_t over an interval $0 < t \leq T$. Note that observability depends upon Ω, T and ϕ, as well as on h.

Putting the two concepts together, we call Σ *canonical* if it is both completely reachable and completely observable. The minimality criterion is now clear: The state-space X of a canonical model is minimal in the sense that there are no elements in X that cannot be reached by use of a suitable stimulus, while no two distinct initial states give rise to the same observed behavior. Thus, a canonical model is characterized by having a state space containing no elements "extraneous" to its input/output behavior $\mathcal{B}_\Sigma$.

Exercises

1. Suppose we have the two-dimensional system

$$x_{t+1}^1 = x_t^2, \qquad x_0^1 = 0,$$
$$x_{t+1}^2 = u_t, \qquad x_0^2 = 0.$$

Assume that $u_t \geq 0$ for all t. Show that the states reachable from the origin are only those for which $x^1 \geq 0$, $x^2 \geq 0$, i.e., states in the first quadrant. Thus, conclude that this system is not completely reachable using positive inputs.

2. Consider the system

$$x_{t+1} = x_t, \qquad x_0 = c,$$

with the observed output being $y_t = x_t^2$. Show that observations of the sequence $\{y_1, y_2, \ldots\}$ will not suffice to determine the value of the initial condition c. Thus, the system is not completely observable.

7. Linear Systems and Canonical Realizations

In order to give concrete mathematical meaning to the sets Ω, Γ and X and the map f forming the "data" of the Realization Problem, we consider now the case when the sets are vector spaces and f is linear.

Assume our information-processing object $\mathcal{I}$ has m independent input channels and p output channels, and that the inputs and outputs are real numbers. Then at each moment $t = 0, 1, 2, \ldots$, we have $u_t \in R^m$ and $y_t \in R^p$. Thus, the input space Ω consists of *sequences* of elements from R^m, whereas the output space Γ is comprised of sequences of vectors from R^p. For technical reasons, we assume only a finite number of inputs are applied, i.e., there exists an $N < \infty$ such that $u_t \equiv 0$ for all $t > N$. Hence, we have

$$\Omega = \left\{(u_0, u_1, u_2, \ldots, u_N) : u_i \in R^m, N < \infty\right\},$$

$$\Gamma = \left\{(y_1, y_2, y_3, \ldots) : y_i \in R^p\right\}.$$

(Note: We impose no finiteness condition on Γ.)

Turn now to the map f. If we assume linearity, the *abstract* map $f : \Omega \to \Gamma$ can be represented *concretely* by the linear input/output relation

$$y_t = \sum_{i=0}^{t-1} A_{t-i}\, u_i, \qquad t = 1, 2, \ldots, \tag{I/O}$$

where the matrices $A_j \in R^{p \times m}$. Another way of looking at this relation is to observe that an input sequence $\omega = (u_0, u_1, u_2, \ldots, u_N)$ is transformed into an output sequence $\gamma = (y_1, y_2, y_3, \ldots)$ by the lower-triangular block Toeplitz matrix

$$\mathcal{F} = \begin{pmatrix} A_1 & 0 & 0 & 0 & \cdots \\ A_2 & A_1 & 0 & 0 & \cdots \\ A_3 & A_2 & A_1 & 0 & \cdots \\ \vdots & \vdots & & & \end{pmatrix}, \qquad A_i \in R^{p \times m}.$$

Thus the abstract map f is represented in concrete terms by the infinite block matrix $\mathcal{F}$. This representation make it evident that when we say that f is given, we mean that we are given the behavior sequence $\mathcal{B}_{\mathcal{I}} = \{A_1, A_2, A_3, \ldots\}$. That is, we have the isomorphism

$$f \cong \{A_1, A_2, A_3, \ldots\}.$$

There are several points to note about the foregoing setup:

1) For each t, the relation (I/O) specifies a set of p equations in the pm unknowns of the matrix A_t. In general, this set of equations is under-determined unless $m = 1$ (a single-input system), and there will be many matrices A_t that will serve to generate the behavior sequence $\mathcal{B}_{\mathcal{I}}$. However, this nonuniqueness disappears under the finiteness hypothesis made above on the input sequence ω (see Chapter 7, Problem 4 for additional details on this point).

2) If there are only a finite number of outputs, i.e., if after some time $T > 1$ we have $y_t = 0$ for all $t \geq T$, then there are only a finite number of nonzero A_i and the matrix $\mathcal{F}$ has only a finite number of nonzero entries.

3) Tacitly assumed in the above setup is that the initial internal state of the system is $x_0 = 0$, i.e., the system starts in an equilibrium state and will not depart from this state unless a nonzero input is applied. If this is not the case, we have a problem of observability, a theme that we'll take up later.

So far we have concentrated upon the behavior sequence $\mathcal{B}$ as the way to describe a system's input/output behavior. (Note: Henceforth, for clarity and ease of writing we shall omit the subscript $\mathcal{I}$ on the behavior sequence when there is no danger of confusion.) But there is another way. Instead of describing a linear system's external behavior by the behavioral sequence $\mathcal{B} = \{A_1, A_2, \dots\}$, it's often characterized by the $p \times m$ *transfer matrix*

$$W(z) = \sum_{i=1}^{\infty} A_i z^{-i},$$

where z is a complex parameter. (Remark: The matrix $W(z)$ can also be thought of as the linear operator relating the discrete Laplace transforms of the system's input to its output. The terminology "transfer matrix" comes from electrical engineering.) We'll see later on that the transfer matrix description often comes in handy as a convenient *algebraic* gadget by which to characterize the properties of the system.

Now we turn to the matter of generating a canonical realization given the input/output description f, i.e., given a behavior sequence B.

In view of the linearity assumption, a realization of f consists of the construction of an n-dimensional vector space X, together with three real matrices $F, G,$ and H of sizes $n \times n, n \times m,$ and $p \times n$, respectively. We *assume* for the moment that $\dim X = n < \infty$. Later we'll return to a discussion of the nature and implications of this assumption. The space X represents the *state space* of our system. It's related to the given input and output spaces via the dynamical equations

$$\begin{aligned} x_{t+1} &= Fx_t + Gu_t, \qquad x_0 = 0, \\ y_t &= Hx_t, \end{aligned} \tag{Σ}$$

$x_t \in X$, $u_t \in R^m$, $y_t \in R^p$. In view of the assumption that X is finite-dimensional, there is no loss of generality in taking $X = R^n$. Given an input sequence $\omega \in \Omega$, it's clear that Σ generates an output $\gamma \in \Gamma$. If the input/output pair $(u_t, y_{t+1})_\Sigma$ from Σ agrees with the pair $(u_t, y_{t+1})_{I/O}$ given by the relation (I/O) above for all t, then we call Σ an *internal* model of the input/output description f. It's a simple exercise to see that this will be the case if and only if

$$A_t = HF^{t-1}G,$$

for all $t = 1, 2, 3, \ldots$. This relation links the input/output description given by the sequence $\{A_1, A_2, \ldots\}$ with the state-variable description Σ given by the matrices F, G and H. The above condition ensures that the external behavior of the two objects f and Σ are identical; however, there may be many systems $\Sigma = (F, G, H)$ satisfying this relation. So we need additional conditions to isolate a canonical model from this set of candidates.

We spoke earlier of the concepts of reachability and observability. These properties of the system involved the state-space X and, hence, are properties of the system Σ. Consequently, the reachability and observability of Σ must be expressible mathematically in terms of the matrices F, G and H. Let's see how this can be done.

From the dynamical relation defining Σ, at time $t = 1$ we have the system state

$$x_1 = Gu_0,$$

i.e., all states that can be "reached" at time $t = 1$ are given as linear combinations of the columns of G. Similarly, at time $t = 2$ the dynamical equations yield the state

$$x_2 = FGu_0 + Gu_1.$$

Thus, all states reachable by time $t = 2$ are linear combinations of the columns of FG and the columns of G. Carrying on this process, we see that at time $t = k$ the reachable states consist of linear combinations of the columns of the matrices $\{G, FG, F^2G, \ldots, F^{k-1}G\}$. But by our finiteness assumption on the state-space X, we have $\dim X = n < \infty$. Thus, the Cayley-Hamilton Theorem tells us that

$$F^r = \sum_{i=0}^{n-1} \alpha_{ri} F^i, \qquad \alpha_{ri} \in R,$$

for all $r \geq n$. Consequently, no states linearly independent of the preceding ones can appear in the above list after the term $F^{n-1}G$. Putting all these observations together, we have the following

REACHABILITY THEOREM. *A state $x \in R^n$ is reachable from the origin for the system Σ if and only if x is a linear combination of the columns of the matrices $G, FG, F^2G, \ldots, F^{n-1}G$.*

COROLLARY 1. *Every state $x \in R^n$ is reachable (i.e., Σ is completely reachable) if and only if the $n \times nm$ matrix*

$$\mathcal{C} = [G \,|\, FG \,|\, F^2G \,|\, \cdots \,|\, F^{n-1}G]$$

has rank n.

COROLLARY 2. *If a state $x \in R^n$ is reachable, then it is reachable in no more than n time steps.*

COROLLARY 3. *The reachable states form a subspace of R^n, i.e., if $x, \bar{x} \in R^n$ are reachable, the states $\alpha x + \beta \bar{x}$ are also reachable for all real α, β.*

Example 1: The Discrete, Controlled Pendulum

Consider the dynamics describing the (small) oscillations of a pendulum of unit length and mass with controllable velocity (in discrete-time)

$$x_{t+1} = Fx_t + Gu_t, \qquad x_0 = 0,$$

where

$$x = \begin{pmatrix} x^1 \\ x^2 \end{pmatrix}, \qquad F = \begin{pmatrix} 0 & 1 \\ -1 & 0 \end{pmatrix}, \qquad G = \begin{pmatrix} 1 \\ 0 \end{pmatrix}.$$

Here x^1 is the position and x^2 is the velocity of the pendulum. Suppose we want to move the pendulum to the zero position, but with unit velocity. Can this be done by controlling on the velocity alone? Intuitively the answer seems to be yes, since we have free control over the velocity. Let's verify this conclusion using the Reachability Theorem.

The state we wish to reach is $x^* = (0\ 1)'$. According to the dynamics, we have

$$x_1 = Gu_0 = u_0 \begin{pmatrix} 1 \\ 0 \end{pmatrix}, \qquad u_0 \in R.$$

Thus, after one time step we can reach any nonzero position, but with zero velocity. Consequently, we can't reach x^* in the first step. At time $t = 2$, however, we have

$$x_2 = FGu_0 + Gu_1 = u_0 \begin{pmatrix} 0 \\ -1 \end{pmatrix} + u_1 \begin{pmatrix} 1 \\ 0 \end{pmatrix}.$$

Here we see that the state x^* can be reached at time $t = 2$ simply by choosing $u_0 = -1, u_1 = 0$. In fact, we have

$$\mathcal{C} = [G \,|\, FG] = \begin{pmatrix} 1 & 0 \\ 0 & -1 \end{pmatrix}.$$

Since $\mathcal{C}$ has rank 2, we conclude from the Reachability Theorem that all states $x \in R^2$ can be reached by time $t = 2$.

Example 2: An Uncontrollable System

Consider the simple two-dimensional system

$$F = \begin{pmatrix} 1 & 1 \\ 0 & 3 \end{pmatrix}, \qquad g = \begin{pmatrix} 1 \\ 2 \end{pmatrix}.$$

In this case, the controllability matrix $\mathcal{C}$ is

$$\mathcal{C} = [g \,|\, Fg] = \begin{pmatrix} 1 & 3 \\ 2 & 6 \end{pmatrix}.$$

Clearly, the rank of $\mathcal{C}$ is 1. Thus, the system is not completely reachable. In fact, the reachable states are those lying in the one-dimensional subspace of R^2 generated by the vector $(1\ 2)'$.

The Reachability Theorem is stated in terms of the matrices F and G, which are part of the internal description of the system Σ. But the problem data $\mathcal{B}$ is given in terms of the behavior sequence $\{A_i\}$. So it's of more than passing interest to ask whether there is a test we can apply to this sequence directly to resolve the matter of system reachability. Under the assumption that the data sequence is *complete*, the answer is yes. Here completeness means essentially that a finite amount of the sequence $\mathcal{B}$ suffices to characterize the entire infinite sequence. We can express this mathematically by requiring that there exist an integer $n < \infty$, such that every matrix A_j is linearly dependent on the matrices A_i, $i \leq n$ for all $j > n$. In short, the first n elements of the sequence $\mathcal{B}$ suffice to determine all the rest. We'll see another version of this condition later. Now let's look at the reachability problem from the external description point of view.

Earlier we noted that the transfer matrix $W(z)$ is an alternate way to describe a system's external behavior. This matrix was defined in terms of the behavior sequence $\mathcal{B}$ as

$$W(z) = \sum_{i=1}^{\infty} A_i z^{-i}.$$

Note that here W is regarded as a purely formal power series (Laurent series, actually), with the indeterminate z serving merely as a time-marker. So there is no question of convergence of the series here. In terms of the matrix $W(z)$ we have the following condition for complete reachability.

TRANSFER MATRIX REACHABILITY THEOREM. *The system described by the transfer matrix $W(z)$ is completely reachable if and only if the rows of W are linearly independent.*

COROLLARY. *If the system has a scalar input $(m = 1)$, the system is completely reachable if and only if the numerator and the denominator of $W(z)$ have no common factor, i.e., W is irreducible.*

Example: Irreducibility of the Transfer Matrix and Reachability

Let the transfer matrix $W(z)$ be

$$W(z) = \frac{1}{(z+1)(z+2)} \begin{pmatrix} 2z+4 \\ z+2 \end{pmatrix}.$$

Since the term $(z+2)$ cancels in the numerator and the denominator in both entries of W, the system is not completely reachable.

Now let's turn to the question of observation. So far we have assumed that the system Σ starts in the state $x_0 = 0$. What if we can't verify this assumption? Do we have any means for identifying the actual initial state x_0 solely from knowledge of the system inputs and the observed output sequence $\{y_1, y_2, y_3, \ldots\}$? If so, then we say that the state x_0 is *observable*. If we can identify the initial state x_0 from the inputs and outputs for any possible initial state x_0, then we call Σ *completely observable*.

At first glance, it would appear from the dynamical equations for Σ that the determination of x_0 from the output would depend on the input sequence, since the inputs generate the outputs. In general, this is indeed the case. However, for a large class of systems, including linear systems, we can skirt around this difficulty by noting that the output is given explicitly by the formula

$$y_t = HF^t x_0 + \sum_{i=0}^{t-1} HF^{t-i-1} Gu_i, \qquad t = 1, 2, \ldots .$$

Here we see linearity saving the day, as the output is decomposed into the sum of the effect of the initial state x_0 and the effect from the input sequence. So without loss of generality we can use the standard input sequence $u_i \equiv 0$ for all $i = 0, 1, 2, \ldots$, since if this is not the case we can consider the outputs

$$\bar{y}_t = y_t - \sum_{i=0}^{t-1} HF^{t-i-1} Gu_i,$$

instead of the given output sequence. So for all questions of observability, we assume $u_t \equiv 0$.

Now consider what it would be like for a state x_0 to be **unobservable**. This would mean that there is another state $\bar{x}_0 \neq x_0$ such that $\bar{x}_0$ and x_0 yield identical outputs for all t, i.e., x_0 and $\bar{x}_0$ are indistinguishable on the basis of the output. Mathematically, this means

$$HF^t(x_0 - \bar{x}_0) = 0, \qquad t = 0, 1, 2, \ldots .$$

However, again we appeal to the Cayley-Hamilton Theorem to see that the above condition need be checked only for a finite number of values of t. More precisely, if $\dim X = n < \infty$, the initial state x_0 is indistinguishable from $\bar{x}_0$ for all t if and only if

$$HF^t(x_0 - \bar{x}_0) = 0, \qquad t = 0, 1, 2, \ldots, n-1.$$

Another way of saying this is that the state $x_0 - \bar{x}_0$ yields the same output as the initial state $x_0 \equiv 0$. Putting all these remarks together, we obtain the following key result.

OBSERVABILITY THEOREM. *An initial state $x_0 \in R^n$ is unobservable for the system Σ if and only if x_0 is contained in the kernel of the matrix*

$$\theta = \begin{pmatrix} H \\ HF \\ HF^2 \\ \vdots \\ HF^{n-1} \end{pmatrix}.$$

(Recall: If A is an $n \times m$ matrix, $\ker A = \{x \in R^m : Ax = 0\}$.)

COROLLARY 1. *Σ is completely observable if and only if rank $\theta = n$, i.e., $\ker \theta = \{0\}$.*

COROLLARY 2. *The <u>unobservable</u> states form a subspace of R^n.*

Example: Satellite Position Determination

The linearized dynamics of a satellite in a near-circular earth orbit can be described by the system matrix

$$F = \begin{pmatrix} 0 & 1 & 0 & 0 \\ 3\omega^2 & 0 & 0 & 2\omega \\ 0 & 0 & 0 & 1 \\ 0 & -2\omega & 0 & 0 \end{pmatrix}.$$

Here the parameter ω represents the angular velocity of the satellite. Suppose we can measure both the radial distance of the satellite y^1 and its angular position y^2. In this case, the observation matrix H becomes

$$H = \begin{pmatrix} 1 & 0 & 0 & 0 \\ 0 & 0 & 1 & 0 \end{pmatrix}.$$

A routine calculation then yields the observability matrix for the system as

$$\theta = \begin{pmatrix} 1 & 0 & 0 & 0 & 3\omega^2 & 0 & 0 & -6\omega^3 \\ 0 & 0 & 1 & 0 & 0 & -2\omega & -\omega^2 & 0 \\ 0 & 1 & 0 & 0 & 0 & 0 & 0 & 0 \\ 0 & 0 & 0 & 1 & 2\omega & 0 & 0 & -4\omega^2 \end{pmatrix}.$$

We see that θ is of rank 4, by inspection; hence, the system is completely observable. So the formal analysis verifies the rather obvious fact that if we know the satellite's radial distance and angular position, then we can pin down its overall position and velocity uniquely.

The Reachability and Observability Theorems enable us to split up the state-space X into four disjoint pieces: states that are reachable/unreachable combined with those that are observable/unobservable. This is the content of the so-called Canonical Decomposition Theorem for linear systems. As noted earlier, our interest is in the piece of X corresponding to those states that are *both* reachable and observable, since they are the only states that can play a role in generating the input/output map f. For this reason, we call an internal model $\Sigma = (F, G, H)$ *canonical* if and only if Σ is both completely reachable and completely observable. Let's summarize our development thus far.

We are given the external behavior

$$f \cong B = \{A_1, A_2, A_3, \dots\}, \qquad A_i \in R^{p \times m}.$$

We then seek to model B with a system $\Sigma = (F, G, H)$ satisfying the two conditions:

1) $A_i = H F^{i-1} G$, $i = 1, 2, 3, \dots$, i.e., Σ agrees with B,

2) Σ is completely reachable and completely observable, i.e., Σ is canonical.

Example: The Fibonacci Numbers

Suppose we apply the input sequence

$$\omega = \{1, 0, 0, \dots, 0\}$$

and observe the output

$$\gamma = \{1, 1, 2, 3, 5, 8, 13, \dots\}.$$

Here we recognize γ as the sequence of Fibonacci numbers generated by the recurrence relation

$$y_t = y_{t-1} + y_{t-2}, \qquad y_1 = y_2 = 1.$$

Since the inputs and outputs are scalars, any linear system with this input/output behavior corresponds to a system with $m = p = 1$, i.e., it's what we call a single-input/single-output system. Furthermore, it's easy to see that the system's behavior sequence is given by

$$B = \{1, 1, 2, 3, 5, 8, 13, \dots\}.$$

Consider the dynamical model

$$x_{t+1} = Fx_t + Gu_t, \qquad x_0 = 0,$$
$$y_t = Hx_t,$$

with $x_t \in R^2$ and

$$F = \begin{pmatrix} 0 & 1 \\ 1 & 1 \end{pmatrix}, \qquad G = \begin{pmatrix} 1 \\ 1 \end{pmatrix}, \qquad H = (\,1 \quad 0\,).$$

A little algebra soon shows that

$$A_i = y_i = HF^{i-1}G,$$

for this system. Furthermore, both the reachability matrix

$$\mathcal{C} = [G\,|\,FG] = \begin{pmatrix} 1 & 1 \\ 1 & 2 \end{pmatrix},$$

and the observability matrix

$$\theta = \begin{pmatrix} H \\ HF \end{pmatrix} = \begin{pmatrix} 1 & 0 \\ 0 & 1 \end{pmatrix},$$

have rank $2 = n = \dim X$. Thus, the system $\Sigma = (F, G, H)$ constitutes a canonical model for the Fibonacci sequence.

This example leaves open the matter of just where the system $\Sigma = (F, G, H)$ came from. How did we *know* that $X = R^2$? And how can we compute n and Σ directly from the behavior sequence $\mathcal{B}$? Once we have a candidate model, it's relatively easy to check whether or not it's canonical. But this is far from having a procedure to calculate the model itself from $\mathcal{B}$ alone. This, in a nutshell, is the central problem of mathematical modeling: How to construct a canonical model from observed data? So let's now turn our attention to the development of systematic procedures for its solution.

Exercises

1. Determine whether or not the following systems $\Sigma = (F, G, H)$ are completely reachable and/or completely observable:

(a) $\quad F = \begin{pmatrix} 0 & 1 & 0 & \cdots & 0 \\ 0 & 0 & 1 & \cdots & 0 \\ 0 & 0 & 0 & \cdots & 0 \\ \vdots & & & & \\ 0 & 0 & 0 & \cdots & 1 \\ \alpha_1 & \alpha_2 & \alpha_3 & \cdots & \alpha_n \end{pmatrix}, \qquad G = \begin{pmatrix} 0 \\ 0 \\ \vdots \\ 0 \\ 1 \end{pmatrix},$

$H = (\, \beta_1 \quad \beta_2 \quad \cdots \quad \beta_n \,).$

(b) $\quad F = \begin{pmatrix} \alpha_1 & 0 & \cdots & 0 \\ 0 & \alpha_2 & \cdots & 0 \\ \vdots & & & \\ 0 & 0 & \cdots & \alpha_n \end{pmatrix}, \qquad G = \begin{pmatrix} 1 \\ 1 \\ \vdots \\ 1 \end{pmatrix},$

$H = (\, \beta_1 \quad \beta_2 \quad \cdots \quad \beta_n \,).$

(c) $\quad F = \begin{pmatrix} 0 & 0 & \cdots & 0 & \alpha_1 \\ 1 & 0 & \cdots & 0 & \alpha_2 \\ 0 & 1 & \cdots & 0 & \alpha_3 \\ \vdots & \vdots & \cdots & \vdots & \vdots \\ 0 & 0 & \cdots & 1 & \alpha_n \end{pmatrix}, \qquad G = \begin{pmatrix} \beta_1 \\ \beta_2 \\ \vdots \\ \beta_n \end{pmatrix},$

$H = (\, 0 \quad \cdots \quad 0 \quad 1 \,).$

(d) $\quad F = \begin{pmatrix} 1 & 3 & 4 \\ 0 & 1 & 1 \\ 2 & 3 & 2 \end{pmatrix}, \qquad G = \begin{pmatrix} 1 & 1 \\ 0 & 1 \\ 2 & 3 \end{pmatrix}, \qquad H = \begin{pmatrix} 1 & 1 & 0 \\ 2 & 2 & 3 \end{pmatrix}.$

(e) $\quad F = \begin{pmatrix} 5 & 9 & 1 & 0 \\ 4 & 12 & 1 & 0 \\ 3 & 0 & 0 & 1 \\ 2 & 1 & 2 & 0 \end{pmatrix}, \qquad G = \begin{pmatrix} 1 & 0 & 0 \\ 1 & 2 & 0 \\ 1 & 5 & 1 \\ 2 & 3 & 2 \end{pmatrix},$

$H = \begin{pmatrix} 5 & 0 & 6 & 7 \\ 1 & 1 & 2 & 3 \\ 4 & 2 & 1 & 2 \end{pmatrix}.$

2. In an attempt to minimize measurements in the satellite position example, we might consider omitting the angular measurement. (a) Determine the observation matrix H in this case. (b) Is the system still completely

observable without the angular measurement? (c) Consider the same two
questions if we eliminate the radial measurement instead of the angular
measurement.

3. Consider the transfer matrix $W(z)$ given in the text example on
reachability. (a) Show that the behavior elements corresponding to this
matrix are $A_1 = (2 \; 1)'$, $A_2 = (-2 \; -1)'$. (b) Deduce from this that the
internal model of this system is given by the matrices

$$F = \begin{pmatrix} 0 & -2 \\ 1 & -3 \end{pmatrix}, \qquad G = \begin{pmatrix} 2 \\ 1 \end{pmatrix}, \qquad H = \begin{pmatrix} 1 & 0 \\ 0 & 1 \end{pmatrix}.$$

(c) Verify that this system is not completely reachable by showing that the
reachability matrix C is not of full rank.

4. A very simplified version of a situation in input/output economics
results in the following linear system:

$$x_{t+1} = Fx_t + Gu_t,$$

where

$$F = \begin{pmatrix} 0 & 0 & \cdots & 0 & a_1 \\ a_2 & 0 & \cdots & 0 & 0 \\ 0 & a_3 & \cdots & 0 & 0 \\ \vdots & \vdots & & \vdots & \vdots \\ 0 & 0 & \cdots & a_n & 0 \end{pmatrix}, \qquad G = \text{diag} \, (g_1, g_2, \ldots, g_n).$$

Here the elements $a_i \geq 0$, while x_t is a vector representing various products of the economic complex, with x_t^n being the finished product and the
quantities x_t^i being intermediate products, $i = 1, 2, \ldots, n-1$. The vector u_t
represents the labor input to the system, while the elements of F and G tell
us how labor affects the products, as well as how one intermediate product
feeds into another. (a) Suppose we can measure only the finished product of
the system. What is the output matrix H in this case? (b) Using this output
matrix H, determine whether the economic complex is completely observable, i.e., can we determine the initial level of the intermediate products if
we know only the final output? (c) Is this system completely reachable?

5. The classical Jacobi method for solving the set of linear algebraic
equations $Ax = b$ is to let D be the diagonal part of A, and compute the
sequence $x_{t+1} = -D^{-1}(A - D)x_t + D^{-1}b$, $x_0 = 0$. Assume, for simplicity,
that D is nonsingular. (a) Relate this iterative scheme to a linear control
system. (b) In the control system, what element of the Jacobi method corresponds to the control? (c) Is the corresponding control system completely
reachable?

6. Compartment models are often used to describe the concentration of a drug in a patient's body. One such model for the kinetics of the heart drug digitoxin involves the dynamics

$$
\begin{aligned}
x_{t+1} &= -(k_1 + k_2 + k_4)x_t, & x_0 &= 0.92D, \\
z_{t+1} &= k_2 x_t - (k_3 + k_5)z_t, & z_0 &= (0.85)(0.08)D, \\
s^1_{t+1} &= k_1 x_t, & s^1_0 &= 0, \\
s^2_{t+1} &= k_3 z_t, & s^2_0 &= 0, \\
s^3_{t+1} &= k_4 x_t, & s^3_0 &= 0, \\
s^4_{t+1} &= k_5 z_t, & s^4_0 &= 0.
\end{aligned}
$$

Here D is a parameter representing the initial dosage of digitoxin administered to the patient, while x_t is the digitoxin concentration in the body, y_t is the concentration of the related drug digoxin and the s^i_t are various bodily sinks for the two drugs. Finally, the parameters k_i are diffusion rate constants between these various compartments. (a) Assume that only the urinary sinks s^1 and s^2 can be measured. Is the system completely observable? (b) Is it possible to determine the initial dosage D on the basis of these measurements of urinary excretion? (Note: These two questions are not the same. Part (a) implies part (b), but not conversely.)

7. Find a canonical realization of the transfer matrix

$$
W(z) = \frac{1}{z^2 - 1}\begin{pmatrix} 2z - 2 & 2 \\ z - 1 & z - 1 \end{pmatrix}.
$$

8. The Ho Realization Algorithm

We begin by writing the infinite behavior sequence $\mathcal{B}$ in the so-called Hankel form

$$
\mathcal{H} = \begin{pmatrix} A_1 & A_2 & A_3 & A_4 & \cdots \\ A_2 & A_3 & A_4 & A_5 & \cdots \\ A_3 & A_4 & A_5 & A_6 & \cdots \\ \vdots & \vdots & & & \end{pmatrix},
$$

where $\mathcal{H}_{ij} = A_{i+j-1}$. Let's *assume* the infinite Hankel array $\mathcal{H}$ has finite rank n, i.e., all $r \times r$ submatrices of $\mathcal{H}$ have determinant zero for $r > n$. We will return below to the question of what to do if this assumption cannot be verified. The simplest case where it is easily verified is when A_k is a fixed constant matrix for all $k \geq N$, for some $N \geq 1$.

Under the finite rank hypothesis, there exists an $r < \infty$ such that

$$
A_{r+j+1} = -\sum_{i=1}^{r} \beta_i A_{i+j},
$$

for real numbers $\beta_1, \beta_2, \ldots, \beta_r$. This also implies that there exist matrices P and Q such that

$$PHQ = \begin{bmatrix} I_n & 0 \\ 0 & 0 \end{bmatrix},$$

where $I_n = n \times n$ identity matrix. A canonical realization of the behavior sequence $\mathcal{B}$ can now be obtained using P and Q according to the following prescription.

HO REALIZATION ALGORITHM. *Let the Hankel array $\mathcal{H}$ have finite rank n. Then a canonical realization of the behavior sequence*

$$\mathcal{B} = \{A_1, A_2, A_3, \ldots\}$$

is given by the system $\Sigma = (F, G, H)$, where the state-space X has dimension n and

$$F = \mathcal{R}_n P\sigma\mathcal{H}Q\mathcal{C}^n, \qquad G = \mathcal{R}_n P\mathcal{H}\mathcal{C}^m, \qquad H = \mathcal{R}_p\mathcal{H}Q\mathcal{C}^n.$$

(Note: Here the notation $\sigma\mathcal{H}$ means "left-shift" the matrix $\mathcal{H}$, i.e., move all columns of $\mathcal{H}$ one position to the left, while the operators $\mathcal{R}_n$ and $\mathcal{C}^m$ have the meanings "keep the first n rows" and "keep the first m columns," respectively.)

Example: Fibonacci Numbers (cont'd.)

To illustrate Ho's algorithm, we return to the behavior sequence given in the preceding section consisting of the Fibonacci numbers $\mathcal{B} = \{1, 1, 2, \ldots\}$. Forming the Hankel array, we have

$$\mathcal{H} = \begin{pmatrix} 1 & 1 & 2 & 3 & 5 & 8 & \ldots \\ 1 & 2 & 3 & 5 & 8 & 13 & \ldots \\ 2 & 3 & 5 & 8 & 13 & 21 & \ldots \\ \vdots & \vdots & \vdots & & & & \end{pmatrix}.$$

Because we *know* that $\mathcal{B}$ consists of the Fibonacci sequence generated by the recurrence relation $A_i = A_{i-1} + A_{i-2}$, it's easy to see that each column of $\mathcal{H}$ is the sum of the two preceding columns; hence, we verify the finite rank assumption, obtaining rank $\mathcal{H} = n = 2$.

To apply Ho's algorithm, all we need to do is find matrices P and Q that reduce $\mathcal{H}$ to the standard form. However, given that rank $\mathcal{H} = 2$, together with the action of the operators $\mathcal{R}_2$ and $\mathcal{C}^2$, it suffices to confine our attention to the 2×2 submatrix of $\mathcal{H}$ given by

$$\mathcal{H}_2 = \begin{pmatrix} 1 & 1 \\ 1 & 2 \end{pmatrix},$$

and reduce $\mathcal{H}_2$ to standard form, i.e., find matrices P and Q such that

$$P\mathcal{H}_2 Q = \begin{pmatrix} 1 & 0 \\ 0 & 1 \end{pmatrix} = I_2.$$

It's easy to see that one such pair is given by

$$P = \mathcal{H}_2^{-1} = \begin{pmatrix} 2 & -1 \\ -1 & 1 \end{pmatrix},$$

$$Q = I_2 = \begin{pmatrix} 1 & 0 \\ 0 & 1 \end{pmatrix}.$$

Furthermore, the left-shift of $\mathcal{H}_2$ is

$$\sigma(\mathcal{H}_2) = \begin{pmatrix} 1 & 2 \\ 2 & 3 \end{pmatrix}.$$

Applying the prescription of the Ho Realization Algorithm yields the canonical system

$$\begin{aligned} F &= \mathcal{R}_2 P \sigma(\mathcal{H}_2) Q \mathcal{C}^2, \\ &= \mathcal{R}_2 \mathcal{H}_2^{-1} \sigma(\mathcal{H}_2) I_2 \mathcal{C}^2, \\ &= \begin{pmatrix} 0 & 1 \\ 1 & 1 \end{pmatrix}, \end{aligned}$$

$$\begin{aligned} G &= \mathcal{R}_2 P \mathcal{H}_2 \mathcal{C}^1, \\ &= \mathcal{R}_2 \mathcal{H}_2^{-1} \mathcal{H}_2 \mathcal{C}^1, \\ &= \begin{pmatrix} 1 \\ 0 \end{pmatrix}, \end{aligned}$$

$$\begin{aligned} H &= \mathcal{R}_1 \mathcal{H}_2 Q \mathcal{C}^2, \\ &= \mathcal{R}_1 \mathcal{H}_2 I_2 \mathcal{C}^2, \\ &= (1 \quad 1). \end{aligned}$$

It's now a simple matter to verify that the system

$$x_{t+1} = \begin{pmatrix} 0 & 1 \\ 1 & 1 \end{pmatrix} x_t + \begin{pmatrix} 1 \\ 0 \end{pmatrix} u_t, \qquad x_0 = 0,$$

$$y_t = (1 \quad 1) x_t,$$

does indeed produce the Fibonacci sequence as its output, using the standard input $u_0 = 1$, $u_t = 0$, $t > 0$. Furthermore, a simple computation shows that the system $\Sigma = (F, G, H)$ is both completely reachable and completely observable.

Remarks

1) Comparing the canonical model for B given in the last section with the model Σ given above, we see they involve different matrices G and H. This is to be expected since the matrices P and Q are not unique. We can always introduce a nonsingular matrix T and use the matrices $\widehat{P} = PT$, $\widehat{Q} = T^{-1}Q$, which yield the new system matrices $\widehat{F} = TFT^{-1}$, $\widehat{G} = TG$, and $\widehat{H} = HT^{-1}$. From elementary linear algebra, we know that such a transformation T represents a linear change of variables in the state-space X. But the choice of coordinate system in X is the only arbitrariness in the canonical realization of B; once the coordinate system is fixed by a choice of T, the realization of B is unique.

2) Even though the actual matrices F, G and H may not be unique, the dimension of X is fixed for *all* canonical realizations of B. Thus, all canonical models for the Fibonacci sequence have state-spaces X of dimension 2.

3) Ho's Algorithm is far from the most computationally efficient method for finding a canonical realization. There are two reasons why we have singled it out for attention here: (i) historically, it was the first such procedure, and (ii) it's easy to explain and see the way the algorithm ties in with the classical invariant factor algorithm of linear algebra. For pointers to its computationally more efficient cousins, the reader should consult the material cited in the Notes and References.

Exercises

1. Consider the behavior sequence

$$B = \{1, 2, 5, 12, 29, 70, \ldots\} = \text{the Pell numbers.}$$

This sequence of numbers is defined by the recurrence relation $A_{i+2} = 2A_{i+1} + A_i$. (a) Show that B has a two-dimensional canonical realization. (b) Compute it using Ho's Algorithm.

2. Show that the behavior sequence

$$B = \{1, 2, 3, 4, 5, \ldots\} = \text{the natural numbers,}$$

also has a two-dimensional canonical realization. (a) Calculate it. (b) Is the canonical system Σ_P of the Pell numbers of Exercise 1 equivalent to the system Σ_N that realizes the natural numbers?

3. Consider the behavior sequence

$$B = \{1, 2, 2, 2, 4, 6, 10, 18, \ldots\},$$

which is generated by the formula

$$A_i = 2^i / i.$$

(a) Determine whether or not B is has a finite-dimensional realization. (b) If it does, use Ho's Algorithm to find it.

9. Partial Realizations

The success of Ho's Algorithm in producing a canonical model of the Fibonacci sequence hinged critically upon the knowledge that dim $X = n < \infty$. Moreover, we made explicit use of the precise value $n = 2$, although it's clear that once n is known to be finite, then a finite, though perhaps tedious, calculation will produce the exact value. But what's to be done if we are denied this *a priori* information? What if we don't know how to "fill-in the dots" in the sequence $\mathcal{B} = \{A_1, A_2, A_3, \dots\}$? In effect, this means that we have only a *finite* amount of data available, and we wish to develop a procedure for canonically realizing it. This is the so-called *partial realization problem*, and leads to some of the deepest and most difficult questions in mathematical system theory. Here we will only touch upon one or two aspects of particular interest.

Define the $n \times m$ principal minor of the Hankel array $\mathcal{H}$ as

$$\mathcal{H}_{n,m} = \begin{pmatrix} A_1 & A_2 & A_3 & \dots & A_m \\ A_2 & A_3 & A_4 & \dots & A_{m+1} \\ \vdots & \vdots & & & \vdots \\ A_n & A_{n+1} & A_{n+2} & \dots & A_{n+m-1} \end{pmatrix}.$$

We say that $\mathcal{H}$ satisfies the Rank Condition if there exist integers n, m such that

$$\text{rank } \mathcal{H}_{n,m} = \text{rank } \mathcal{H}_{n+1,m} = \text{rank } \mathcal{H}_{n,m+1}, \qquad \text{(RC)}$$

i.e., the rank $\mathcal{H}_{n,m}$ remains unchanged by the addition of one block row or one block column. Whenever the Rank Condition holds, we can appeal to the following result to justify use of Ho's Algorithm.

PARTIAL REALIZATION THEOREM. *Let n, m be integers such that the Rank Condition (RC) is satisfied. Then a system Σ, generated by Ho's Algorithm applied to $\mathcal{H}_{n,m}$, is a canonical model of the behavior sequence $\mathcal{B}_{n,m} = \{A_1, \dots, A_{n+m}\}$.*

Furthermore, if the Rank Condition is not satisfied for n, m, then every partial realization of $\mathcal{B}_{n,m}$ has dimension greater than rank $\mathcal{H}_{n,m}$.

Example: The Fibonacci Numbers (cont'd.)

For the Fibonacci sequence $\mathcal{B} = \{1, 1, 2, 3, \dots\}$, it's easily checked that the Rank Condition is satisfied for $n = m = 2$, i.e.,

$$\text{rank } \mathcal{H}_{22} = \text{rank } \mathcal{H}_{32} = \text{rank } \mathcal{H}_{23} = 2.$$

Thus, using Ho's Algorithm with $\mathcal{H} = \mathcal{H}_{22}$, we obtain the same canonical realization as in the last section. The Partial Realization Theorem then ensures that this model will canonically realize the finite sequence

$$\mathcal{B}_{2,2} = \{A_1, A_2, A_3, A_4\} = \{1, 1, 2, 3\}.$$

Remark

It can be shown that for *any* infinite sequence $\{A_1, A_2, A_3, \ldots\}$, the Rank Condition will be satisfied infinitely often. Hence, every finite sequence of data can be canonically realized as a subsequence of some finite behavior sequence.

Exercises

1. Consider the partial behavior sequence

$$\mathcal{B}_6 = \{1, 1, 1, 2, 1, 3\}.$$

(a) Show that the system $\Sigma = (1, 1, 1)$, i.e., $F = [1]$, $G = [1]$, $H = [1]$, realizes the first three terms of $\mathcal{B}_6$. (b) Using the Partial Realization Theorem, show that the "natural" continuation of $\mathcal{B}_6$ is

$$\mathcal{B}_{11} = \{1, 1, 1, 2, 1, 3, 2, 3, 5, 2, 9\},$$

which is the input/output behavior of the canonical system $\Sigma = (F, G, H)$, where

$$F = \begin{pmatrix} 1 & 1 & 0 \\ 0 & -1 & 1 \\ 1 & 0 & -1 \end{pmatrix}, \qquad G = \begin{pmatrix} 1 \\ 0 \\ 0 \end{pmatrix}, \qquad H = (1 \quad 0 \quad 0).$$

2. The behavior sequence

$$\mathcal{B} = \{14, 23, 28, 34, 42, 50, 59, 66, 72, 79, 86, 96, 103, 110, 116,$$
$$125, 137, 145, 157, 168, 181, 191, 207, 215, 225, 231, 238, 242\}$$

represents the local stops on the East Side IRT subway line in Manhattan. (a) Is this sequence finitely realizable? (b) Use Ho's Algorithm to construct a canonical realization of the conductor's litany as s/he calls out these stops. (c) In the old days this line also had a stop at 18th Street. How would the dimension of the canonical realization change if this stop were to be reinstituted on the IRT line?

3. The sequence of numbers $\{3, 8, 8, 4, 89, 75, 32, 30, 2\}$ represents the planetary diameters (to the nearest thousand statute miles) for our solar system (starting with Mercury). If another planet were to be discovered outside the orbit of Pluto, use a partial realization of the above sequence to predict the diameter of the newcomer.

10. *Approximate Modeling and Inexact Data*

Under the tacit assumption that the behavioral data $\mathcal{B}$ is exact, we have seen that there is a minimal explanation of $\mathcal{B}$ that is essentially unique. Our realization results demonstrate this fact within the context of data arising from a linear input/output structure. But the general result has much broader currency, which we can express as:

The Uniqueness Principle

If the data $\mathcal{B}$ are exact and complete, there is one and only one canonical system that explains (reproduces) the data.

The term "complete" in the statement of The Uniqueness Principle means simply that the data $\mathcal{B}$ did indeed arise from a system belonging to the class within which we seek the explanation. In this chapter that means the class of linear systems.

We see by The Uniqueness Principle that all minimal explanations of $\mathcal{B}$ are isomorphic. And this remains true independent of whether the data came from a linear or a nonlinear model. But everything hinges crucially on the data being exact, as the following classical example shows.

Example: Lagrange Interpolation

Suppose that the data $\mathcal{B}$ we are given consists of the pairs of real numbers $(t, x_t), t = 1, 2, \ldots, N$. We seek an explanation for this data within the class of polynomials. That is, we seek a polynomial p such that

$$p(t) = x_t, \qquad t = 1, 2, \ldots, N.$$

In this class of models, a minimal model is simply the polynomial p of least degree satisfying the above condition.

In classical numerical analysis, the "realization theorem" for this problem is called the *Lagrange interpolation formula* for the data $\mathcal{B}$. This formula yields the model

$$p^*(t) = \sum_{t=1}^{N} x_t \prod_{j \neq t} \frac{z - x_t}{x_t - x_j}.$$

This formula clearly provides a realization of the data $\mathcal{B}$. While it's not immediately evident that p^* is the minimal realization, it's a fairly straightforward matter to verify that it is. Note here that the condition that $\mathcal{B}$ be complete is satisfied trivially, since for any N we are assuming at the outset that the data is indeed generated by a polynomial. Finally, the isomorphism condition is also satisfied, since the requirement that p^* agree with the given data fixes the leading coefficient of any candidate polynomial. Thus, there is exactly one minimal polynomial agreeing with $\mathcal{B}$.

To see how things go wrong with the Lagrange scheme when the data is not known exactly, suppose we take a million points that lie exactly on a quartic curve—but not on any curve of lower degree. Then using the Lagrange scheme to "explain" this data, we will have $\deg p^* = 4$. Now let's perturb the data by adding a small amount of noise. Then generically the Lagrange formula will lead to a polynomial p such that $\deg p^* = 10^6$ for the new model. In plain English, the Lagrange formula is highly unstable with respect to small perturbations in the data. And, in fact, the complexity of the model p^* in this noisy case depends on the number of data points N. This is what people mean when they say that the Lagrange formula simply fits the noise. This example illustrates the only general statement that can be made about noisy data:

The Uncertainty Principle

Inexact (noisy, uncertain) data gives rise to non-unique (uncertain) explanations (models).

From what has been said up to now, the task of noisy realization theory is fairly clear: To describe how the uncertainty in the data $\mathcal{B}$ is converted into uncertainty in the model we use to explain that data. But, as has been noted by R. E. Kalman, this epistemological problem has often been merged with the purely psychological fact that we humans have a seemingly deep-seated need for unique answers to our questions. Enter classical statistics.

The standard procedures of statistics arose earlier this century in an attempt to deal with both these mathematical and psychological problems in one fell swoop. The assumptions underlying the statistical approach go as follows:

    i. All uncertainty is generated by a fixed probabilistic mechanism;

    ii. The data is obtained as an independent sample of the population of objects described by the fixed probability mechanism.

So, statistically speaking, any noisy data set $\mathcal{B}$ is a finite, independent sample from a fixed infinite population governed by a fixed probability law. So, as Kalman has noted, the classical statistician is engaging in a "prejudice" when s/he analyzes data on the basis of this standard sampling model. To illustrate the point, the standard "least-squares" method falls into this prejudicial trap, since it provides a unique answer to the problem of fitting a straight line to noisy data—in direct violation of The Uncertainty Principle.

While we have no space here to enter further into these deep philosophical and mathematical waters, the point we wish to emphasize is that when it comes to identification of "good" model from noisy data, it's very easy to fall into prejudices that can completely invalidate the conclusions of the analysis. By The Uncertainty Principle, it's simply not possible to obtain a unique model from such data; hence, prejudice of one type or another must always be present in any scheme that identifies a single model from the data. What's needed is to, first of all, be on the lookout for such invalidating prejudices, and, secondly, to think about new mathematical approaches to the overall question of extracting information about the underlying system from noisy data. We encourage the reader to consult the papers cited in the Notes and References, especially those by Kalman and Jan Willems, for a number of thought-provoking ideas about how this might be done.

Exercises

1. Newton's inverse-square law of gravitation can be thought of as a "realization" of Kepler's laws of planetary motion, i.e., that the planets move in elliptical orbits—provided that we regard Kepler's laws as being exact. Consider what would be the situation if (as is indeed the case) Kepler's laws were not exact. In particular, discuss why both Kepler and Newton were "lucky."

2. Consider a hyperplane H in R^n of dimension r and codimension q, where $q + r = n$. Suppose that we have exact measurements of the data, which consists of points $x_i \in R^n$, $i = 1, 2, \ldots, N$ lying on H. (a) What does it take for this set of data to be complete? (b) If the data is complete, in a suitable coordinate system we can write down a real matrix

$$\begin{pmatrix} I_r \\ G \end{pmatrix},$$

where I_r is the $r \times r$ identity matrix and the entries of the $q \times r$ matrix G are the so-called Plücker coordinates of H. Again assuming exact data, how would you use this representation to describe H by a set of linear equations? (Answer: $(G - I_q)x = 0$.) (c) Suppose now the data is noisy, i.e., $x_i = \hat{x}_i + \bar{x}_i$, where $\hat{x}_i$ denotes the *exact* or *true* component of x_i, while $\bar{x}_i$ is the noise. Describe in everyday terms the problem of identifying H in this situation. In particular, discuss the role played by the part $\hat{x}_i$ in any such identification procedure.

11. The State-Space X

For our goal of using mathematical system theory as a basis for modeling certain aspects of cognitive processes in the brain, it turns out to be useful to give a somewhat more explicit characterization of the elements of the canonical state-space X. Up to now we have just thought of the elements of X, the states, as being points of the space R^n. In the advanced theory of linear systems, we can attach a much deeper interpretation to the states:

A state $x \in X$ represents an "encoding" of the input ω in the most compact form that's consistent with the production of the output γ from the input via the input/output map f. In more technical terms, each state is an *equivalence class of inputs,* where we regard two inputs ω, ω' as being equivalent if they generate the same output under f, i.e., $\omega \approx \omega'$ if and only if $f(\omega) = f(\omega')$. Thus, the encoding of $\omega \rightarrow x = [\omega]_f$ represents the way the system Σ "remembers" the input ω. But what does such an encoding operation look like?

The simplest way to see explicitly how x codes the input ω is to *formally* associate each input ω with a vector of polynomials. Since, by definition,

$$\omega = (u_0, u_1, u_2, \ldots, u_N),$$

where each $u_i \in R^m$, $N < \infty$, we can formally identify ω with the vector polynomial

$$\omega \leftrightarrow u_0 + u_1 z + u_2 z^2 + \cdots + u_N z^N \doteq \omega(z),$$

where z is an indeterminate symbol. It's important to note here that we are *not* thinking of z as being any sort of number; z is just an indeterminate symbol whose only role is to act as a "time-marker" for the input ω. So the symbol z^t corresponds to the "time t" at which the input u_t is applied to the system.

Now we turn to the internal dynamics matrix F. Using the foregoing identification of input sequences with polynomials, we have the isomorphism $\Omega \approx R^m[z]$, the set of polynomials with coefficients in R^m. Let $\psi_F(z)$ be the minimal polynomial of F, i.e., $\psi_F(z)$ is the nonzero polynomial of least degree such that $\psi_F(F) = 0$. As is well-known from matrix theory, $\psi_F(z)$ is a divisor of the *characteristic polynomial* of F, which is always of degree n if $F \in R^{n \times n}$. Hence, $\deg \psi_F(z) \leq n < \infty$.

For a variety of technical reasons that would take us too far afield to elaborate here, it turns out that

$$x \doteq [\omega]_f \in \Omega \mod \psi_F(z),$$

i.e., x is the remainder after division of the vector polynomial ω by the scalar polynomial $\psi_F(z)$.

Example: The Fibonacci Sequence (cont'd.)

Let us return to the Fibonacci sequence realization given earlier. Just as dim $X = 2$ for any canonical realization, it's also true that the polynomial $\psi_F(z)$ is the same for every such model. So let's take F to be

$$F = \begin{pmatrix} 0 & 1 \\ 1 & 1 \end{pmatrix},$$

which gives $\psi_F(z) = z^2 - z - 1$. For a general scalar input

$$\omega = u_0 + u_1 z + u_2 z^2 + \cdots + u_N z^N,$$

we find that

$$x = \text{remainder } \left[\omega / \psi_F(z) \right].$$

As ω ranges over all of $R[z]$, it's an easy exercise to see that x ranges over the set of all *linear* polynomials, i.e.,

$$X \approx \{\text{polynomials } p(z) : \deg p \leq 1\}.$$

Hence, $\dim X = 2 = \deg \psi_F(z)$, a result that follows immediately since here the minimal and characteristic polynomials of F coincide. In general,

$$\dim X = \deg \{\text{minimal polynomial of } F\}.$$

To see how a *specific* input is coded into an element of X, suppose we apply the standard input $u_0 = 1$, $u_t = 0$, $t > 0$. Then ω is isomorphic to the constant polynomial 1, and we have

$$x = \frac{\omega}{\psi_F(z)} = \frac{1}{(z^2 - z - 1)},$$

implying that $x = 1$, i.e., $[\omega]_f = 1$. If instead we had chosen the input

$$u_0 = 1, \ u_1 = 1, \ u_2 = 1, \ u_t = 0, \qquad t > 2,$$

then $\omega \approx 1 + z + z^2$. Thus, we would have

$$x = [\omega]_f = \text{remainder } [\omega / \psi_F(z)],$$

$$= \text{remainder } \left[\frac{(z^2 + z + 1)}{(z^2 - z - 1)} \right],$$

$$= 2z + 2.$$

The Problems section for this chapter gives additional examples of how various systems code inputs as states. The important point to note is that the coding is determined by the matrix F alone. As a result, different realizations will encode the inputs differently and, hence, we will see different state-spaces X. This observation confirms our remarks in Chapter 1 to the effect that the state space is **not** an intrinsic part of the system's observed

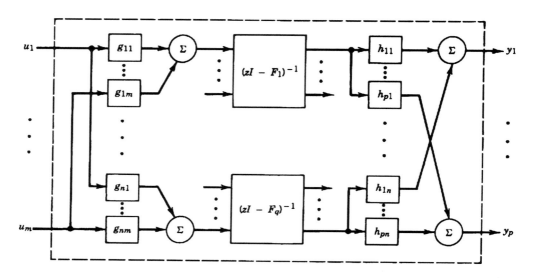

Figure 6.2. Structure of a Canonical Realization

behavior; rather, it's a mathematical construct whose sole purpose is to provide a way of "remembering" the inputs in a form convenient for generating the corresponding outputs.

To conclude our discussion of the Realization Problem, it's instructive to examine pictorially the structure of a canonical realization. In Fig. 6.2 above, we see the general picture of a canonical model. The blocks labeled $(zI - F_i)^{-1}$ correspond to a decomposition of F into Jordan normal form, i.e.,

$$F \cong \text{diag}\,(F_1, F_2, \ldots, F_q),$$

while the boxes labeled g_{ij} and h_{ij} represent the components of the matrices G and H, respectively. What's significant about this picture is the high degree of internal connectivity between the various elements.

The collection of results given up to now can compactly be summarized in the following result:

REALIZATION THEOREM. *Given an input/output map* $f \colon \Omega \to \Gamma$, *there always exists a canonical model* $\Sigma = (F, G, H)$ *such that* $\mathcal{B}_I = \mathcal{B}_\Sigma$. *Furthermore, the model* Σ *is unique up to a change of coordinates in the state space.*

(Note: For the Realization Theorem to hold, it may be necessary to have an infinite-dimensional state space.)

We saw a moment ago that for a canonical realization there are a large number of small pieces linked together in a tightly woven web of interconnections. This is a point we shall return to below. But now we temporarily leave linear system theory and return to the brain-mind question.

Exercises

1. The discussion of this section amounts to showing that the state-space X is isomorphic to the set $\Omega/\ker f$, where f is the system input/output map. (a) Prove this fact. (b) Show that the set $\Omega/\ker f$ is isomorphic to the set $f(\Omega)$. (Remark: This Exercise shows that instead of characterizing states as equivalence classes of inputs, we could just as easily have taken them to be classes of outputs.)

2. Suppose we write a state as $x = \zeta \cdot g$, where g is a fixed polynomial vector and ζ is a polynomial of degree n. (a) Describe in algebraic language the problem of finding an input $\omega \in \Omega$ that transfers the state x to 0. (b) Characterize algebraically what such an input looks like.

12. The Cognitive Theorem

In the context of the behaviorist-cognitivist debate, we can re-state the above Realization Theorem in psychological terms as

THE COGNITIVE THEOREM. *Given any stimulus-response pattern $\mathcal{B}_\mathcal{I}$, there always exists a cognitive model with behavior $\mathcal{B}_\Sigma$ such that $\mathcal{B}_\Sigma = \mathcal{B}_\mathcal{I}$. Further, this cognitive model is essentially unique.*

Remarks

1) The Cognitive Theorem states only that associated with any *physically observable* stimulus-response pattern $\mathcal{B}_\mathcal{I}$, there is an *abstract* set X, a point $x_0 \in X$ and *abstract* maps ϕ and h, such that $\Sigma = (X, \phi, h, x_0)$ forms a canonical model with $\mathcal{B}_\mathcal{I} = \mathcal{B}_\Sigma$. In order for the Cognitive Theorem to form the basis for a cognitive (or materialist) theory of behavior, it's necessary for these abstract objects to somehow be related to *actual* mental states. Later we shall examine just how this might be done.

2) At one level, the Cognitive Theorem says that there is no essential difference between the behaviorist and the cognitive schools of psychology— they are *abstractly equivalent*. On another level, the two theories are worlds apart; it all depends upon your point of view. The cognitive model offers an *explanatory mechanism* (the states X and the maps g and h) for behavior. It also has a built-in *predictive capacity* (the dynamics $x_{t+1} = \phi(x_t, u_t)$). The behaviorist model provides neither. It offers only a record of the experimental observations; the raw data, so to speak.

Exercises

1. Discuss what kind of experimental work would be needed to transform the abstract theory of brains outlined here into a *concrete* theory of cognitive processes.

2. The Cognitive Theorem assumes the behavioral data is exact. Consider how to re-state the Theorem when the data is noisy.

13. *Memory and Behavior*

The diagrammatic representation of the realization problem given in Section 2 makes it clear that the role of the state-space X is to "mediate" between the external inputs from Ω and the observed outputs from Γ. But what kind of psychological interpretation can we attach to this process of mediation? Put more precisely, what functional interpretation can we attach to the maps g and h? As already discussed, the only consistent answer to this question is to assert that the role of g is to "encode" an external stimuli ω as an internal state, whereas the role of h is to "decode" a state, thereby producing an observable output γ. We have already seen that this encoding operation can be explicitly represented by operations involving polynomials (at least in the case of linear systems), so we might think of Σ as a kind of "pattern recognition" device: the input pattern ω is "remembered" as the state $[\omega]_f$, which is represented by any polynomial ω^* such that $\omega - \omega^* \equiv 0 \mod \psi_F$. The simplest such polynomial ω^* is found by dividing ω by ψ_F and taking ω^* to be the remainder.

Suppose the stimulus-response pattern $f: \Omega \to \Gamma$ corresponds to some sort of elementary behavior. For example, it might be the raising of your arm or the scratching of your nose. The Cognitive Theorem then tells us that however complicated this behavior may be, it's composed of a combination of elementary behavioral "atoms," interconnected as in Fig. 6.2. It is the high level of interconnectivity that enables extremely complicated behaviors f to be put together from such elementary atoms. From an evolutionary standpoint, we wouldn't expect to see a noncanonical realization of f because it's just too large and unwieldy (too many complicated components substituting for the high level of connectivity). In short, it's more efficient and reliable to interconnect many simple behavioral modes than to rely on fewer, more complicated types. And this is exactly what we do see when we look at a brain. The elementary components, the neurons, have extremely simple behavior; but the density of interconnections is overwhelmingly large. Such experimental observations strongly argue for the view of a brain as a canonical realization of external behaviors.

Now let's turn to the problem of *pattern recognition*. It's clear that one of the characteristic features of intelligence is the ability to learn and respond to a wide variety of external stimuli (patterns). Once a pattern is learned, in some fashion the brain must be able to recognize the pattern again among the myriad patterns presented by the outside world. The setup outlined above provides a simple criterion for how this might be accomplished.

Suppose the pattern we want to recognize is represented by the input $\Phi \in \Omega$, and we want to build a system that fails to react to any input $\pi \neq \Phi$. By what has gone before, we know that the system is totally characterized by its minimal polynomial ψ, so our problem is to find a system Σ that recognizes Φ, but fails to respond to an input $\pi \neq \Phi$. In the language of polynomials, the solution is trivial: we need to find a ψ such that $\psi | \pi$ but $\psi \nmid \Phi$. A solution is possible if and only if $\pi \nmid \Phi$. So the pattern discrimination problem can be solved as long as the pattern Φ that's to be recognized is not a multiple of the recognition circuit *and* every pattern we want to reject is such a multiple. In order to carry out such a discrimination, a brain must clearly have many such elementary "circuits" wired-up in various series-parallel combinations. Referring again to Fig. 6.2, we think of each of the blocks in the canonical realization as being one such elementary circuit, the entire circuit being devoted to recognition of a *single* such pattern Φ. Then a brain would consist of an unimaginably large number of copies of Fig. 6.2.)

To summarize our position so far, each compound behavior $(\omega, f(\omega))$ is remembered as a state of the "machine" Σ that cognitively represents f. The system Σ is explicitly constructed in the form of a brain, so that it will produce the "response" $\gamma = f(\omega)$ when the input ω is presented. However, if some other input $\pi \neq \omega$ is given, then Σ will, in general, produce a response $f(\pi) \neq \gamma$. Hence, Σ is capable of generating many behaviors besides the specific pattern $(\omega, f(\omega))$ from which it originally arose. This observation allows us to present a theory for how *internal* (i.e., unobservable) stimulus-response patterns may arise, and their connection to the issue of thoughts, feelings and emotions. This theory is based upon the idea that unobservable "private" concepts like thoughts and feelings appear as a result of "parasitic" impulses arising as a side-effect of the brain's *normal* functions, which are to process sensory inputs into observable behaviors. In the next section we examine how such an idea can be given mathematical form within the framework of our dynamical system metaphor.

14. Thoughts and Group Invariants

So far we have created a mechanism at the functional level by which the external behavioral modes of an organism can be coded and decoded via the internal mental states of some kind of a "brain." But any decent model for a brain must account for subjective emotional experiences like pain, love, jealousy and pleasure, and not just externally observed actions like motion, sleep and talking. So we want to explore the manner in which this mechanism might give rise to what we ordinarily consider to be private, personal, internal "thoughts," as distinct from externally observable behavior.

According to recent work in brain physiology, the central cortex of the human brain consists of around four million neuronal modules, each composed of a few thousand nerve cells. Each module is a column that is vertically oriented across the cerebral cortex, and is about 0.25 mm across and between 2 and 3 mm long. These modules are now believed to be the functional units of communication throughout the association cortex, which forms about 95% of the human neocortex. Thus, a human brain can be thought of as something like a piano with four million keys. Carrying this musical analogy a step further, let's postulate the existence of four parameters that the cortical modules utilize in generating the virtually infinite number of spatiotemporal patterns constituting our conscious experiences. These parameters are *intensity* (the integral of the impulse firing in the particular module's output lines), the *duration* of the impulse firing from the module, the *rhythm* (or temporal pattern of modular firings) and the *simultaneity* of activation of several modules.

As a working hypothesis, we associate each neuronal module with a state-variable model $\Sigma_f = (F, G, H)$ of a particular behavioral pattern f. Thus, even though a module Σ_f is *originally* needed to account for the pattern f, once the mechanism (wiring diagram) corresponding to Fig. 6.2 is physically implemented in the neuronal hardware, the module Σ_f may generate many other behavioral responses, as well. We have already seen that the output from such a system Σ_f is

$$ y_t = \sum_{i=0}^{t-1} H F^{t-i-1} G u_i. $$

Thus, Σ_f will reproduce the input/output behavior f as long as the input sequence ω is the one originally given as part of the description of f. However, if a different input sequence ω^* is given, then Σ_f will, in general, produce an output sequence $\gamma^* \neq \gamma$. As a consequence, each of our neuronal modules Σ_f corresponds to a particular "learned" behavioral pattern f. But it can also produce an infinite variety of other actions $f^* \neq f$—once the neuronal pathways (essentially the connective structure of F, together with the connections G and H linking the inputs and outputs to the states) have been laid down.

The foregoing type of ambiguity (or lack of one-to-one correspondence) between f and Σ_f can be eliminated by employing the tacit assumption that a standard input sequence is used, generally $u_0 = 1, u_t = 0, t \neq 0$. It's tempting to conjecture that much of the processing of stimuli carried out by the body's receptor organs is arranged to implement such a normalization prior to the input reaching the neuronal module. We shall assume that this is the case, and that there is a one-to-one match-up between cortical modules and behaviors.

Under the foregoing hypotheses, there are on the order of four million or so "elementary" behaviors, one for each cortical module. These "atoms" of cognitive life correspond to the keys on the piano. The intermodular connections, coupled with the four parameters of intensity, duration, rhythm and simultaneity, then generate all behavioral modes. Let's now take a look at how these elementary behavioral modules could be stored in a real-life human brain.

First of all, each module consists of about 2,500 neurons capable, therefore, of storing 2,500 bits of information. If we assume that a single real number requires 25 bits, then a given module can store around 100 real numbers. If the system $\Sigma_f = (F, G, H)$ corresponding to the module has a state space of dimension n, and the number of input channels m and output channels p are such that $p, m \leq n$, then to store Σ_f requires $n(p + m + n)$, i.e., $O(n^2)$, numbers. With a brute force storage arrangement of this type, each module can only correspond to a system Σ_f of dimension ≤ 10. But this seems much too small to be able to account for even reasonably complex "elementary" behaviors.

A way out of this problem is to recall that the canonical model Σ_f is determined only up to a change of coordinates in the state space. Thus, any other model $\widehat{\Sigma}_f = (\widehat{F}, \widehat{G}, \widehat{H}) = (TFT^{-1}, TG, HT^{-1})$, $\det T \neq 0$, will display exactly the same elementary behavior. By standard arguments in linear system theory, it can be shown that as T ranges over the group of nonsingular $n \times n$ matrices, there exists a representative of the behavior class of f, call it $\widetilde{\Sigma}_f = (\widetilde{F}, \widetilde{G}, \widetilde{H})$, such that the number of nonfixed elements in $\widetilde{\Sigma}_f$ is $O(n)$. That is, by viewing the states in an appropriate basis, it's possible to represent the behavior f by storing only $O(n)$ numbers. These numbers form what are called the *Kronecker invariants* of the group action defined by letting the nonsingular $n \times n$ matrices act upon the state-space X as indicated above.

A reasonable conjecture is that evolutionary adaptation has arranged matters so that the "hard-wired" neuronal connections in the cortex are such that the brain represents each learned behavior in something close to this optimal "Kronecker" coordinate system. As a result, with the same set of 2,500 neurons in each module, it's possible to accommodate elementary behavioral modes f requiring canonical realizations Σ_f whose dimension can be on the order of 100 or so, an order of magnitude increase over the brute-force storage scheme.

Up to now we have considered each cortical module Σ_f as a representation of a given externally-observed behavioral pattern f. But what about internal thoughts? How can we account for aspects of consciousness involving things like hope, fear, pain, hunger, thirst and other nonbehavioral—but no less real—mental phenomena? Is there any way to accommodate these aspects of consciousness within the framework developed above? A theory of the brain should be able to account for these phenomena, too, so let's see how it might be done using the formal scheme outlined above.

To make progress on the matter of emotional states and thoughts, let's reconsider the diagram of Σ given in Fig. 6.2, looking a bit harder at the meaning of the blocks denoted there by g_{ij}, h_{ij} and $(zI - F_i)^{-1}$. Our view is that the elements g_{ij} and h_{ij} are just pre- and post-processors that link the cortical module to sensory effectors/affectors, *as well as to parts of the brain and to other modules.* The elements $(zI - F_i)^{-1}$ and the lines into and out of these blocks then represent the internal workings of the cortical module itself. With this picture in mind, we consider separately the question of emotional states and cognitive states.

There is now a great deal of experimental evidence suggesting that most emotional states like hunger, pain, and taste have their origin in the *limbic system,* that collection of nuclei and connecting pathways at the base of the brain. If this is indeed the case, then as far as cortical modules are concerned it makes little difference whether the inputs come from external sensory stimuli or from another part of the brain like the limbic system, the cerebellum or the optical cortex.

From the perspective of the cerebral cortex, where our modules Σ_f "live," inputs from the sense organs and inputs from the limbic system are treated equally, and appropriate cortical modules are developed early on to handle each. In terms of Fig. 6.2, some of the input channels to the g_{ij} come from sensory receptors, while others come from the limbic system. The emotional states arising in the limbic system may or may not evoke observable outputs. This will depend upon the post-processors h_{ij}, since, as we know, sometimes emotional states generate observable responses— crying, hunger pangs, violent movements—and sometimes not. In any case, in our set-up there is no need to distinguish emotional states from sensory stimuli, other than to note that one comes from the outside world while the other comes only from outside the neocortex. To the "inner world" of the neocortex, they are completely indistinguishable and are processed in exactly the same way (at least insofar as our theory goes).

Accounting for cognitive thoughts poses a somewhat more delicate task, since such thoughts are assumed to be self-generated within the cortex itself, quite independently of stimuli from the sense organs or other parts of the brain. Our somewhat speculative approach to this problem is to regard these thoughts as by-products of the primary cortical stimulation coming through the external input channels g_{ij}. We have already assumed that each module Σ_f is established by a particular behavioral mode f, with the g_{ij} conditioned to pre-process the appropriate stimuli in order to transform it into standard form. But it's also the case that each such cortical module shares connections with ten or so neighboring modules, which may generate stimuli feeding *directly* into the internal blocks $(zI - F_i)^{-1}$ and bypassing the pre-processors. Such inputs would, in general, cause the behavioral module Σ_f to emit outputs to the h_{ij}. This, in turn, may even result in a behavioral output different from the "design" behavior f if the threshold of the h_{ij} is exceeded.

In general, we may assume that such direct stimuli from the other modules is weak compared to that from the pre-processors. So when the "right" input signal for f is present, the "noise" from the other modules is too feeble to influence the output of Σ_f. Note also that in order for Σ_f to be ready to function properly when the design stimuli for f are applied, it must be the case that the matrix F is stable, generating a rather quick damping of any disturbance back to the zero state. Otherwise Σ_f would not be in a position to respond appropriately to rapid repetition of the same stimuli. Just how this condition is translated into the actual physical structure and chemical properties of the neurons, axons and synapses of the brain is a matter for experimental investigation.

So we conclude that thoughts are generated only when the module Σ_f is in its quiescent state waiting to perform its main function. Moreover, such thoughts are generated by the noise present in Σ_f as a result of its connections to other modules. At first hearing, this may seem like a very bizarre notion of how thoughts arise, since conventional wisdom dictates that thoughts are somehow voluntary creations of the human mind and, as a result, stand above the merely involuntary activities associated with various bodily activities involving pure survival. However, upon further examination the basic idea seems not so outlandish after all, since there is no *a priori* reason to imagine that the brain itself can distinguish between a creative thought and the impulses that generate a hunger pang or the bat of an eyelash—they are all electrochemical impulses in the neural circuitry. And it's only in the external world that the different *interpretations* appear. It's at exactly this stage that the idea of *consciousness* enters the picture, an idea that we will not consider further here in this very introductory and speculative exposition.

To summarize, the brain's cortical modules correspond to elementary behaviors f that are represented internally by the objects Σ_f. For compactness and efficiency, we further assume that nature has arranged things so that the objects Σ_f are stored by the Kronecker invariants of Σ_f, which are obtained from the canonical state coordinate change,

$$F \to TFT^{-1}, \qquad G \to TG, \qquad H \to HT^{-1}, \qquad \det T \neq 0.$$

Each such collection of numbers characterizes an entire *class* of systems Σ_f, each member of which represents the same external behavior f. The simplest such element $\widetilde{\Sigma}_f$ of each class contains $O(n)$ parameters, enabling the brain to efficiently reproduce elementary behaviors involving state spaces of dimension on the order of 100. Since there are around four million such cortical modules in the human brain, various series-parallel connections of those elementary behavioral/cognitive "atoms" provide ample material for the almost unlimited variety of thoughts, emotions and experiences of human life. This view of the brain is entirely *functional,* of course, omitting all material aspects associated with the actual physicochemical activities of any real brain. A big experimental challenge is to bring the functional view presented above into congruence with the known experimental facts surrounding the brain's actual physical structure and behavior.

Schematically, we can represent this system-theoretic view of the brain as displayed in Fig. 6.3, which shows the relationship between the Behavioral and Cognitive (Structural) approaches to brain modeling. This figure emphasizes the central role played by the Realization Theorem as a bridge between the two psychological schools of thought. Here we see also the important function of the group invariants in determining the activities of the individual cortical modules, as well as the functional role of the canonical state space in its relation to the generation of thoughts and emotions in the brain. Of course, Fig. 6.3 is only a sketch for a system-theoretic view of the brain, and virtually everything remains to be done to turn it into an actual theory.

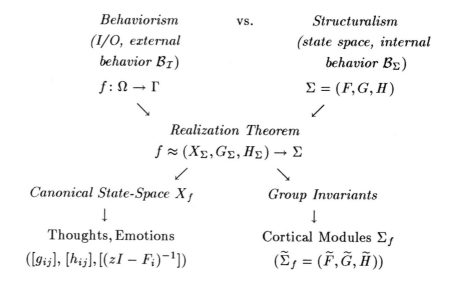

Figure 6.3. A System-Theoretic View of the Brain

Exercises

1. Consider the completely reachable system $\Sigma = (F, G, -)$, where the columns of G are denoted $g_1, g_2, \ldots, g_m$. Consider the ordered set of vectors

$$g_1, g_2, \ldots, g_m, Fg_1, Fg_2, \ldots, Fg_m, \ldots, F^{n-1}g_1, \ldots, F^{n-1}g_m.$$

Now consider the *Young's diagram*

	I	F	F^2	$\cdots$	F^{n-1}	
g_1	$\times$	$\times$	$\times$	$\cdots$	$\times$	k_1
g_2	$\times$			$\cdots$		k_2
g_3	$\times$					k_3
$\vdots$						$\vdots$

Here the rule for placing the crosses in the diagram is as follows: Begin with column 1 and place a cross if element (i, j) is linearly independent of its predecessors in the ordered list of vectors. Then go to column 2 and repeat the process. The integer k_i is just the number of crosses placed in row i. (a) Show that this procedure leads to exactly n crosses being placed in the diagram. Thus, $\sum_i k_i = n$. (b) Show that the vectors selected by the above process constitute a basis for R^n, and that in this basis the system matrix F assumes the form $F^\sharp = \left[F^\sharp_{ji} \right]$, where

$$F^\sharp_{ii} = \begin{pmatrix} 0 & 1 & 0 & \cdots & 0 \\ 0 & 0 & 1 & \cdots & 0 \\ \vdots & & & & \\ -a_{ii1} & \cdot & \cdot & \cdot & -a_{iik_i} \end{pmatrix},$$

$$F^\sharp_{ji} = \begin{pmatrix} & & 0 & & \\ & & \vdots & & \\ -a_{ji1} & \cdots & -a_{jik_j} & 0 & \cdots & 0 \end{pmatrix}, \qquad j \neq i, \qquad k_j \le k_i.$$

$$F^\sharp_{ji} = \begin{pmatrix} & 0 & \\ & \vdots & \\ -a_{ji1} & \cdots & -a_{jik_i} \end{pmatrix}, \qquad k_j > k_i.$$

Here the a's are constants and $F^\sharp_{ji}$ is a matrix of size $k_j \times k_i$. (b) What is the form of the input matrix G in this basis? (Remark: The integers $\{k_i\}$ are called the *Kronecker invariants* for the system Σ.)

2. The Kronecker form $\Sigma^\sharp = (F^\sharp, G^\sharp)$ has been obtained using only a coordinate transformation T in the state-space X. This change acts on F and G as $F \to TFT^{-1}$, $G \to TG$. Suppose now that we also admit a coordinate change V in the input-space Ω, i.e., we allow the input change

$G \rightarrow GV^{-1}$, as well as admitting a linear feedback law L that acts on F and G as $F \rightarrow F - GL, G \rightarrow G$. These three changes T, V and L constitute what is called the *feedback group* acting on the system $\Sigma = (F, G, -)$. Prove that under the feedback group, all the a's can be eliminated from $F^\sharp$ and $G^\sharp$, leaving only the Kronecker invariants to characterize the completely reachable pair (F, G). In other words, the Kronecker invariants constitute a complete, independent set of invariants completely characterizing the orbit of the pair (F, G) under the feedback group.

15. Linearity?

Let's conclude our speculations with a few remarks concerning the emphasis upon *linear* structures in this chapter. After all, given the enormous complexity of the functions the brain somehow carries out, on what grounds can we justify the arguments given here, which seem to be highly dependent upon linearity? There are several answers to this question, depending upon the level at which it is considered. Let's examine a number of them in turn.

• *Realization Level*—Our main tool has been the Canonical Realization Theorem. And, as already noted, the equivalence between a behavior B and a canonical model Σ_f in no way depends on the linearity of the input/output map f. The theorem is true under very weak hypotheses on f, Ω and Γ. So at this level, anyway, there is no problem at all with the linearity assumption.

We have focused most of our specifics upon the case of linear behaviors f because this is the setting in which the algebraic ideas can most easily be made explicit and accessible to nonmathematicians. An important aspect of our development was the description of the canonical state space as an equivalence class of inputs. For linear processes, this space can be described explicitly by simple mathematical gadgets—polynomials. For more general behaviors f, an explicit characterization is either impossible, or is at least algebraically much more complicated as, for instance, when f is bilinear, in which case the state space is an algebraic variety.

So we don't necessarily claim that the brain modules are actually linear; only that they are based upon the same *concepts* that have been given here explicitly for the linear case.

• *Dynamical Level*—In another direction, one might object that the dynamical processes of observed neural phenomena are so complicated that there *must* be complicated nonlinearities at work. Perhaps so. But the recent work in cellular automata theory outlined in Chapter 3 and in chaos theory and fractals as detailed in Chapter 4, as well as other recent work in dynamical system theory, seems to indicate otherwise. The main message of all of this work is that very complicated behavior can (and does) emerge from simple (sometimes even linear) local interactions when the number of interconnected subsystems is large enough. And with at least four million or so cortical modules to work with, it's not unreasonable to suppose that almost arbitrarily complicated patterns might arise in the brain from linear or almost linear building blocks.

   • *Approximation Level*—As discussed, we make no claims that the cortical modules are necessarily linear; however, if they are truly nonlinear, we have the comforting system-theoretic fact that *any* reasonably smooth behavior f can be arbitrarily closely approximated by a *bi*linear process. And such processes are amenable to the same sort of algebraic treatment we have presented for linear processes.

   So, in summary, it's not the linearity of f that is important; it's the concept of a canonical realization and the algebraic structure of its associated state space. These are the ingredients that make the magic work. Now let's add some mathematical muscle to these brave remarks. Since the approximation result above shows the importance of bilinear processes in the overall scheme of the nonlinear, we conclude the chapter by showing how to extend the notion of reachability to the bilinear setting.

16. Reachability of Bilinear Systems

We consider the bilinear system

$$x_{t+1} = \left(F + \sum_{i=1}^{m} u_t^i N_i \right) x_t + Gu_t, \qquad x_0 = 0,$$

where the matrices $F, N_i \in R^{n \times n}$, while $G \in R^{n \times m}$. Note that here u_t^i denotes the ith component of the control vector u_t. Our interest is in determining the states reachable from the initial state $x_0 = 0$.

   To address the reachability problem, it's convenient to introduce the object

$$\langle F, N_1, \ldots, N_m \rangle (G) \doteq \text{the smallest subspace of } R^n \text{ containing the range}$$
$$\text{of } G \text{ that is invariant under } F, N_1, N_2, \ldots, N_m.$$

Now we define the matrix sequence $\{P_i\}$ by the rule

$$P_{i+1} = [FP_i \,|\, N_1 P_i \,|\, \cdots \,|\, N_m P_i],$$
$$P_1 = G.$$

It's a straightforward exercise in linear algebra to see that if we set $\mathcal{C}_B \doteq [P_1 \,|\, P_2 \,|\, \cdots \,|\, P_n]$, then

$$\langle F, N_1, N_2, \ldots, N_m \rangle (G) = \text{ range } \mathcal{C}_B.$$

This simple result leads to the following

BILINEAR REACHABILITY THEOREM. *For a bilinear system with $G \neq 0$, the states reachable from the origin span the subspace of R^n given by* range $\mathcal{C}_B$. *In other words, the reachable states are generated by the columns of the* **bilinear** *reachability matrix*

$$\mathcal{C}_B = [P_1 \,|\, P_2 \,|\, \cdots \,|\, P_n].$$

So we see that by the mathematical artifice of introducing the matrix sequence $\{P_i\}$, we can express the reachability condition for bilinear systems in *exactly* the same form as for the purely linear case. The only difference is that we substitute the bilinear reachability matrix $\mathcal{C}_B$ for its linear counterpart $\mathcal{C}$. This is an illustration of what amounts to a "folk theorem" in the mathematical system theory world: Every result for linear systems can be transferred directly to bilinear processes by suitably "souping up" the corresponding linear result.

The foregoing result is nothing more than a whiff of what can be done by way of extending the concepts of reachability, observability, realization, stability, and control from the linear to the nonlinear. Some of these extensions will be brought out in the next chapter within the context of optimal control. The rest are covered in all their gory detail in the books and papers cited in the chapter Notes and References.

Exercise

1. Consider the bilinear system described by the matrices

$$F = \begin{pmatrix} 1 & 0 \\ -1 & -1 \end{pmatrix}, \qquad N = \begin{pmatrix} 1 & 4 \\ 3 & -2 \end{pmatrix}, \qquad g = \begin{pmatrix} 1 \\ 1 \end{pmatrix}.$$

(a) Is this system completely reachable? (b) Characterize the reachable states.

Discussion Questions

1. Discuss the claim made by some that it's possible for devices like computers to have actual mental states. To what degree do you believe that mental states are possible only in human brains? (The *central-state identity* theory asserts that they are. That is, mental events, states and processes are identical with neurophysiological events taking place in a human-type of brain.)

2. *Functionalism* is based upon the idea that a mental state can be defined by its causal relations to other mental states and that such states can be realized by many types of systems. What do you think about the prospects for a workable theory of cognitive behavior based upon this type of "bootstrapping" idea?

3. One often sees the analogy between the brain/mind and the hardware/software of a digital computer, with the brain corresponding to hardware and mind associated with the software. Does this analogy make any sense to you? If so, how do you think the software gets written? Does the brain's hardware offer any insight into the design of new computer architectures? Can computer software advances give us any help in better understanding of the mind? Do you think the human "operating system" is wired-in at birth?

4. We have associated the *abstract* states X of a canonical realization of a behavior sequence B with the *physical* states of a brain. In electrical engineering, the state variables constituting the space X are associated with the resistors, capacitors and/or inductors of an RLC-type electrical circuit. Consider whether or not the elements of such a circuit and clusters of neurons in a brain bear any relationship to each other.

5. Can you think of any experiments that would test our contention that elementary behaviors f are "coded" into the connectivity pattern of a neuronal module?

6. Behavior patterns in real life are generally learned responses. That is, the behavior f is the result of positive and negative reinforcements (at least this is what many behavioral psychologists claim). How would you extend our system-theoretic model to include learning?

7. The state space of a canonical realization contains no unreachable or unobservable states. If the abstract states are identified with physical brain states, how would you interpret unreachable and/or unobservable states of the brain?

8. The philosopher Karl Popper and the neurophysiologist John C. Eccles have postulated a brain-mind theory termed *dual-interactionism*, in which brain and mind are two separate entities interacting through some sort of "liaison" modules in the physical brain. Discuss the degree to which such a picture is compatible (or can be made compatible) with the system-theoretic framework developed in this chapter.

9. In the *holographic memory* theory of brain operation as developed by Karl Pribram and David Bohm, the memory function of the brain is assumed to function like a hologram in the sense that memory is not localized in the brain. Roughly speaking, the physical brain "tunes in" or "reads" somehow

a holographic universe that exists on some plane transcending space and time. Consider the manner in which the "cortical module" model developed here can account for this holographic aspect of brain function.

10. The discussion of this chapter carefully omits any considerations of intelligence. What additions, if any, do you think would be needed to our cognitive model of behavior in order to have it shed light on the question of *intelligent* behavior? Could a computer display intelligent behavior, at least in principle? Why?

11. One of the original and still bench-mark operational tests for intelligence is the *Turing test,* in which an interrogator poses questions of any sort to two respondents whose physical nature is hidden from the interrogator. One of the respondents is a computer, the other a human. If, after an extended series of questions and answers to and from both respondents, the interrogator is unable to distinguish the computer from the human, then the computer is deemed to be intelligent.

Discuss the adequacies of this sort of behavioristic view of intelligence. In particular, consider whether passing a Turing test constitutes "true" intelligence or is only a *simulation* of an intelligent being. Is there any real difference between the two?

12. The computer metaphor for cognition assumes that we are speaking of a *digital* computer. Would matters be any different if we were to consider an *analog* computer instead?

13. Gödel's Incompleteness Theorem says that in any formal axiomatic system rich enough to express all possible statements about the natural numbers, there is a statement we cannot prove using the inferential rules of this system. Nevertheless, this statement can be seen to be true by "jumping outside" the system. In computer science, it is established that any type of computing machine can be represented by a particular type of primitive computer, a *Turing machine.* Virtually all scientific arguments against the mechanization of intelligent thought have the following schemata: Joe cannot be represented by a Turing machine (computer), since if Joe were a realization of some Turing machine T, then (by Gödel) there would be something A that Joe could not do. But Joe *can* do A. Therefore, Joe is not a realization of T. And this is true of all possible computers. Hence, Joe transcends the limits of mechanism. (Remark: Here the activity A could be to prove the special Gödel sentence for the Turing machine T. For computer aficionados, this result is equivalent to what's called the Halting Problem. We'll discuss all of these matters in painstaking detail in Chapter 9.)

Does this antimechanism argument seem convincing to you? Can you offer any counter arguments or pick out any hidden assumptions in the argument?

14. Suppose a friend told you he had constructed a machine that could feel pain, love, jealousy, fear, and so forth. Suppose further that the machine can communicate to you in plain, everyday language. Now you ask it to describe what its pains, loves, etc., *feel* like. What do you suppose the response will be?

Let's now assume you're a mischievous type of person, and you want to cause this machine a lot of grief. For example, you want to make it feel jealous of your new girlfriend, who has just been introduced to the machine. What kind of actions/statements would you take to elicit this emotion from the machine?

15. The Turing test introduced in Discussion Question 11 focuses entirely upon the computer's input/output behavior, deeming a machine to be intelligent if it *behaves* like an intelligent human. Philosopher John Searle has advanced what has come to be called the *Chinese Room Test* as a critique of the adequacy of Turing's idea. Here is the essence of Searle's argument:

Assume you are locked into a closed room having no windows, the only entrance and exit being a single door having a large mailslot in it. In the room are a number of flashcards carrying Chinese ideographs (characters), together with a large dictionary-like book whose entries are just pairs of these Chinese symbols. This book has an instruction manual (in English) telling you that whenever you see the first character of a pair come in through the mailslot, your job is to pass back out the flashcard showing the second symbol of the pair.

Outside the room stand a group of native speakers of Chinese, who periodically pass cards carrying Chinese symbols into the room via the mailslot. In accordance with the book's instruction manual, you receive these cards, look up the symbol in the book and then pass back out the card carrying the corresponding symbol called for by the book. So, for example, if the incoming card carried the symbol "squiggle," the book might have the entry (squiggle, squoggle), in which case you would send the card showing the character "squoggle" through the slot.

The point of Searle's argument is that since you understand not one word (or symbol) of Chinese, this game is purely an exercise in abstract symbol manipulation for you. But for the Chinese speakers outside the room, the symbols they pass in and the "responses" you pass back out make perfectly good sense. For example, the incoming cards may be asking your opinion of the current state of the world economy, with your outgoing cards comprising perfectly sensible replies, like "it stinks" or "interest rates are too high."

Searle's claim is that your actions inside the Chinese Room duplicate perfectly what's going on inside a digital computer as it processes input data into outputs. So no matter how intelligent and clever the machine looks from the outsider's, third-person perspective, from the insider's, first-person point of view, there is no understanding whatsoever of what the symbols actually **mean.** The machine is merely shuffling symbols in accordance with a set of rules embodied in its program. So whatever intelligence there is in the machine is there solely in virtue of the intelligence that has been placed there by the program. And in this way Searle claims not only that the Turing test fails as a way to discriminate machine thinking from human thought, but also that no digital computer capable of mere symbol-processing is ever going to think like you and me. In short, strong AI is impossible. Discuss the pros and cons of this "anti-behavioristic" argument.

16. Proponents of strong AI come in two basic flavors: Top-Down and Bottom-Up. The first group holds to the view that it's possible somehow to, in effect, ignore the material structure of the brain and "skim-off" the rules of thought as a basis for their programs. The efforts of Top-Down'ers like Herbert Simon, Alan Newell and Roger Schank are displayed in various types of expert systems programs, as well as in many well-chronicled attempts to fence-in the supposed rules of thought, such as frames, templates and artificial worlds.

Bottom-Up'ers, on the other hand, argue that since thought processes are intimately connected with the human brain, it just may be that there's something about the physical structure of that brain that's important, essential even, for such thoughts to occur. Thus, they claim that the neurophysiological structure of the brain, embodied in the neurons and their connective pattern, cannot be ignored in the search for the Holy Grail of strong AI. People like Douglas Hofstadter, David Rumelhart and Gregory Hinton are leaders of this group, who go under the general rubric of the "new connectionists."

Discuss the strengths and weaknesses of these two competing positions on strong AI.

17. Discussion Question 13 gives an argument against strong AI based upon an appeal to Gödel's Incompleteness Theorem. But Gödel's result is true only if the formal logical system (i.e., the program) is consistent. This means that it's not possible for the same statement to be proved both true and false within the framework of the rules of inference of the system. Does this consistency requirement offer an opening to argue against the relevance of Gödel's Theorem as a refutation of strong AI?

18. The theory of *input/output economics* was initiated by Wassily Leontief in the 1930s. To illustrate the theory, suppose we have a vector x whose components describe the output of the industrial sector of an economy. We also have a vector y whose elements are the rate at which consumers absorb the industrial output. Leontief argued that we should have the linear

relation $x = Ax + y$ linking these two vectors, where A is a matrix of "technological coefficients." The primary task of this theory is to identify these coefficients on the basis of measurements of x and y. Does Leontief's overall program seem reasonable to you? Can you reformulate the Leontief idea in system-theoretic terms using the ideas of this chapter? In particular, what is the internal model corresponding to Leontief's input/output setup?

19. We saw that any nonlinear process can be arbitrarily closely approximated by a bilinear system of sufficiently high dimension. This is the system-theoretic analog to the famous Weierstrass Approximation Theorem from real analysis stating that any continuous function can be arbitrarily closely approximated by a polynomial. Discuss the merits and demerits of the claim that this bilinear approximation result implies that it's unnecessary to devote our time and energies to the study of anything other than bilinear processes.

Problems

1. Consider the transfer matrix

$$W(z) = \sum_{i=1}^{\infty} A_i z^{-i},$$

where the behavior sequence $\mathcal{B} = \{A_i\}$ is that associated with the system $\Sigma = (F, G, H)$.

a) Establish the relationship $W(z) = H(zI - F)^{-1}G$.

b) Show that the blocks $(zI - F_i)^{-1}$ in Fig. 6.2 correspond to a decomposition of the proper rational matrix $W(z)$ into matrix partial fractions.

c) Prove that complete reachability of Σ corresponds to the condition that the matrix $W(z)$ be *irreducible;* i.e., there exist polynomial matrices $S(z)$ and $T(z)$ of sizes $p \times m$ and $m \times m$, respectively, such that $W(z) = S(z)T^{-1}(z)$, where $\det T(z)$ is a polynomial of degree n and

$$\text{rank } \begin{pmatrix} S(z) \\ T(z) \end{pmatrix} = m,$$

for every complex z.

d) Suppose the matrices F, G and H are given by

$$F = \begin{pmatrix} 1 & 1 & 0 \\ 0 & -1 & 1 \\ 1 & 0 & -1 \end{pmatrix}, \qquad G = \begin{pmatrix} 1 \\ 0 \\ 0 \end{pmatrix}, \qquad H = (1 \quad 0 \quad 0).$$

Show that the transfer matrix for the above system is the rational function

$$W(z) = \frac{z^2 + 2z + 1}{z^3 + z^2 - z - 2}.$$

2. Let $F \in R^{n \times n}, g \in R^n$ and assume that (F, g) is a completely reachable pair, i.e., the matrix $[g \mid Fg \mid \cdots \mid F^{n-1}g]$ has rank n.

a) Show that there exists a nonsingular $n \times n$ matrix T such that the transformed system

$$\hat{F} = TFT^{-1}, \qquad \hat{g} = Tg,$$

has the form

$$\hat{F} = \begin{bmatrix} 0 & 1 & 0 & 0 & \cdots & 0 \\ 0 & 0 & 1 & 0 & \cdots & 0 \\ 0 & 0 & 0 & 1 & \cdots & 0 \\ \vdots & \vdots & \vdots & \vdots & \ddots & \vdots \\ 0 & 0 & 0 & 0 & \cdots & 1 \\ -\alpha_n & -\alpha_{n-1} & -\alpha_{n-2} & -\alpha_{n-3} & \cdots & -\alpha_1 \end{bmatrix}, \qquad g = \begin{bmatrix} 0 \\ 0 \\ 0 \\ \vdots \\ 0 \\ 1 \end{bmatrix},$$

where the constants $\{\alpha_i\}$ are the coefficients of the characteristic polynomial of F. That is, the polynomial

$$\det(zI - F) = z^n + \alpha_1 z^{n-1} + \alpha_2 z^{n-2} + \cdots + \alpha_{n-1} z + \alpha_n.$$

(Note: This Problem shows that by a suitable change of coordinate system in the state space, any single-input, completely reachable system can always be brought to the above *control canonical form*. See Problem 12 below for an illustration of the mathematical advantages of this structure.)

b) Extend this result to the case of $m > 1$ inputs.

3. Prove that the sequence of *prime numbers*

$$\mathcal{B} = \{2, 3, 5, 7, 11, \dots\},$$

has no finite-dimensional realization (difficult!).

4. Consider the input sequence $\omega = \alpha_0 + \alpha_1 z + \alpha_2 z^2 + \cdots + \alpha_N z^N$. Compute the "remembered" state $[\omega]_f$ for the following systems:

a) $\psi_F(z) = z$,

b) $\psi_F(z) = z - 1$ ("integrator"),

c) $\psi_F(z) = z^k$ ("truncator"),

d) $\psi_F(z) = z^k - \alpha$ ("spectrum analyzer"),

e) $\psi_F(z) = z^k + \sum_{i=0}^{k-1} \beta_i z^i$, $\qquad \beta_i \in R.$

5. Let Σ_i, $i = 1, 2, 3, 4$ be linear systems connected in the following series-parallel network

Compute the input/output relation of this system in terms of the Σ_i. (Remark: According to Fig. 6.2, the brain consists of a staggeringly large number of copies of this kind of network.)

6. Prove that the properties of complete reachability and complete observability are *generic*, i.e., in the space of all linear systems, the reachable and observable systems form an open, dense set. (Remark: This means that for each reachable and observable system, there is an open neighborhood of the system containing only reachable and observable systems. Further, every neighborhood of an unreachable and/or unobservable system contains a reachable and observable system.)

7. *(Duality Principle)* Given the **control** system $\Sigma_C = (F, G, -)$, show that any properties of Σ_C can be expressed as equivalent properties of the **observation** system $\Sigma_O = (F', -, G')$, i.e., the two systems Σ_C and Σ_O are *dual* under the transformation

$$F \longrightarrow F', \qquad G \longrightarrow H'.$$

(Remark: This pivotal result enables us to halve the work in analyzing system structure by considering *only* reachability or *only* observability properties of the system. The Duality Principle also holds for nonlinear systems, at the expense of a more elaborate definition of Σ_C and Σ_O.)

8. Let W_1 and W_2 be two transfer matrices. We say that W_1 *divides* W_2 if there exist polynomial matrices V and U such that $W_1 = V W_2 U$. We say that the system Σ_2 *simulates* Σ_1 if the state-space X_1 of Σ_1 is isomorphic to a subspace of X_2.

a) If $\{\psi_i(W)\}$ denote the *invariant factors* of W, $i = 1, 2, \ldots r$, show that Σ_1 can be simulated by Σ_2 if and only if $\psi_i(W_1)$ divides $\psi_i(W_2)$ for all i. (Remark: This means that a computer with transfer matrix W_2 can simulate a brain with transfer matrix W_1 if and only if each invariant factor of the brain is a divisor of the corresponding invariant factor of the computer.)

b) Show that from a dynamical point of view, the condition $W_1 | W_2$ means that the inputs and outputs of the machine having transfer matrix W_2 are re-coded by replacing the original input ω_2 by $U(z)\omega_2$, whereas the output γ_2 is replaced by $V(z)\gamma_2$. Such a change involves a delay $d = \deg \psi_{w_2}$, where ψ_{w_2} is the denominator polynomial of W_2.

9. Let $\mathcal{K}$ be the subspace of R^n defined by $\mathcal{K} = \{x \in R^n : Kx = 0\}$, where $K \in R^{n \times n}$. We say that the system $\Sigma = (F, G, -)$ is *controllable relative to* $\mathcal{K}$ if there exists a time $N \leq n$ such that $Kx_N = 0$. In other words, every state in $\mathcal{K}$ is controllable.

We call Σ *conditionally controllable* from a subspace $\mathcal{M}$ if and only if for every $x_0 = My$, $y \in R^n$, there is a control input that drives x_0 to the origin, i.e., there is a time $N \leq n$ such that $x_N = 0$.

a) Prove that Σ is controllable relative to $\mathcal{K}$ if and only if

$$\text{rank } \left[KG \,|\, KFG \,|\, KF^2G \,|\, \cdots \,|\, KF^{n-1}G \right] = \text{rank } K.$$

b) Prove that Σ is conditionally controllable from $\mathcal{M}$ if and only if

$$\text{rank } \left[M \,|\, G \,|\, FG \,|\, \cdots \,|\, F^{n-1}G \right] = \text{rank } \left[G \,|\, FG \,|\, F^2G \,|\, \cdots \,|\, F^{n-1}G \right].$$

10. Prove that it's possible for the constant linear system $\Sigma = (F, G, -)$ to be completely reachable if and only if the number of inputs (the number of columns of the matrix G) is greater than or equal to the number of nontrivial invariant factors of F.

11. *(Pole-Shifting Theorem)* Suppose $\Lambda = \{\lambda_1, \lambda_2, \ldots, \lambda_n\}$ is an arbitrary symmetric set of complex numbers, i.e., if $\lambda \in \Lambda$, then $\bar{\lambda} \in \Lambda$. Assume that the single-input system $\Sigma = (F, g, -)$ is completely reachable. Show that there exists a unique *feedback matrix* $k \in R^{1 \times n}$, such that the closed-loop system matrix $F - gk'$ has Λ as its set of characteristic values. (Note: This result shows that the stability characteristics of a completely reachable system can be arbitrarily altered by application of a suitable feedback control law. See Chapter 7 for more about feedback control and its relations to optimality and stability.)

12. Consider the scalar transfer matrix

$$W(z) = \frac{c_{n-1}z^{n-1} + \cdots + c_1 z + c_0}{z^n + d_{n-1}z^{n-1} + \cdots + d_1 z + d_0}.$$

a) Show that the system $\Sigma = (F, G, H)$ given by

$$F = \begin{pmatrix} 0 & 1 & \cdots & 0 \\ 0 & 0 & \cdots & 0 \\ \vdots & & & \\ 0 & 0 & \cdots & 1 \\ -d_0 & -d_1 & \cdots & -d_{n-1} \end{pmatrix}, \quad G = \begin{pmatrix} 0 \\ \vdots \\ 0 \\ 1 \end{pmatrix}, \quad H = (\, c_0 \; \cdots \; c_{n-1} \,),$$

realizes $W(z)$, i.e., $W(z) = H(zI - F)^{-1}G$.

b) Is this realization canonical, in general?

c) If not, how could you make it so?

13. Consider the transfer matrix
$$W(z) = \frac{z^2 + 2z + 1}{z^3 + z^2 - z - 2} \, .$$
a) Show that the behavior sequence associated with W is
$$\mathcal{B} = \{1, 1, 1, 2, 1, 3, \dots\}.$$
b) Show that a canonical realization of $W(z)$ has dimension $n = 3$.

14. Consider the two scalar behavior sequences
$$\mathcal{B}_1 = \{1, 1, 1, 2, 1, 3, \dots\},$$
$$\mathcal{B}_2 = \{1, 2, 3, 4, 5, 6, \dots\}.$$

a) In the preceding Problem we saw that $\mathcal{B}_1$ has a minimal realization of dimension $n_1 = 3$. Show that the sequence of natural numbers $\mathcal{B}_2$ has a canonical realization of dimension $n_2 = 2$.

b) Now consider two-input, single-output system $(m = 2, p = 1)$ whose behavior sequence is
$$\mathcal{B} = \{(1 \ 1), (1 \ 2), (1 \ 3), (2 \ 4), (1 \ 5), (3 \ 6), \dots\},$$
$$= \{A_1, A_2, A_3, \dots\}.$$
This behavior sequence is formed by combining the behaviors $\mathcal{B}_1$ and $\mathcal{B}_2$ componentwise, which suggests that a minimal realization of $\mathcal{B}$ might have dimension $n_1 + n_2 = 5$. Show by direct calculation or otherwise that this conjecture is false. Specifically, show that a canonical realization of $\mathcal{B}$ has dimension $n_B = 4$.

15. Prove that a transfer matrix $W(z)$ has a finite-dimensional canonical realization if and only if W is a strictly proper rational matrix, i.e., the numerator of every component of W is of lower degree than the denominator.

16. Let $W(z)$ be a strictly proper rational transfer matrix of degree n. Assume that the denominator of W has only simple zeros $z_1, z_2, \dots, z_n$. Define the matrices
$$K_i = \lim_{z \to z_i} (z - z_i) W(z),$$
and let $r_i = \operatorname{rank} K_i$, $i = 1, 2, \dots, n$. By definition of rank, there exist matrices L_i and M_i of sizes $p \times r_i$ and $r_i \times m$, respectively, such that $K_i = L_i M_i$. Prove that the following system $\Sigma_W = (F_w, G_w, H_w)$ canonically realizes $W(z)$:
$$F_W = \operatorname{diag}(z_1 I_{r_1}, z_2 I_{r_2}, \dots, z_n I_{r_n}),$$
$$G_W = \begin{bmatrix} M_1 \\ M_2 \\ \vdots M_n \end{bmatrix}, \qquad H_W = \begin{bmatrix} L_1 & L_2 & \cdots & L_n \end{bmatrix}.$$

(Note: Here "diag $(\cdot)$" denotes the diagonal matrix whose elements are $z_i I_{r_i}$, where I_{r_i} is the $r_i \times r_i$ identity matrix.)

Notes and References

§1. The brains-minds-machines issue has never been more actively pursued than today, providing the central focus for the academic discipline now called the "cognitive sciences." A sampling of the various views on these matters is available in

Flanagan, O., *The Science of the Mind,* revised edition, MIT Press, Cambridge, 1991,

Churchland, P. S., *Neurophilosophy,* MIT Press, Cambridge, MA, 1986,

Churchland, P. M., *Matter and Consciousness,* MIT Press, Cambridge, 1984,

Gunderson, K., *Mentality and Machines,* 2d ed., University of Minnesota Press, Minneapolis, MN, 1985,

Aleksander, I. and P. Burnett, *Thinking Machines,* Oxford University Press, Oxford, 1987.

For an account of the brain-mind issue by two of the twentieth century's most prominent thinkers, see

Popper, K., and J. Eccles, *The Self and Its Brain,* Springer, Berlin, 1977.

§2. Mathematical system theory appears to have gotten its initial push from the problem of how to design an electrical circuit with specific performance characteristics using a minimal number of components. The generalization of this problem has led to the mathematical theory of linear systems considered in this chapter.

§3. An excellent summary of the development of the behaviorist and cognitive schools of psychology is given in the Flanagan book cited under §1. Further consideration of some of the issues involved, especially from the philosophical standpoint, is given in

Fodor, J., "The Mind-Body Problem," *Scientific American,* 244 (1981), 114–124,

Mind and Cognition: A Reader, W. G. Lycan, ed., Blackwell, Cambridge, MA, 1990,

The Mind and the Machine, S. Torrance, ed., Ellis Horwood, Chichester, UK, 1984.

Important critiques of the principles and programs of behavioral psychology are

Chomsky, N., "Review of Skinner's *Verbal Behavior*," in *Readings in the Philosophy of Language*, L. Jakobovits and M. Miron, eds., Prentice-Hall, Englewood Cliffs, NJ, 1967,

Dennett, D., "Skinner Skinned," in Dennett, D., *Brainstorms: Philosophical Essays on Mind and Psychology*, Bradford Books, Montgomery, Vermont, 1978.

The point of view espoused in this chapter is very close to that of the "Bottom-Up" school of thought in artificial intelligence (AI), especially as it pertains to the mind-body problem. For introductory accounts of AI, see

McCorduck, P., *Machines Who Think*, Freeman, San Francisco, 1979,

Boden, M., *Artificial Intelligence and Natural Man*, Basic Books, New York, 1977,

Hunt, M., *The Universe Within: A New Science Explores the Human Mind*, Simon & Schuster, New York, 1982,

Haugeland, J., *Artificial Intelligence: The Very Idea*, MIT Press, Cambridge, MA, 1985,

Casti, J., *Paradigms Lost: Images of Man in the Mirror of Science*, Chapter 5, Morrow, New York, 1989 (paperback edition: Avon, New York, 1990).

For some adverse views on AI, see

Searle, J., "Minds, Brains and Programs," *The Behavioral and Brain Sciences*, 3 (1982), 417–457,

Searle, J., *Minds, Brains and Science*, Harvard University Press, Cambridge, MA, 1984,

Weizenbaum, J., *Computer Power and Human Reason*, Freeman, San Francisco, 1976,

Dreyfus, H., and S. Dreyfus, *Mind Over Machine*, Free Press, New York, 1986,

Winograd, T. and F. Flores, *Understanding Computers and Cognition*, Addison-Wesley, Reading, MA, 1987,

Lucas, J., "Minds, Machines and Gödel," in *Minds and Machines*, A. Anderson, ed., Prentice-Hall, Englewood Cliffs, NJ, 1964,

Penrose, R., *The Emperor's New Mind*, Oxford University Press, Oxford, 1989.

§4–5. Textbook accounts of the external versus internal way of looking at dynamical processes are

Casti, J., *Linear Dynamical Systems,* Academic Press, Orlando, 1987,

Brockett, R., *Finite-Dimensional Linear Systems,* Wiley, New York, 1970,

Fortmann, T., and K. Hitz, *An Introduction to Linear Control Systems,* Dekker, New York, 1977,

Rosenbrock, H. *State-Space and Multivariable Theory,* Wiley, London, 1970,

Kailath, T., *Linear Systems,* Prentice-Hall, Englewood Cliffs, NJ, 1980.

§6. The recognition of reachability and observability as the key system-theoretic properties needed to determine canonical models is traceable to the work of Rudolf Kalman in the 1960s. The basic references are

Kalman, R., "On the General Theory of Control Systems," *Proc. 1st IFAC Congress, Moscow,* Butterworths, London, 1960,

Kalman, R., "Canonical Structure of Linear Systems," *Proc. Nat. Acad. Sci. USA,* 48 (1962), 596–600.

It's also of interest to note the relationship between the reachability of a single-input linear system and the nth-order differential equation

$$\frac{d^n x}{dt^n} + a_1 \frac{d^{n-1} x}{dt^{n-1}} + \cdots + a_n x = u(t),$$

as customarily seen in classical engineering and physics texts. Rewriting this nth-order system in matrix form, we arrive directly at the *control canonical* form given in Problem 2(a). It's then an easy matter to verify that every such nth-order, constant coefficient, linear system is completely reachable. Hence, this critical system-theoretic property is "built in" by the manner in which the equation is expressed. This observation underscores the point that the first step to a good theory is often a good notation!

§7. For many more details and examples on reachability and observability for linear systems, see the texts cited under §4–5 above.

§8. The Ho Realization Algorithm was first presented in

Ho, B. L., and R. E. Kalman, "Effective Construction of Linear State-Variable Models from Input/Output Functions," *Regelungstech.,* 14 (1966), 545–548.

Other algorithms can be found in the books by Casti and Brockett noted under §4–5 above. Of special interest are the computational aspects of the various realization methods, some of which are notoriously unstable from a numerical standpoint. An especially detailed study of the numerical aspects of a well-known recursive algorithm for computing minimal realizations is presented in

de Jong, L. S., "Numerical Aspects of Recursive Realization Algorithms," *SIAM J. Cont. Optim.*, 16 (1978), 646–659.

§9. The partial realization question has been extensively explored by the engineer R. E. Kalman in work summarized in

Kalman, R., "On Partial Realizations, Transfer Functions and Canonical Forms," *Acta Poly. Scand.*, 31 (1979), 9–32.

For other important work along the same lines, see

Kalman, R., "On Minimal Partial Realizations of a Linear Input/Output Map," in *Aspects of Network and System Theory*, R. Kalman and N. de Claris, eds., Holt, New York, 1971,

Rissanen, J., "Recursive Identification of Linear Systems," *SIAM J. Cont. Optim.*, 9 (1971), 420–430.

An account of some of the relationships between the Partial Realization Problem for linear systems and the properties of Hankel and Toeplitz matrices is given in

Fuhrmann, P., "On the Partial Realization Problem and the Recursive Inversion of Hankel and Toeplitz Matrices," in *Linear Algebra and Its Role in Systems Theory*, R. Brualdi, D. Carlson, B. Datta, C. Johnson, and R. Plemmons, eds., Amer. Math. Soc., Providence, RI, 1985.

§10. As we have defined it, the Realization Problem is a noise-free version of the *statistical* problem of parameter identification for a linear model, when observations of the output are made in the presence of noise. A simple version of the statistical problem is as follows. We are given the scalar input/output relation

$$y_t = \sum_{i=1}^n A_{t-i} u_i + \epsilon_t, \qquad t = 1, 2, \dots ,$$

where the observations $\{y_t\}$ and the inputs $\{u_t\}$ are known exactly. The quantities $\{\epsilon_t\}$ are assumed to be normally-distributed random variables representing the uncertainty in the model. The problem is to develop a "good" internal model of the form

$$x_{t+1} = Fx_t + Gu_t,$$
$$y_t = Hx_t + \epsilon_t,$$

so that the behavior of the model agrees with that of the input/output relation in some agreed-upon statistical sense. For various approaches to this kind of "stochastic" realization problem, see the works

Hazewinkel, M., and A. H. G. Rinnooy Kan, eds., *Current Developments in the Interface: Economics, Econometrics, Mathematics,* Reidel, Dordrecht, 1982,

Kalman, R., "Identification of Linear Relations from Noisy Data," in *Developments in Statistics,* P. R. Krishnaiah, ed., Vol. 4, Academic Press, New York, 1982,

Kalman, R., "Identification of Noisy Systems," *Russian Math. Surveys,* 40 (1985), 25-42,

Deistler, M., and B. Pötscher, eds., *Modeling Problems in Econometrics, Applied Mathematics & Computation,* Vol. 20, Nos. 3 and 4, 1986,

Hannan, E. and M. Deistler, *The Statistical Theory of Linear Systems,* Wiley, New York, 1988.

Of special interest in connection with the problem of approximate modeling is work by Jan Willems, which outlines a theory for measuring the tradeoff between the power of a model and the amount of error one is willing to tolerate in the model's predictions. This work is summarized in

Willems, J., "An Approach to Exact and Approximate Modeling of Time Series," *Contemporary Math.,* 47 (1985), 479-490.

Willems has also looked at the question of how to bring system-theoretic thinking to bear on classical problems of physics like Newton's law of gravitation. For an account of these ideas, see

Willems, J., "System-Theoretic Models for the Analysis of Physical Systems," *Richerche de Auto.,* 10 (1979), 71-106.

§11. The treatment given here of the concept of state may seem needlessly abstract, or even eccentric, to those schooled along more traditional lines of classical physics, where the state is usually thought of as a *phase* space whose elements are the position and velocity vectors of the various components of the system. This classical, almost Newtonian, view is a very special case of the far more general notion of state outlined here, and it's a major conceptual mistake to think of the state as being something to which we can attach any sort of intuitive, physical interpretation. Our presentation makes clear the primary role of the state space as a purely mathematical construction, introduced to mediate between the measured inputs and observed outputs. Any physical interpretation that one can attach to these states is purely coincidental and quite independent of their primary function. The modern algebraic view of linear systems further amplifies and clarifies this point. For an account of this algebraic point of view, see the Casti book cited under §4–5 or

Kalman, R., P. Falb and M. Arbib, *Topics in Mathematical System Theory,* McGraw-Hill, New York, 1969,

Tannenbaum, A., *Invariance and System Theory: Algebraic and Geometric Aspects,* Springer Lect. Notes in Math., Vol. 845, Berlin, 1981.

§12–14. A much more detailed account of the arguments presented in these sections is found in

Casti, J., "Behaviorism to Cognition: A System-Theoretic Inquiry into Brains, Minds and Mechanisms," in *Real Brains, Artificial Minds,* J. Casti and A. Karlqvist, eds., Elsevier, New York, 1987, pp. 47–75.

Mathematical models of various brain and mind functions have appeared with great regularity in the literature over the past couple of decades. For instance, William Hoffman has developed a model of brain function that rests on the premise that a neuron is an infinitesimal generator of perceptions, cognition and emotion. These ideas make extensive use of the correspondence between a Lie group germ and neuron morphology to give a very thought-provoking account of many aspects of form memory and vision. Hoffman employs the usual mathematical structure governing invariance in the presence of an infinitesimal generator, namely, Lie transformation groups together with their prolongations, to establish higher-order differential invariants. These invariants then show how the memory engram is stored within the brain. For an account of these ideas, see

Hoffman, W., "The Neuron as a Lie Group Germ and a Lie Product," *Q. Applied Math.,* 25 (1968), 423–441,

Hoffman, W., "Memory Grows," *Kybernetik,* 8 (1971), 151–157,

Hoffman, W., "Subjective Geometry and Geometric Psychology," *Math. Modelling,* 3 (1981), 349–367,

Hoffman, W., "The Visual Cortex is a Contact Bundle," *Appl. Math. & Comp.,* 32 (1989), 137–167.

In a quite different direction, the neurophysiologist Karl Pribram and the theoretical physicist David Bohm have jointly proposed a model of holographic memory storage in the brain, which was briefly considered in Discussion Question 9. Their holographic view has the great merit that it enables one to account for the observed fact that patients with brain damage do not seem to lose their memory function, suggesting that memory is stored in a distributed manner throughout the brain rather than being localized to particular regions. The Bohm-Pribram theory is discussed more fully in

Bohm, D., *Wholeness and the Implicate Order,* Routledge and Kegan Paul, London, 1980,

Pribram, K., "Towards a Holonomic Theory of Perception," in *Gestalttheorie in der modernen Psychologie,* S. Ertel, ed., Steinkopff, Durnstadt, 1975.

A more popular account of many of the same ideas is available in

Talbot, M., *The Holographic Universe,* HarperCollins, New York, 1991.

In a long series of papers over more than two decades, Stephen Grossberg has proposed a theory of learning, perception, cognition, development and motor control involving a rather elaborate theory of nonlinear processes. This theory emphasizes the role of "adaptive resonances" in neural circuitry to explain behavioral phenomena. The distilled essence of this work is summarized in the volume

Grossberg, S., *Studies of Mind and Brain,* Reidel, Dordrecht, 1982.

§15–16. Most of the results presented here within the context of linear systems can be extended to a wide class of nonlinear systems—but at the price of a somewhat more elaborate mathematical formalism and machinery. For an account of these matters, see

Casti, J., *Nonlinear System Theory,* Academic Press, Orlando, FL, 1985,

Sontag, E., *Mathematical Control Theory,* Springer, New York, 1990,

Nijmeijer, H. and A. J. van der Schaft, *Nonlinear Dynamical Control Systems,* Springer, New York, 1990,

Isidori, A., *Nonlinear Control Systems,* 2nd Edition, Springer, Berlin, 1989.

Just to see how linear realization procedures extend to the bilinear case, the reader should have a look at

d'Alessandro, P., A. Isidori, and A. Ruberti, "Realization and Structure Theory of Bilinear Dynamical Systems," *SIAM J. Control,* 12 (1974), 517-535.

DQ #11. The Turing test was first proposed by Alan Turing in the article

Turing, A., "Computing Machinery and Intelligence," *Mind,* 59 (1950) (Reprinted in the Anderson volume cited under §3 above).

The Turing test has been subjected to many critiques in the intervening 40 years. Many of them are discussed in the anti-AI volumes also noted earlier under §3.

DQ #13. A discussion of the pros and cons of Gödel's Theorem as an anti-AI argument are given in the well-known volume

Hofstadter, D., *Gödel, Escher, Bach: An Eternal Golden Braid,* Basic Books, New York, 1979.

In this connections, see also Chapter 5 of the Casti book cited under §3.

DQ #15. The Chinese Room argument is brought out in great detail in the Searle material cited under §3. Of special interest in this connection is the article from *Behavioral and Brain Sciences,* since it contains peer commentary from leading philosophers and computer scientists on the adequacy of Searle's claims.

DQ #16. For an account of connectionism as an AI strategy, see

Connections and Symbols, S. Pinker and J. Mehler, eds., MIT Press, Cambridge, MA, 1988,

Parallel Models of Associative Memory, G. Hinton and J. Anderson, eds., Lawrence Erlbaum Associates, Hillsdale, NJ, 1981.

DQ #18. For an introductory account of Leontief's theory, see

Leontief, W., "Structure of the World Economy: Outline of a Simple Input-Output Formulation," *Amer. Econ. Rev.,* 64 (1974), 823-834.

PR #8. A more extensive discussion of this problem is found in

Kalman, R., "Lectures on Controllability and Observability," Centro Internazionale Matematico Estivo Summer Course 1968, Cremonese, Rome, 1968.

PR #9. For more details about relative and conditional reachability, see the Casti text on linear systems cited under §4–5 above.

CHAPTER SEVEN

TAMING NATURE AND MAN: CONTROL, ANTICIPATION AND ADAPTATION IN SOCIAL AND BIOLOGICAL PROCESSES

1. Classical Optimal Control Theory

At various points in our narrative, we've touched upon the matter of influencing the behavior of a dynamical process by regulating its input. In the last chapter we saw how the input to a system could be considered as a pattern of stimuli generating a behavioral response, while other chapters have dealt with variations in system parameters that can, under appropriate circumstances, cause a system to radically change its mode of behavior. However, in most of these cases it was at least tacitly assumed that the variation in the inputs or parameters was brought about more by the vagaries of nature than by human design. This is a pretty passive version of control, and hardly agrees with what we usually mean when we think of a process as being "controlled." Consequently, in this chapter we'll take a more activist point of view, looking at how we can regulate a dynamical process in order to make it behave in accordance with the wishes of a controller rather than the whims of nature. Such a perspective carries along with it the idea that the benefits of control don't come for free, and that we have to weigh those benefits against the cost of exerting the controlling actions. This trade-off leads directly to the idea of an *optimal control process,* in which optimality is defined relative to some criterion that measures both the cost of control and the benefits of the desired system behavior.

Classical optimal control theory arose out of the so-called "simplest problem" of the calculus of variations. The prototypical control process is that of finding a function (input) $u(t)$, $0 \leq t \leq T$, such that the integral

$$J(u) = \int_0^T g(x, u) \, dt,$$

is minimized. Here the system state $x(t)$ is connected to the control input $u(t)$ through the differential equation

$$\dot{x} = f(x, u), \qquad x(0) = x_0.$$

There are many variations upon the basic theme of this problem, variations involving the assumptions we make about the space of admissible control inputs Ω, the constraints imposed upon the system state, the degree of smoothness in the functions f and g, whether T is finite or infinite, other types of nonintegral criteria and so forth. To illustrate a few of the many ways control processes arise, let's look at some examples.

Example 1: Estimation and Control

Suppose we have a dynamical process governed by the equation

$$\dot{x} = f(x,t) + \text{dynamical error}, \qquad x(0) = c,$$

giving rise to the observed output

$$y(t) = h(x,t) + \text{ observational error.}$$

Here, for simplicity, let's assume that f and h are scalar-valued functions, and that the initial state c is known exactly. The task is to estimate the state of the process over some finite time period $0 \leq t \leq T$.

Suppose we agree to let the error term in the dynamics be represented by a function $u(t)$, and agree to select this function in some fashion so as to produce the optimal estimate of the state. This function accounts for both our uncertainty about the true dynamics, as well as whatever stochastic effects may be influencing the state. It's clear that, on the one hand, we want to choose u so that the system state follow a trajectory reasonably close to that dictated by the vector field f. On the other hand, we don't want our estimate to depart too much from what we have actually observed. In short, the control function u is to be chosen to optimally trade off these two costs.

Measuring the costs by a least-squares criterion, we are led to the control problem:

$$\min_{z} \int_0^T \left[k_1(t)\{y(t) - h(x(t))\}^2 + k_2(t)\{\dot{z}(t) - f(z(t),t)\}^2 \right] dt.$$

Here $k_1(t)$ and $k_2(t)$ are nonnegative weighting functions representing our relative confidence in the dynamics f and the observations y. We can recast this formulation into more familiar terms by recalling that the control function u was introduced to account for dynamical error, i.e., we have assumed that the *true* dynamics are

$$\dot{x} = f(x,u,t).$$

Thus, the optimal control problem becomes

$$\min_{u} \int_0^T \left[k_1\{y(t) - h(x(t))\}^2 + k_2 u^2(t) \right] dt,$$

subject to the constraint

$$\dot{x} = f(x,t) + u, \qquad x(0 = c.$$

In subsequent sections we will see how to find such a minimizing control function $u(t)$.

Example 2: Allocation Processes

A type of problem that crops up regularly in decisionmaking environments involves the allocation of some resource, like money, materials or time, among several competing activities. Suppose we have an amount X of a single resource, and there are N activities to which we can parcel out the resource. Assume that if we allocate an amount x_i to activity i, we receive a return $g_i(x_i)$, $i = 1, 2, \ldots, N$. The task is to decide how to expend the resource X in order to maximize the overall return.

This simple allocation process can be easily formulated as a discrete-stage process. We wish to maximize the quantity

$$\sum_{i=1}^{N} g_i(x_i),$$

subject to the constraints

$$x_1 + x_2 + \cdots + x_n \leq X, \qquad x_i \geq 0, \qquad i = 1, 2, \ldots, N.$$

Sometimes it's even possible to obtain a closed-form solution to such multistage allocation processes by making various sorts of structural assumptions about the nature of the return functions $\{g_i\}$. We'll take up this question again in the Problems section.

Example 3: Stochastic Control

In Example 1 we considered a situation in which there is uncertainty in both the system dynamics and the system state. But we didn't specify the nature of the uncertainty. Often we are able to say that the uncertainty is due solely to random disturbances, and that there is no error in our choice of the system's vector field f. So let's assume that we know the probability distribution governing the disturbance. In this case, we can write the dynamics in discrete time as

$$x_{t+1} = f(x_t, u_t, w_t), \qquad x_0 = c, \tag{$\ddagger$}$$

where u_t is the controlling action, and w_t represents the random disturbance, both of which are acting on the state at time t.

In this situation, of course, the state trajectory is not fixed by the choice of the control inputs, but depends on how the random quantities $\{w_t\}$ happen to "come out." So it makes no sense to speak of **the** system trajectory. Rather, we can speak only about an *ensemble* of trajectories, the likelihood of any particular member of the ensemble appearing being dictated by the probability distribution function governing the random variables $\{w_t\}$. For simplicity, let's suppose that these random quantities are independent and identically distributed, so that there is just a single probability distribution function $p(w)$ to worry about.

With these stochastic considerations in mind, let's suppose we wish to minimize the sum

$$J = \sum_{t=0}^{N} g(x_t, u_t, w_t),$$

over all possible choices of the control inputs $\{u_t\}$. Since the state trajectory depends on how the random variables happen to come out, so does the value of the criterion J. So let's agree that we'll choose our controls so as to minimize the expected value of J, where the expectation is taken over the probability density function p. This finally leads to the well-defined control problem:

$$\min_{u} \sum_{i=1}^{M} p(w_t^i) \sum_{t=0}^{N} g(x_t, u_t, w_t^i),$$

where w_t^i is the ith possible value for the random variable w_t and $p(w_t^i)$ is the probability of its occurrence. The state and control are related through the dynamics (‡).

Sometimes it happens that we don't know the probability density function p, in which case we are faced with a so-called *adaptive control problem*. In this event, we often try to use measurements of the system's behavior in order to uncover the nature of the probability law p. We'll see examples of this also as we go along. But for now let's return to a consideration of the methods for actually finding the optimal controls for these kinds of problems.

Two complementary lines of attack have been developed to deal with such questions: the *open-loop* approach, which emphasizes use of the Pontryagin Minimum (or Maximum) Principle, and the *closed-loop* approach using Bellman's Principle of Optimality and dynamic programming. The next two sections look briefly at each of these approaches.

Exercises

1. Suppose you are the head of the US Federal Reserve System, and it's your task to set the discount rate, i.e., the interest rate that the Fed charges banks around the country for the money that's supplied to them. In coming to your decision as to what the "right" discount rate should be, you must take into consideration things like the inflation rate, the rate of money growth, unemployment, the dollar/yen/D-mark exchange rates, as well as the political climate. Consider how you would formulate this decisionmaking task as an optimal control problem.

2. Assume that you are a refinery manager and have X barrels of oil that's to be used over a one-year period. Part of this oil is to be sold at monthly intervals, while the other part is to be retained for use in the production of more oil. Suppose that d_k is the number of barrels you sell in month k, for which you receive a return of $r(d_k)$. Moreover, let's suppose you use z_k barrels in month k, which produces a yield of az_k barrels of oil at the end of the month, $a < 1$. Your problem is to decide how much oil should be sold each month in order to maximize your return over the 12-month period. (a) Formulate this problem as an optimal allocation process. (b) Assume the return function has the form $r(d_k) = bd_k$, $b > 0$. Show that in this case, the optimal allocation policy is to sell nothing during the first eleven months, and then sell everything at the end.

2. The Minimum Principle and Open-Loop Control

For the sake of exposition and to avoid annoying and distracting technical caveats, let's assume that the functions g and f, as well as the space of admissible inputs and the constraint space, possess whatever mathematical properties we need in order to have our formal manipulations make sense. To find the function $u^*(t)$ minimizing the integral J under the differential equation side constraint, the Pontryagin approach involves introducing a Lagrange multiplier function $\lambda(t)$ and then using this function to define a Hamiltonian $H(x, u, \lambda, t)$ for the system. The Pontryagin Minimum Principle then asserts that the optimal control $u^*(t)$ is the function u that minimizes H. Let's take a look at how this method works.

Let the control input $u(t) \in R^m$ with the state $x(t) \in R^n$ for each $t \in [0, T]$. Using the *co-state* vector $\lambda(t)$, we form the Hamiltonian function H as

$$H(x, u, \lambda, t) = g(x, u, t) + \lambda' f(x, u, t),$$

where "$'$" denotes the transpose operation. Thus, H is a scalar-valued function of x, u and λ. According to the Minimum Principle, the control function $u^*(t)$ that minimizes the criterion J is exactly that element $u \in \Omega$ that minimizes H pointwise in t. Thus,

$$u^*(t) = \arg \min_{u \in \Omega} H(x, u, \lambda, t).$$

Of course, the function H contains the unknown co-state function λ. So in order to minimize H and find u^*, we must develop a relation between u and λ. Following an approach very similar to that employed in the calculus of variations to obtain the classical Euler-Lagrange equation, it can be shown that the functions x, u and λ are related to each other through satisfaction of the following two-point boundary value problem:

$$\dot{x} = \frac{\partial H}{\partial \lambda}(x, u, \lambda, t), \qquad x(0) = x_0,$$

$$-\dot{\lambda} = \frac{\partial H}{\partial x}(x, u, \lambda, t), \qquad \lambda(T) = 0, \qquad 0 \leq t \leq T.$$

Assuming there are no constraints on the allowable variation in the control u, a necessary condition for the optimizing input function u^* is that it satisfy the equation

$$\frac{\partial H}{\partial u}(x, u, \lambda, t) = 0,$$

where the above boundary-value problem is used to express both x and λ as functions of u.

We have assumed that the final time T is fixed, but that there are no conditions imposed on the final state $x(T)$. This choice led to the boundary condition given above on the co-state function $\lambda(t)$. If the final time is not fixed, determination of what this boundary condition should be is a bit more complicated, leading to what are termed *transversality conditions*. For details the reader is referred to the texts cited in the chapter Notes and References. Now let's see how the Minimum Principle works in action.

Example 1: Linear Dynamics, Quadratic Criterion

To illustrate the foregoing approach, consider the case when the criterion function g is quadratic and the state dynamics f are linear. We then have

$$g(x, u, t) = \tfrac{1}{2}[(x, Qx) + (u, Ru)],$$

$$f(x, u, t) = Fx + Gu,$$

where we impose the conditions $Q \geq 0$, $R > 0$ to ensure that a solution to the problem exists and that it is unique. In this case, the Hamiltonian function is

$$H(x, u, \lambda, t) = \tfrac{1}{2}[(x, Qx) + (u, Ru)] + (\lambda, Fx + Gu),$$

so the corresponding two-point boundary-value problem becomes

$$\dot{x} = Fx + Gu, \qquad x(0) = x_0,$$

$$-\dot{\lambda} = Qx + F'\lambda, \qquad \lambda(T) = 0.$$

The condition for the minimizing control is

$$Ru^* + G'\lambda = 0.$$

This, in turn, leads to the optimal control law

$$u^*(t) = -R^{-1}G'\lambda(t).$$

Thus we see that in order to compute the optimizing control function $u^*(t)$, we must solve the boundary-value problem for the co-state function $\lambda(t)$.

Example 2: Constrained States and Controls

The relatively simple form of the solution in the preceding example is a consequence of both the linearity of the dynamics and the absence of any constraints other than the differential equation specifying the dynamics. But even if we drop one or another of these simplifying assumptions, the Minimum Principle can still be effectively employed. To illustrate, in this example we retain the linearity but impose restrictions on both the states and controls.

Suppose we wish to minimize the integral

$$J(u) = \frac{1}{2} \int_0^T \left[x_1^2(t) + u^2(t) \right] dt,$$

where the system dynamics are

$$\dot{x}_1(t) = x_2(t), \qquad x_1(0) = c_1,$$
$$\dot{x}_2(t) = -x_2(t) + u(t), \qquad x_2(0) = c_2.$$

In addition, we restrict the admissible controls by requiring that

$$-1 \le u(t) \le 1, \qquad 0 \le t \le T.$$

To use the Minimum Principle, we must form the relevant Hamiltonian for the system. This function is easily found to be

$$H(x, u, \lambda) = \frac{1}{2} \left[x_1^2(t) + u^2(t) \right] + \lambda_1(t) x_2(t) - \lambda_2(t) x_2(t) + \lambda_2(t) u(t).$$

Using the above Hamiltonian, the necessary conditions for optimality are

$$\dot{x}_1^*(t) = x_2^*(t), \qquad x_1^*(0) = c_1,$$
$$\dot{x}_2^*(t) = -x_2^* + u^*(t), \qquad x_2^*(0) = c_2,$$
$$\dot{\lambda}_1^*(t) = -x_1^*(t), \qquad \lambda_1^*(T) = 0,$$
$$\dot{\lambda}_2^*(t) = -\lambda_1^*(t) + \lambda_2^*(t), \qquad \lambda_2^*(T) = 0.$$

To find the control $u^*(t)$ minimizing H, we first separate all of the terms containing $u(t)$ from the Hamiltonian. This yields

$$\frac{1}{2} u^2(t) + \lambda_2^* u(t).$$

At times t when the optimal control is unsaturated, we have

$$u^*(t) = -\lambda_2^*(t).$$

Clearly, this occurs whenever $|\lambda_2^*(t)| \leq 1$. But when $|\lambda_2^*(t)| > 1$, then the minimizing control is

$$u^*(t) = \begin{cases} -1, & \text{for } \lambda_2^*(t) > 1, \\ +1, & \text{for } \lambda_2^*(t) < -1. \end{cases}$$

To obtain $u^*(t)$, we must solve the state and co-state equations. What's important to note, however, is that the optimal control that emerges is generally **not** the same as what we would obtain if we merely computed the minimizing control for the unconstrained problem and allowed it to saturate whenever the stipulated control constraints are violated.

As the above discussion shows, use of the Minimum Principle always yields the optimal control law $u^*(t)$ as a function of the time t. Intuitively speaking, this kind of solution demands that we look at a clock and determine the current time t. We then compute the quantity $u^*(t)$ and apply this control input at that moment. This is what's termed *open-loop* control, since determination of what action to take at any moment is a function solely of the current time and doesn't involve what the system is actually doing, i.e., what the current system state $x(t)$ happens to be. Note also that from a computational point of view, determination of $u^*(t)$ may not always be an entirely straightforward affair, since it involves the solution of a two-point boundary-value problem for the co-state λ. This is often a nontrivial computational problem, especially if the interval length T is large. On the other hand, the Minimum Principle approach is very flexible in regard to the kind of constraints that can be imposed, especially constraints on the control law u. And the determination of $u^*(t)$ involves only the solution of a set of *ordinary* differential equations, although generally of a nonlinear nature. Now let's look at the alternate approach based upon dynamic programming.

Exercises

1. Consider the problem of finding the control law $u(t)$ that minimizes the integral

$$J(u) = \frac{1}{2} \int_0^T [(x, Qx) + (u, Ru)] \, dt,$$

where x and u are related via the *nonlinear* dynamical equations

$$\dot{x} = f(x, u), \qquad x(0) = c.$$

(a) Formulate the Hamiltonian function for this problem. (b) Use the Minimum Principle to determine the equations that the minimizing function $u^*(t)$ must satisfy. (c) Apply this general result to the case when $x(t) \in R^2$, $u(t) \in R$ and

$$f(x,\, u) = \begin{pmatrix} x_2(t) \\ -x_1(t) + [1 - x_1^2(t)]\, x_2(t) + u(t) \end{pmatrix},$$

$$Q = \begin{pmatrix} 1 & \frac{1}{2} \\ \frac{1}{2} & 1 \end{pmatrix}, \qquad R = (\,1\,).$$

2. The problem of finding the curve $y(x)$ of minimal length between two points $(0,0)$ and $(1,1$ involves minimizing the integral

$$J = \int_0^1 \sqrt{1 + \dot{y}^2(s)}\, ds, \qquad y(0) = 0, \qquad y(1) = 1.$$

(a) Formulate this as an optimal control problem. (b) Using the Minimum Principle, verify that the solution is a straight line, i.e., a curve whose slope is constant.

3. Consider the problem of finding the curve $x(t)$ that minimizes the integral

$$J(x) = \int_0^{4\pi} [\dot{x}^2(s) - x^2(s)]\, ds, \qquad x(0) = 0.$$

(a) Show that this problem has no solution. Specifically, show that the curve $x(t) = 0$ satisfies the necessary conditions arising from the Minimum Principle. But the minimum value of J is **not** zero. (b) Find a function $x(t)$ such that $J(x)$ is arbitrarily large and negative. (Hint: Consider a trigonometric function.)

3. Feedback Control and Dynamic Programming

The key element in the dynamic programming approach to optimal control is captured by Bellman's Principle of Optimality. Assume that $u^*(t)$ is the optimal control function and that $x^*(t)$ is the associated optimal state trajectory. Let $v = u^*(0)$ denote the optimal action to be taken at the initial moment, with the initial state being $x(0) = c = x^*(0)$. Then the Principle of Optimality states that the part of the optimal trajectory starting at time Δ from the state $c + f(c, v, 0)\Delta$ is also the optimal trajectory for a problem that begins not at time $t = 0$ in the state c, but at time $t = \Delta$ in the state $c + f(c, v, 0)\Delta$. In other words, any part of an optimal state trajectory is also an optimal trajectory. The general idea is depicted in Fig. 7.1.

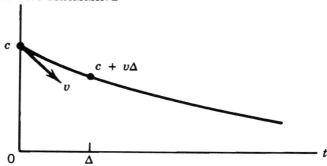

Figure 7.1. The Principle of Optimality

Let's now use the Principle of Optimality to develop an equation for the control that minimizes

$$J = \int_t^T g(x, u) \, dt,$$

where

$$\dot{x} = f(x, u), \qquad x(t) = c.$$

Note that this formulation of the problem agrees with that given earlier if we set $t = 0$, $c = x_0$. So what we are doing here is imbedding the original problem within a *family* of problems parametrized by the initial time t and the initial state c. The minimal value of J, as well as the associated optimal control law, is determined by two quantities: the initial state c and the length of time of the process, which is characterized by $T - t$ since T is assumed to be fixed.

We introduce the function

$I(c, t) = $ the value of J obtained for a process of duration $T - t$
that starts in state c when an optimal control is used.

Then regardless of what initial control action $v = u(0)$ is applied, the Principle of Optimality, together with the definition of the function $I(c, t)$, means that we must have the inequality

$$I(c, t) \le \{g(c, v)\Delta + I(c + f(c, v)\Delta, \, t + \Delta)\} + o(\Delta). \qquad (\dagger)$$

Here we have made use of the additivity property of integrals

$$\int_t^T = \int_t^{t+\Delta} + \int_{t+\Delta}^T,$$

as well as the Mean-Value Theorem for integrals. We clearly want to make the right side of (†) as small as possible. So we choose v to minimize, leading to the recurrence relation for I as

$$I(c,t) = \min_{v}\{g(c,v)\Delta + I(c + f(c,v)\Delta, \, t + \Delta)\} + o(\Delta). \qquad (*)$$

For a process of length $T - t = 0$, we have the trivial starting condition

$$I(c,T) = 0. \qquad (**)$$

The relation $(*)$ links the solution to a problem starting at time t with one starting at time $t + \Delta$. Thus Eq. $(*)$, together with the initial condition $(**)$, enables us to calculate the *optimal-value function* $I(c,t)$ as well as the optimal decision function $v(c,t)$ for each possible initial state c and every interval length $T - t$. So by this approach we solve not only the original problem involving some *fixed* values of c and t, but at the same time solve also a host of additional problems having different initial states and various interval lengths. Let's look for a moment at what's involved in actually computing the functions $I(c,t)$ and $v(c,t)$.

Let's agree to quantize each component of the state into K discrete levels. Further, suppose that each component of the control is quantized into L levels. Then if the state $c \in R^n$, there are K^n distinct values of the state that must be considered at each time moment. Furthermore, if the control $v \in R^m$, we must test each of the L^m control levels against each of these states in order to compute the function $I(c,t)$ for a single value of t. In addition, to calculate $I(c,t)$ it's necessary to have in memory the function $I(\cdot, t + \Delta)$ for *every* possible state value. This means that we need at least K^n locations of high-speed memory, not counting what may be needed for other purposes. Let's say $K = 100$. Then if n is small, say, $n = 1$ or 2, the memory required is modest, on the order of a few thousand locations; but real problems have $n = 10, 20$ or even up to a few hundred. In such cases no brute force use of $(*)$ can be contemplated, even with the biggest and fastest of modern computers, and more sophisticated techniques are required in order to cut this computational requirement down to size. Some of these methods are discussed in the material cited in the Notes and References so we won't go into them here. Note also that though not as severe a restriction as the foregoing memory difficulty, the computational burden also increases markedly as we increase the number of control components and/or refine their discretization. However, the computational requirements only increase linearly with the control discretization rather than displaying the geometric growth seen when we increase the state dimensionality.

If we pass to the limit letting $\Delta \to 0$ in Eq. (*), we obtain the famous Bellman-Hamilton-Jacobi equation for the function I as

$$-\frac{\partial I}{\partial t} = \min_v \left\{ g(c, v) + \left(f(c, v), \frac{\partial I}{\partial c} \right) \right\}, \qquad t < T,$$

$$I(c, T) = 0.$$

With these ideas about how dynamic programming works in principle, let's look at a couple of examples and see how it works in practice.

Example 1: The Linear-Quadratic Problem (cont'd.)

In the last section we showed how to solve a quadratic optimization problem with linear dynamics using the Minimum Principle. Here we look at the same problem, showing how the solution looks when we use the Bellman-Hamilton-Jacobi equation and dynamic programming. Recall that the functions defining the cost criterion and the dynamics of the problem are

$$g(x, u) = \tfrac{1}{2}[(x, Qx) + (u, Ru)],$$

$$f(x, u) = Fx + Gu.$$

The Bellman-Hamilton-Jacobi equation in this case becomes

$$-\frac{\partial I}{\partial t} = \min_v \left\{ \tfrac{1}{2}[(c, Qc) + (v, Rv)] + \left(Fc + Gv, \frac{\partial I}{\partial c} \right) \right\}, \qquad t < T,$$

$$I(c, T) = 0.$$

It's a simple matter to verify that the function $I(c, t)$ is a pure quadratic form in c. That is, there exists a matrix function $P(t)$ such that

$$I(c, t) = (c, P(t)c),$$

which, after some tedious although straightforward algebra, yields

$$v_{\min}(c, t) = v^*(c, t) = -R^{-1}G'P(t)c, \tag{$\ddagger$}$$

where the function $P(t)$ satisfies the matrix Riccati equation

$$-\dot{P}(t) = Q + PF + F'P - PGR^{-1}G'P, \qquad P(T) = 0.$$

The optimal state trajectory is the solution of the *closed-loop* dynamics

$$\dot{x}^*(t) = [F - GR^{-1}G'P(t)]x^*(t), \qquad x^*(t) = c.$$

The special structure of the criterion function and the linear dynamics allow us to obtain a very neat, closed-form representation for the optimal feedback control law in terms of the solution to a matrix Riccati equation. Unfortunately, in most cases these nice analytic features are absent, forcing us to take recourse to the use of numerical methods. Our next example illustrates this point.

Example 2: A Numerical Example

Suppose we want to choose the sequence $\{u_t\}$ to minimize the sum

$$J = \sum_{k=0}^{4} (2 + u_k) 3^{x_k} + |x_5 - 1|,$$

where the system dynamics are

$$x_{t+1} = x_t + \left(2 - 2x_t + \frac{5}{4}x_t^2 - \frac{1}{4}x_t^3\right) u_t, \qquad x_0 = c.$$

Assume that the state and controls are constrained by the relations

$$0 \le x \le 3, \qquad -1 \le u \le 1.$$

The nonlinearities in the dynamics, coupled with the constraints, preclude any simple, closed-form solution. So we compute the optimal control numerically.

Let's agree to quantize the states and controls uniformly, using the discretizations $\Delta x = 1$, $\Delta u = 1$. Under these conditions, the set of admissible states is $X = \{0, 1, 2, 3\}$, while the set of admissible controls is $U = \{-1, 0, 1\}$.

We begin by defining the optimal value function $I(c, t)$ as before. Using the Principle of Optimality, it's straightforward to derive the recurrence relation

$$I(c, t) = \min_{v = \{-1,0,1\}} \left\{ (2 + v)3^{-c} + I\left[c + \left(2 - 2c + \frac{5}{4}c^2 - \frac{1}{4}c^3\right)v, t + 1\right] \right\}$$

$$I(c, 5) = |c - 1|, \qquad c = 0, 1, 2, 3, \qquad t \le 4.$$

In computing the optimal value function, non-quantized values of the state may occur on the right-hand side. If so, we use linear interpolation between values of $I(c, t)$ at the two nearest quantized states. The complete solution for this problem is shown in Table 7.1, where the upper number at each grid point is the optimal control. The lower figure at each grid point represents the optimal cost in starting the system at that state and time.

The most important point to note about the above examples is that Eq. (‡) expresses the optimal control as a function of the system state c, rather than as a function of the current time t. This means that we must *measure* what the system is actually doing and apply the control that is appropriate for that state, rather than just looking at a clock and determining the control as a function of the time alone. Basically, the state measurement is "fed back" from the system output to its input, giving rise to the concept of *feedback*, or *closed-loop*, control as shown in Fig. 7.2.

Table 7.1 Solution of the Dynamic Programming Problem

x \ t	0	−1	−1	−1	−1	
3	0.83720	0.73762	0.69898	0.89236	0.54979	2
2	1	0	0	0	−1	
	1.14363	0.94735	0.67669	0.40601	0.13534	1
1	1	1	1	1	0	
	2.05099	1.78033	1.50965	1.23898	0.73576	0
x = 0	1	1	1	1	0	
	3.94735	3.67669	3.40601	3.13534	3.00000	1
t = 0		1	2	3	4	5

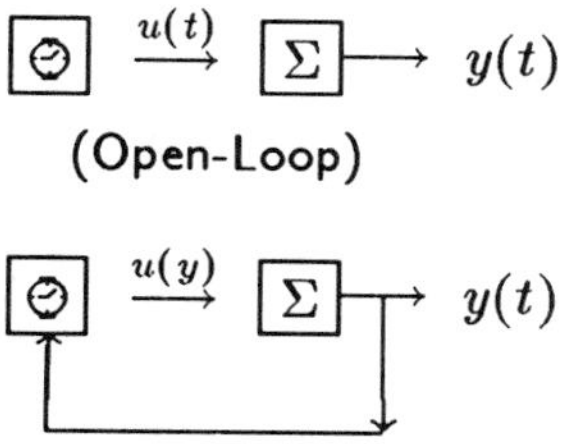

$$\boxed{\odot} \xrightarrow{\ u(t)\ } \boxed{\Sigma} \longmapsto y(t)$$

(Open-Loop)

$$\boxed{\odot} \xrightarrow{\ u(y)\ } \boxed{\Sigma} \longmapsto y(t)$$

(Closed-Loop)

Figure 7.2. Open- versus Closed-Loop Control

We have already seen the "curse of dimensionality" that greatly complicates the computational solution of high-dimensional control problems by dynamic programming. It should also be noted that the basic equation for the optimal-value function I is a partial differential equation rather than the ordinary differential equation obtained using the Minimum Principle. However, instead of being a boundary-value problem, the dynamic programming equation for I is an *initial-value* problem which is generally far easier to deal with computationally than a boundary-value problem. Thus, we can trade off ordinary differential equations and boundary-value problems for partial differential equations and initial-value problems. In general, which method we use depends upon a whole host of circumstances that space constraints prevent us from entering into here. The interested reader is invited to consult the many excellent references on this circle of questions cited in the chapter Notes and References.

In Section 1 we discussed briefly the problem of stochastic control, where there is a known random disturbance affecting either the system dynamics and/or the cost function. Far more realistic, of course, is to assume that the probability law governing these disturbances is **unknown,** and that we learn about it as a result of observing the system's response to our controlling actions. So to conclude our discussion of dynamic programming, let's look at a fairly extensive example showing one way to address this *adaptive control problem.*

Example 3: Adaptive Control

Suppose we are faced with a random variable that can take on the value 0 or 1 with probabilities p and $1 - p$, respectively. Since we don't know p, we decide to conduct a set of experiments involving the random variable, record the outcomes and then estimate p on the basis of these experiments. But there are costs associated with conducting each experiment, as well as costs incurred by making a wrong estimate of p. Our task is to determine when to stop the experiments, as well as to decide what will then be our best estimate of p. Let's specify the problem in more technical detail.

Assume that before the process begins we have *a priori* information that n 1's will occur in N trials. So before the experiments begin, our estimate of the quantity p is $\hat{p} = n/N$. But as far as *actual observations* are concerned, let's assume that m 1's have been seen in M trials. Since p is unknown, we regard it as a random variable with a distribution function that changes during the course of our experiments as we make observations on the underlying process generating the 0's and 1's. Solely on the basis of the a priori information, let's take this distribution function to be

$$dG(p) = \frac{p^{n-1}(1-p)^{N-n-1}}{B(n, N-n)} \, dp,$$

where B is the beta function

$$B(r, s) = \frac{r! \, s!}{(r+s+1)!}.$$

After m 1's have been observed in M trials, we modify dG by the rule

$$dG_{M,m}(p) = \frac{p^m(1-p)^{M-m} \, dG(p)}{\int_0^1 p^m(1-p)^{M-m} \, dG(p)}.$$

Let the quantity $c_{M,m}$ represent the expected mean-square error of incorrect estimation after M observations have resulted in m 1's. We set

$$c_{m,M} = \int_0^1 (p_{M,m} - p)^2 \, dG_{M,m}(p),$$

where $p_{M,m}$ is the estimate that minimizes $c_{M,m}$. The value of $p_{M,m}$ is easily seen to be

$$
\begin{aligned}
p_{M,m} &= \int_0^1 p\,dG(p), \\
&= \frac{m+n}{M+N}.
\end{aligned}
$$

A straightforward calculation then yields

$$
c_{M,m} = \alpha\left(\frac{m+n}{M+N}\right)\left(\frac{m+n+1}{M+N+1} - \frac{m+n}{M+N}\right).
$$

Here the parameter α represents the normalization constant needed to ensure that the quantity $f(p) = p^m(1-p)^n$ is indeed a probability density function for nonnegative integers n and m. In our case, it turns out that

$$
\alpha = \frac{1}{\int_0^1 p^m(1-p)^n\,dp}.
$$

The calculations up to now have dealt with the cost of making bad estimates of p. Let's now turn to the cost of actually performing the experiments.

We assume that if M experiments have been made, the price we have to pay for these experiments is $k(M)$. Thus, the cost of experimentation is allowed to vary during the process. In addition, we shall assume that in the absence of information to the contrary, estimated probabilities are to be regarded as true probabilities. Finally, we set an upper bound R on the number of experiments that can be performed. With these considerations in mind, let's see how to formulate the optimal experimentation policy.

Define the function $I(m, M)$ to be the expected cost of a sequence of experiments that begins with m 1's having been observed in M trials when we use an optimal decision policy. Here, of course, the decision to be made at each stage is whether to stop the process or carry on with another experiment. Applying the Principle of Optimality, we obtain the following functional equation for I:

$$
I(m, M) = \min_{C,S}
\begin{cases}
C: & k(M) + p_{m,M}I(m+1, M+1) + \\
& (1 - p_{m,M})I(m, M+1), \\
S: & \alpha\left(\frac{m+n}{M+N}\right)\left(\frac{m+n+1}{M+N+1} - \frac{m+n}{M+N}\right),
\end{cases}
$$

for $m = 0, 1, \ldots, M$, $M = R-1, R-2, \ldots, 0$. In this equation "C" stands for the decision to continue the experiments, while "S" represents the decision to stop the process. In view of the truncation assumption on the maximum number of experiments, we have

$$I(m, R) = \alpha \left(\frac{m+n}{R+N} \right) \left(\frac{m+n+1}{R+M+1} - \frac{m+1}{R+M} \right).$$

These relations enable us to calculate the function $I(m, M)$ and to determine the optimal stopping rule, as well as to optimally estimate p at the end of the process.

The above equations were investigated numerically for a wide range of values of the parameter R and for several cost functions $k(M)$. The results were very illuminating. For instance, when the cost of experimentation $k(M)$ is constant or when it increases with M, with a priori distribution function for p being

$$dG(p) = \begin{cases} 0, & \text{for } -\infty \leq p < 0, \\ \frac{1}{2}, & \text{for } 0 \leq p < 1, \\ 1, & \text{for } 1 \leq p \leq \infty, \end{cases}$$

the optimal policy has the following structure:

   a. Continue the experiments if M is small (not enough information available upon which to base an estimate of p).
   b. Stop the process if M is sufficiently large.
   c. Continue the experiments for intermediate values of M, unless extreme runs of either 0's or 1's occur.

And, of course, the optimal estimate of p is just the last estimated value when the process stops.

The foregoing example, while a bit complex, shows the great power and flexibility of dynamic programming to attack almost any type of process involving control and decisionmaking. Now let's turn our attention for the balance of the chapter to processes of a novel nature, in which the decisions are based not only upon what has happened in the past, but also on what we expect to see in the future. These are the so-called *anticipatory control processes*.

Exercises

1. Let $a_1, a_2, \ldots, a_n$ be a set of nonnegative real number. Use dynamic programming to prove the arithmetic-geometric mean inequality

$$\frac{1}{n} \sum_{i=1}^{n} a_i \geq \left(\prod_{i=1}^{n} a_i \right)^{1/n}.$$

2. The standard linear programming problem involves maximizing the sum

$$\sum_{i=1}^{n} c_i x_i,$$

over all x_i satisfying the constraints

$$x_j \geq 0, \qquad \sum_{j=1}^{n} a_{ij} x_j \leq b_i, \qquad i = 1, 2, \ldots, m; \ j = 1, 2, \ldots, n.$$

Here the quantities c_i, a_{ij} and b_i are all real numbers. Show how to formulate this as a dynamic programming problem.

3. Consider the linear-quadratic control problem with $x(t), u(t) \in R^2$. Suppose we want to minimize

$$J = \sum_{k=0}^{4} \left[x_1^2(k) + x_2^2(k) + u_1^2(k) + u_2^2(k) \right] + \frac{5}{2} \left[(x_1(5) - 2)^2 + x_2(5) - 2)^2 \right].$$

The system dynamics are

$$x_1(k + 1) = x_1(k) + x_2(k) + u_1(k), \qquad x_1(0) = 2,$$
$$x_2(k + 1) = x_2(k) + u_2(k), \qquad x_2(0) = 1.$$

Let the constraints on the states and controls be

$$0 \leq x_1 \leq 2, \qquad -1 \leq x_2 \leq 1,$$
$$-1 \leq u_1 \leq 1, \qquad -1 \leq u_2 \leq 1.$$

Quantize the state variables as $\Delta x_1 = \frac{1}{2}$, $\Delta x_2 = \frac{1}{4}$ but do not quantize the controls. Rather, let the controls take on any admissible value such that for any given quantized present state, the next state is also a quantized state. (This will eliminate the need for any interpolation.) (a) Determine the functional equation for the optimal value function $I(c, k)$. (b) Compute the optimal control policy $u^*(k)$ and the optimal state trajectory $x^*(k)$.

4. Suppose we have a scalar control process governed by the dynamics

$$\dot{x} = u(t) + w(t), \qquad x(0) = -3,$$

where $w(t)$ is a random forcing term with probability density function

$$p(w) = \frac{1}{\sqrt{\pi}} e^{-w^2}.$$

We choose the control function $u(t)$ to minimize the *expected value* of

$$J(u) = \int_0^{10} [x^2(s) + u^2(s)] \, ds + \frac{5}{2} x^2(10),$$

subject to the constraints

$$-3 \leq x \leq 3, \qquad -1 \leq u \leq 1.$$

Quantize the state, control and time as follows: $\Delta x = 1$, $\Delta u = 1$, $\Delta t = 1$. In addition, quantize the probability density function p by the rule

$$p(w = 1) = \frac{1}{4}, \qquad p(w = 0) = \frac{1}{2}, \qquad p(w = -1) = \frac{1}{4}.$$

(Note that this discrete distribution has the same mean and variance as the original continuous density function p.) Using this discrete version of the original problem, compute the optimal control law and the minimal expected cost by dynamic programming. (Answer: The minimal expected cost from the initial state $x(0) = -3$ is 27.36.)

5. A gambler receives advance information over a noisy communication channel about the outcome of certain independent sporting events. Assume that each contest is between two evenly-matched teams, and that p is the probability that the correct result will be transmitted. Assume that the gambler begins with X units of capital and bets so as to maximize his expected capital after making K wagers. Use dynamic programming to determine the gambler's optimal betting policy. (Hint: Define the optimal value function $I(X, k)$ to be the gambler's expected capital after k bets when he uses an optimal betting policy and starts with X units of capital.)

4. Anticipatory Systems

In either open-loop or feedback mode, the foregoing classical control processes are inherently *reactive* in character; i.e., the control action is always taken in response to the past and present behavior of the system itself. Here we want to consider another class of control processes in which the controlling action depends not only upon what the system is (and has been) doing, but also upon what we *think* the system will be doing in the future. We term such processes *anticipatory*.

When we speak of anticipatory processes, we appear to edge dangerously close to ignoring one of the pillars upon which modern theoretical science rests, namely, that effects should not precede their causes. However, as we shall see below, there is no cause for alarm. The type of process we have in mind is perfectly consistent with traditional notions of causality, since we don't claim that the system behavior is determined by *actual* future events, but rather by our current *beliefs* about the nature of those events. As a trivial example, suppose we are standing on a corner waiting to cross to the other side of the street. We look up the street and see a car coming at some speed in our direction. Using the personal computer between our ears, we quickly examine our current position, the position and velocity of the car, and the width of the street and *predict* that if we step off the curb into the street, it may turn out to be our last living act. As a result of this *prediction,* we take the decision (controlling action) to remain where we are and let the car pass by before crossing. In light of this simple example, we see that such anticipatory processes abound in modern life, and it's seems odd that they have not been more extensively studied. Much of the remainder of this chapter is devoted towards rectifying this puzzling state of affairs.

Consideration of both the general idea of an anticipatory process, as well as the simple example above, shows that the crucial ingredient separating the type of classical process discussed earlier from an anticipatory process is the notion of *self-reference.* The anticipatory process contains an internal model of itself, and this model is used on a time scale faster than real time to influence the actions taken at any given moment. Furthermore, as the process unfolds and more information is obtained about the operating environment, this internal model is continually updated according to one scheme or another, thereby providing a means for the system to both adapt and evolve. Before entering into a detailed consideration of how such anticipatory systems arise in nature, let's first take a longer look at the basic components of this kind of process.

For the type of anticipatory system we have in mind, three components are needed: (1) the system Σ itself, (2) a controller C, and (3) a model M of Σ. These three objects are connected together as shown in Fig. 7.3.

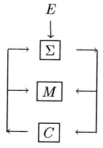

Figure 7.3. The Components of an Anticipatory System

In the diagram, the model M sees the behavior of Σ as well as the results of past control actions from C. As a result of this information, *M predicts* what the future states of Σ will be and transmits this information to the controller C, which then integrates the prediction with its measurements of Σ and the environment E in order to generate the *current* controlling action. This decision is then input to Σ and the cycle repeats itself.

It's of fundamental importance to note two aspects of this setup: (1) the model M must operate on a time scale faster than that of Σ, and (2) the model must utilize information about past behaviors of Σ and past actions of C to improve its predictions about what Σ will be doing in the future; i.e., M must possess some sort of learning and adaptation mechanism. How much faster the time scale of M needs to be in relation to that of Σ is determined by the speed of information processing in C, as well as by the speed of signal transmission from C to Σ and from Σ to C and to M. Basically, what's needed is that M be able to predict far enough into the future of Σ that the predictions can effectively be used to influence the actual behavior of Σ at that future time.

At this point, a traditional control engineer would be likely to comment that the type of anticipatory scheme outlined above seems not to differ much in character from the standard adaptive control procedures already well-scouted in the control engineering literature. It's worthwhile to pause for a moment to address this point.

The standard adaptive control problem is structured along the following lines. We are given a dynamical process

$$\dot{x} = f(x, u, a),$$

where $a \in R^k$ is some vector of unknown parameters. Furthermore, we have the observations

$$y = h(x, u),$$

and our job is to choose a control law both to control the system and to learn about the values of a. In general, the control is then a function not only of the state x, but also of our current best estimate of the parameters a, i.e., $u = u(x, \hat{a})$, where we have written $\hat{a}$ to represent the estimated value of the parameters. As the process unfolds, we obtain new observations y and we process them according to one learning scheme or another to update our estimate of a. The important point to note here is that the actual control input is determined only as a function of the current (and, perhaps, past) state and our current best estimate of the parameters.

By way of contrast, in an anticipatory system, besides the above dynamics, observations, and learning scheme for a, we have a model M that predicts future states of the system based upon all this past information. Thus, the control that is actually applied at the current moment is a function not only of the current and past states and the best estimate of the parameters, but also of the predicted future states. Thus,

$$u = u(x(t), \hat{a}, \hat{x}(t + T_i)), \qquad T_i > 0, \qquad i = 1, 2, \ldots, s,$$

where $\hat{x}(\cdot)$ represents the predicted future states. So we see that there is an additional factor here that enters into the computation of the control law used at each time, namely, the predicted future states. One might object that since the predicted future states $\hat{x}(t + T)$ are created by the model M from past states and controls, the anticipatory scheme is basically just a special type of classical feedback control process in which the control law is a somewhat more complicated function of the state than usual. Formally speaking, this is indeed the case. But it's far from clear that it constitutes a fatal flaw. Quite the contrary, actually.

First of all, the fact that the predicted states are functions of the past enables us to preserve the causal relationship between the system state and its inputs; that is, the next state depends only upon the current and past states and controls. Second, if it's an objection that the predicted states used in generating the control law are functions of the past states and controls, then the same objection applies to the standard adaptive control problem involving the estimated parameters $\hat{a}$, which themselves are determined solely from past states and inputs. Needless to say, there are good reasons to retain the distinction between the states and estimated parameters in the classical setup, and we shall attempt to show in the remainder of the chapter that the same kind of distinction is crucial when we come to consider anticipatory processes. But there really is something new that's added when we pass to anticipatory systems: the model M. And it's this "something" that dramatically changes the character of the kind of behavior we can expect and the ways in which we can interact with the underlying system Σ. But rather than belabor this point in the abstract, let's look at a specific example.

Example: A Chemical Reaction Network

To illustrate some of the ideas associated with anticipatory processes, we consider an example due to Robert Rosen of a biosynthetic reaction network. If we let A_i represent a chemical substrate at stage i, while E_i is the enzyme that catalyzes it, we can graphically depict the network as shown in Fig. 7.4.

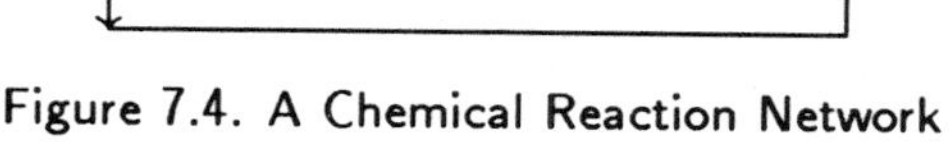

$$A_0 \overset{k_1}{\underset{}{\rightrightarrows}} A_1 \xrightarrow{k_2} \cdots \xrightarrow{k_{n-1}} A_{n-1} \xrightarrow{k_n} A_n$$

Figure 7.4. A Chemical Reaction Network

Here the quantity k_i is the rate of the ith reaction, $i = 1, 2, \ldots, n$. Further, we have assumed there is a forward activation step in the network so that the concentration of substrate A_0 serves to activate the production of A_n. That is, the concentration of A_0 at time t *predicts* the concentration of A_n at some *future* time $t + T$. So we choose $k_n = k_n(A_0)$ to embody this forward activation step, leaving all other reaction rates k_i constant, $i = 1, 2, \ldots, n - 1$.

Under the foregoing hypotheses, the rate equations for the system are

$$\frac{dA_i}{dt} = k_i A_{i-1} - k_{i+1} A_i,$$

$$\frac{dA_n}{dt} = k_n(A_0) A_{n-1}, \qquad i = 1, 2, \ldots, n - 1.$$

Let's assume that the purpose of the forward activation step is to stabilize the level of substrate A_{n-1} in the face of ambient fluctuations in the initial substance A_0. We can represent this requirement mathematically by the condition

$$\frac{dA_{n-1}}{dt} = 0,$$

which is independent of A_0. Thus, from the rate equations this condition means that we must have

$$k_{n-1} A_{n-2} = k_n(A_0) A_{n-1}.$$

We shall satisfy this condition by choosing the functional form of $k_n(A_0)$, which will then embody the predictive model implicit in the forward loop of the network.

From the linear dynamics, it's easy to see that

$$A_{n-2}(t) = \int_0^t K_1(t - s) A_0(s) \, ds,$$

$$A_{n-1}(t) = \int_0^t K_2(t - s) A_0(s) \, ds,$$

where K_1 and K_2 are functions determined entirely by the chemical reaction rates k_i, $i = 1, 2, \ldots, n - 1$. These expressions show explicitly that the value of A_0 at a given moment determines the values of A_{n-1} and A_{n-2} at a *later* time.

The control condition on $k_n(A_0)$ now becomes

$$k_n(A_0) = k_{n-1} \frac{\int_0^t K_1(t - s) A_0(s) \, ds}{\int_0^t K_2(t - s) A_0(s) \, ds}, \qquad (*)$$

i.e., the reaction rate k_n at any time t is determined by the value of A_0 at a prior instant $t - T$. Or, equivalently, the value of $A_0(t)$ determines k_n at a *future* time $t + T$. Consequently, we see the manner in which the initial substrate A_0 serves to *adapt* the pathway so as to stabilize the condition

$$\frac{dA_{n-1}}{dt} = 0.$$

Finally, we note that the homeostasis maintained in the pathway is obtained entirely through the modeling relation between A_0 and A_{n-1}, i.e., by virtue of the relation (∗) linking the prediction of the model to the actual rate k_n. Thus, the homeostasis is preserved entirely through adaptation generated on the basis of a *predicted* value of A_{n-1}. In particular, there is no feedback in the system and no mechanism available for the system to "see" the value of the quantity that's being controlled.

In the above scheme it's important to note that the anticipatory linkage between A_0 and A_{n-1} will be adaptive only as long as the relation (∗) holds, i.e., only as long as the linkage "wired-in" by the forward activation step and the actual linkage dictated by the chemical kinetics remain the same. If there should be a deviation in this linkage, then the rate change given by (∗) will become *maladaptive* to a degree measured by the magnitude of the deviation. Since such deviations can be expected to occur for *all* real processes, we conclude that a forward activation step of the above type can retain its adaptive character only for a characteristic period of time that is dependent upon the nature of the larger system within which it is embedded, as well as upon the character of the system's interactions with that larger system. This phenomenon, which is termed *temporal spanning,* has no analogue in nonanticipatory systems. In the next section we demonstrate how this idea of temporal spanning can give rise to another kind of feature that is also absent in classical reactive systems: global system failure.

Exercises

1. Many primitive organisms in nature are negative phototropic, i.e., they move toward darkness. Show how this observed behavior can be described as an anticipatory control process. Do the same for the wintering behavior of deciduous trees, by explaining how the shedding of leaves and other changes that occur in the autumn are the result of the tree making use of an anticipatory model.

2. Describe in system-theoretic terms how a system like a living cell or a corporation generates an internal predictive model of itself.

3. In the chemical reaction network of the text, the initial substrate A_0 plays two quite distinct roles. What are they? (Hint: One of the roles is purely symbolic or linguistic.)

4. In economics and finance, the theory of *rational expectations* is often used in order to account for the beliefs of economic agents about the future

and their effect on the actions that these agents take today. Roughly speaking, this theory says that the an agent's estimate of a price is rational if the price he anticipates leads to an actual price that will, *on the average,* coincide with his expectation. Thus, the rationally-expected price will diverge from the true price only on account of some unpredictable uncertainty in the price-setting mechanism. So, for instance, if you are trying to predict the Dow Jones Industrial Average, the rational expectations hypothesis states that, on the average,

the subjective estimate of the DJIA = the true value of the DJIA.

Discuss the rational expectations hypothesis within the framework of anticipatory processes.

5. Perfect Subsystems and Global Failures

In today's world we are all too depressingly aware of the phenomenon of global system failure without any apparent local cause(s) to account for it. National economies limp along like war-tattered refugees as politicians, economists, and others attribute the problems to inflation, labor unions, interest rates, foreign competition and a plethora of other local "causes." Other examples of this inability to attribute system failure to local causes abound in both the social and biological spheres, so it's of interest to show that this is exactly the type of behavior that can be *expected* from any anticipatory system.

Suppose we have a system Σ composed of N subsystems. Assume that the behavior of the ith subsystem S_i is given by the input/output relation

$$\phi_i(u_i(t), y_i(t+h)) = 0, \qquad u_i \in R^m, \qquad y_i \in R^p, \qquad i = 1, 2, \ldots, N,$$

where $u_i(t)$ and $y_i(t+h)$ are the inputs and outputs of subsystem S_i at times t and $t + h$, respectively. Thus, the subsystems receive inputs that may be from either the external environment or the outputs of other subsystems, producing outputs at some later time in accordance with the rule ϕ_i. We assume that as far as the subsystem S_i is concerned, everything is operating as it should whenever the input/output pair $(u_i, y_i) \in \Omega_i$, where Ω_i is some specified subset of $R^m \times R^p$. In addition, the overall system Σ has its own input/output relation

$$\Phi(v, w) = 0, \qquad v \in R^n, \qquad w \in R^q,$$

where both the input v and the output w are from and to the external environment of Σ, respectively. Let's agree that the global system Σ is operating properly whenever the input/output pair $(v, w) \in \Omega$, where Ω is a given subset of $R^n \times R^q$.

We now examine the logical possibilities between the following propositions:

A. Each subsystem S_i is operating properly.

B. The overall system Σ is operating properly.

There are four cases to consider:

• $A \Rightarrow B$: In this case, proper operation of the subsystems implies proper operation of Σ. Note that this leaves open the possibility that the system Σ could function properly even if some of the subsystems were operating incorrectly. Considering the contrapositive $\sim B \Rightarrow \sim A$, we see that failure of Σ implies failure of at least one subsystem. So in this case every global failure of Σ arises out of the failure of at least one local subsystem.

• $B \Rightarrow A$: Here the correct functioning of the overall system implies that each subsystem is functioning correctly. The contrapositive implies that if a component subsystem functions incorrectly, then so does the complete system. Note, however, that this conclusion leaves open the possibility that the overall system may have failure modes that do not stem from the failure of any individual subsystem. It is this possibility that we will explore in more detail in a moment.

• A and B are logically equivalent: This means that every global failure gives rise to the failure of some subsystem, and conversely. In this case there can be no global failures that do not arise from the failure of some subsystem.

• A and B are logically independent: In this situation there is no causal connection between the conditions under which Σ functions correctly and those conditions under which the subsystems operate properly. Thus, the set Ω cannot be decomposed into properties of the local component sets Ω_i and, hence, we cannot define Ω without reference to the overall system.

Now let's show how any system possessing a feedforward loop of the type described earlier leads to the second type of logical implication discussed above.

Consider the anticipatory system displayed in Fig. 7.5.

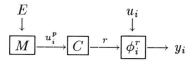

Figure 7.5. An Anticipatory Loop for Subsystem S_i

Here we have the model M of subsystem S_i processing the environmental input E (as well as other information) in order to produce the quantity u_i^p, the *predicted* value of the input to S_i at some future time $t + h$. This predicted value is transmitted to the local controller C, who then uses it to compute the value of a parameter vector r employed in the input/output law ϕ_i^r. This law, in turn, governs how subsystem S_i will process its *actual* input at $t + h$. Call this actual input $u_i(t + h)$. Now let's see how this setup can lead to a global system failure.

First of all, we have assumed that the correct behavior of the overall system Σ means that the system output lies in the acceptable region Ω. Assume now that all components of Σ, aside from those involved in the above loop for S_i, are functioning properly. Under these circumstances, proper functioning of Σ reduces to the adaptability of the component ϕ_i^r in the feedforward loop above. As long as the predicted input u_i^p agrees with the actual input u_i, we necessarily have the overall state of Σ belonging to Ω. In other words, proper behavior of the global system Σ is now directly related to the fidelity of the model M in the feedforward loop.

Since the predictive model is, in general, closed to interactions that the subsystem S_i is open to, there will necessarily be an increasing deviation between the predictions of the model and the actual inputs experienced by the subsystem S_i. As a result of these deviations between $u_i(t + h)$ and $u_i^p(t + h)$, the preset value of the parameter vector r will render the rule ϕ_i^r increasingly less competent to deal with the actual inputs. So, in general, there will be some characteristic time T at which the deviation becomes so great that ϕ_i^r is actually *incompetent*. Consequently, because of the direct relation between the competence of the subsystem S_i and the maintenance of the state of the global system Σ in Ω, the overall system will fail at this time T. We can identify this moment of failure T as the intrinsic life span of the system Σ. So we see that the global failure of Σ is not the result of a local failure since, by hypothesis, every unmodeled component of Σ (including those involved in the feedforward loop) is functioning correctly *according to any local criterion.*

The heart of this argument resides in the observation that there is no local criterion that will enable the controller of subsystem S_i to recognize that anything has gone wrong. As far as C is concerned, it receives the prediction u_i^p and processes it *correctly* into the parameter r. Similarly, the operator of subsystem S_i receives the instructions r from the controller and *correctly* executes the rule ϕ_i^r specifying how the actual input u_i is to be transformed into the output. The basic problem is back upstream at the level of the model M, a problem that **cannot be seen** by either the controller C or the operator of the subsystem. The loss of competence of ϕ_i^r to process the input to S_i can only be expressed in terms of the behavior of the overall system Σ, and not by reference to just the feedforward loop considered in isolation. So we see that any system with a feedforward loop possesses the possibility for a global mode of failure that cannot be identified with any type of local failure. Now let's show how this general scheme relates to the biochemical example discussed earlier.

Example: A Biosynthetic Pathway

We consider the chemical reaction network displayed earlier in Fig. 7.4. For this system, the substrate A_0 acted as a *predictor* for future values of the substrate A_n. Thus our model M is $M = \{A_0\}$. The subsystem S_i is just the substrate A_n, so we take $i = n$ and $S_n = \{A_n\}$. Finally, the parameter r generated by the "controller" of S_n is simply the reaction rate k_n involved in the dynamical processing of substrate A_{n-1}. So we set $r = k_n$, the dynamics ϕ_n^r being the relation linking A_n with A_{n-1}. The crucial point here is that the adaptation parameter k_n is determined solely by the *predicted* value of the subsystem input.

Having now seen some of the major differences between feedback and feedforward controllers, it's perhaps useful to pause briefly and devote a short section to summarizing what we have learned so far.

Exercise

1. In the text, we considered the following two propositions:

   A. Each subsystem S_i is operating properly.

   B. The overall system Σ is operating properly.

We saw that there are four mutually exclusive logical possibilities between these two propositions: $A \Rightarrow B$, $B \Rightarrow A$, $A \equiv B$, $A \not\Leftrightarrow B$. Give real-life examples of each of these cases.

6. A Taxonomy of Adaptive Systems

In the preceding pages we have seen a variety of properties of both feedback and feedforward control systems. One of the most important distinctions that can be drawn between these two classes of systems is the type of information that they employ in carrying out their regulatory activity. Since the utilization of information lies at the heart of any type of adaptive control procedure, it's evident that differences in the manner in which information is used to control the system will generally result in radically different types of control and, hence, different types of system behavior. Here we briefly outline a crude taxonomy of adaptive control mechanisms to show the fundamental differences between those based upon feedback and those that employ feedforward loops of the type discussed above.

Roughly speaking, there are four types of qualitatively distinct adaptive mechanisms, depending upon whether the adaptive loop is feedback or feedforward, with or without memory. Let's take a brief look at each of these classes of systems and their characteristic behaviors.

• *Class I: Feedback Regulators, No Memory*—The distinguishing feature of feedback controllers is that they operate solely on the basis of measurements of the controlled system; there is no sensing of the environment. Typically the system has some desired "set-point," and the controller monitors the deviation of the system from this level, exerting control based upon the deviation of the actual system behavior from its desired level. It should be emphasized that this is error-actuated control, and that the controller does nothing if the current state of the system is in the acceptable region. The simplest types of feedback homeostats have no memory, the present activity of the controller being independent of its past behavior.

• *Class II: Feedback Regulators with Memory*—In this class we lump together all feedback controllers that can make a memory trace of their past actions, and that employ this memory to change various aspects of the system, such as the set-point or parameters like the time constants in the control loop. Any such system can be thought of as a *learning* system. But just as with feedback regulators without memory, such systems are still special-purpose devices. That is, once fabricated and operating, their behavior is fixed; the special-purpose hardware needed for their operation is difficult to modify for other purposes.

• *Class III: Feedforward Regulators, No Memory*—Whereas feedback regulators are characterized by using only the state of the controlled system for their operation, feedforward controllers have a quite different set of properties. First of all, the controller can sense aspects of its environment that can then be correlated with future states of the controlled system. This correlation between the environment at time t and the state at some future time $t + \Delta t$ is "hard-wired" into the controller. Further, such a controller can modify the dynamics of the controlled system in accordance with the values of both the current state of the environment and the *predicted* future state(s) of the system. Thus, feedforward controllers contain a *model* of the external world embodied in the correlation between the controlled system state and the state of the environment.

It's important to note that the operation of the feedforward controller is no longer determined primarily by an error signal, as it was for feedback control systems. Now the operation feeds off of regularities in the environment rather than from the randomness that generates the error signal used by feedback controllers. This difference is significant, because in a feedback system the behavior of the system has already deteriorated before the controller can begin to act. Since any feedback loop has an intrinsic time constant, it can happen that performance has irreversibly deteriorated before the controller can take action. In addition, in environments that fluctuate too rapidly, the controller may end up tracking only the fluctuations rather than stabilize the system. Feedforward controllers avoid these difficulties altogether.

The model of the external world built into a feedforward controller introduces the concept of "software" into adaptive mechanisms, since by changing the model we can modify the entire system without completely redesigning it. However, the behavior of such a controller is still primarily hardware-dependent.

- *Class IV: Feedforward Regulators with Memory*—Feedforward controllers with memory can employ their past behavior to modify their internal models. Thus, in some sense these systems are capable of re-programming themselves. Perhaps ironically, such types of adaptive systems must employ features of the Class I feedback regulators in their operation, since they must employ an error signal in order to modify their internal model. So they suffer from the same limitations as all feedback systems. Nevertheless, there is an enormous gain in adaptive power with systems that can modify their models of the world on the basis of past experience.

Keeping the above properties of controllers in mind, we can begin to address a host of questions involving both biological and social organization pertaining to the theme of adaptation. In particular, we can consider the degree to which the adaptive capability of individual units interacts with the adaptive capacity of the overall system of which they are a part. In the last section we saw that if the control mechanism were of either Class III or IV, such local adaptivity could nevertheless lead to global maladaptation. On the other hand, it's through defects in the adaptability of local units that higher-level associations with new properties can be created and maintained, the corollary being that too much capacity in the units precludes or destroys such global associations.

Much of our interest in studying adaptive processes in a social context is derived from the feeling that many of the social organizations of modern life are malfunctioning, and that we must take steps to make them more "adaptive," in some sense. The preceding discussion suggests that at least some of the maladaptation arises from the fact that our systems are *too adaptive.* They have, in effect, too comprehensive a view of what constitutes their own interests, too much access to environmental variables through technology and the media, too many models correlating behavior and the environment from a multitude of special interest groups—in short, too much adaptation at too low a level generated by too much special-purpose hardware. Utilizing the more technical aspects of the above feedback and feedforward systems,

together with the lessons from biological organization, we now have at least
the raw materials needed to make a concerted attack on these sorts of questions. But for now let's set these cosmic themes aside, lowering our level
of rhetoric a bit to consider an example of an anticipatory process in which
the role of the predictive model can be more explicitly displayed—the case
of a living cell.

Exercises

1. Give examples illustrating the four types of adaptive systems.

2. The Darwinian theory of evolution involves the process of adaptation
brought about by the "law" of natural selection. Which of the types of
adaptive systems does Darwinian evolution represent? Does the type change
if you think of selection acting at the level of an organism's genotype rather
than at the Darwinian level of the phenotype?

7. Metabolism-Repair Systems

The quintessential example of a system containing an internal model of itself is a living cell. Functionally, we can regard the cell as possessing two
components: a *metabolic* part representing the basic chemical activity characterizing the cell, and a *maintenance* or *repair* component that ensures
continued viability of the cell in the face of external disturbances of many
sorts. Thus, we may think of the cell as a small chemical factory receiving
inputs from its external environment and processing these inputs into outputs in accordance with instructions coded into the cellular DNA. The DNA
also contains the necessary instructions for not only how the cell should operate, but also for how it should be built. This genetic material provides a
complete internal model of all aspects of the cell. Our goal is to distill the
essence of this situation into a mathematical representation that will show
us each of these functional activities and their mutual interrelations.

Let's assume that Ω represents the set of environmental inputs available
to the cell, while Γ is the set of output products that a cell might be capable
of producing. These sets are determined primarily by the kind of operating
environment within which the cell can exist, as well as by the laws of chem-
istry and physics constraining the nature and type of chemical reactions
that can take place in that environment. Further, let $H(\Omega, \Gamma)$ be the set of
physically realizable metabolisms that the cell can display. Again, $H(\Omega, \Gamma)$
is determined by the particular operating environment and the laws of nature. With these definitions and notations, we can represent the metabolic
activity of the cell as a map

$$f : \Omega \to \Gamma,$$
$$\omega \mapsto \gamma.$$

At this stage, we immediately recognize cellular metabolism as an illustration of the type of input/output relation discussed earlier in another context in Chapter 6. Consequently, if the metabolic map f were linear, we would have at our disposal all the machinery developed there to bring to bear upon our study of cellular metabolic behavior. But a cell does not live by metabolism alone. So we must think about how to add the maintenance/repair function to this metabolic activity.

We begin by noting that the functional role of the repair component is to restore the metabolism f when there is some fluctuation in either the ambient environment ω or a disturbance to the metabolic machinery f itself. In this connection, it's crucial to state explicitly the way in which the cell's metabolic component acts. The external input ω, which, is actually a *time-series* of chemical quantities, is processed into the metabolic output γ, also a time-series of output products. At this point the cellular metabolism f disappears. It is then the job of the repair function to employ some of the cellular output to produce a new metabolic map that is then used to process the next input sequence, call it $\hat{\omega}$. Let's suppose that during the course of evolutionary time the cell has evolved in such a way that its design, or basal, metabolism has been set to produce the output γ when the input is ω. So if the next input $\hat{\omega}$ happens to be the design input ω, a properly functioning cellular repair mechanism should produce the metabolism f. In other words, if there has been no departure from the "design specs," then the repair operation doesn't have to do any real "repair." It only needs to produce more of the same, so to speak. Putting all these remarks together, we see that the repair function can be abstractly represented by the map

$$P_f : \Gamma \to H(\Omega, \Gamma),$$
$$\gamma \mapsto f,$$

where we have written the subscript f to indicate explicitly the "natural" boundary condition

$$P_f(f(\omega)) = P_f(\gamma) = f,$$

for the repair machinery P_f. So we see that the role of the repair operation is to counteract, or at least stabilize, the cell's metabolic behavior in the face of inevitable fluctuations in the outside world and/or inherent errors in the metabolic machinery f. In short, P_f is an error-correcting mechanism. But who (or what) corrects errors made by P_f? Who repairs the repairers? It does no good to introduce another level of repair into the situation, as the incipient infinite regress is obvious. But what's the alternative?

Nature's solution to the repair problem is simplicity itself: Throw the old repair mechanism away and build a new one! In other words, *replication*. The basic scheme is very straightforward. The cell's genetic component (essentially the DNA) receives information about the cell's metabolic processing apparatus f and then transforms this information into a copy of the repair map P_f. We can represent the reproductive machinery of the cell by the map

$$\beta_f \colon H(\Omega, \Gamma) \to H(\Gamma, H(\Omega, \Gamma)),$$
$$f \mapsto P_f.$$

Just as with the repair map P_f, the replication operation is subject to the boundary condition that "if it ain't broken, don't fix it." This means that if the design metabolism f is presented, the map β_f should transform it into the associated repair map P_f. Thus, $\beta_f(f) = P_f$. We can summarize our discussion so far by the diagram shown in Fig. 7.6.

$$\Omega \xrightarrow{\ f\ } \Gamma \xrightarrow{\ P_f\ } H(\Omega, \Gamma) \xrightarrow{\ \beta_f\ } H(\Gamma, H(\Omega, \Gamma))$$

Figure 7.6. An Abstract Metabolism-Repair System

The preceding development provides only the bare bones minimum insofar as describing the operation of a living cell. Let's now consider a number of aspects surrounding Fig. 7.6, both in its relation to the ideas of anticipation considered earlier, as well as its relationship to the behavior of *real* cells.

• *Implicit Control*—Classical control systems involve a direct intervention in the system dynamics by the controller. In the (M, R) setup above, there are no direct control inputs being injected into the system at any stage of the operation; in fact, there is no controller at all, at least in the sense that that term is usually employed in control engineering. The only signals coming into the cell from the outside are those generated by the environment. Thus, the control is *implicit,* arising solely from the boundary conditions $P_f(f(\omega)) = f$ and $\beta_f(f) = P_f$, together with the feedforward paths from the metabolism component to the repairer and from the repair component to the replication operation. This distinction between implicit versus explicit control makes it clear that the kind of system we are speaking about here differs radically from the sort of adaptive schemes generally discussed in the control engineering literature.

• *Translation and Transcription*—In our discussion of cellular processes in Chapter 3, we saw that the information coded into the cellular DNA is used in two quite different ways. First, it's *translated* into instructions for assembling RNA and various types of amino acids, which in turn generate the proteins. But it's also *transcribed* in order to replicate the DNA during the process of cellular reproduction. Since the translation operation involves

the production of chemical compounds needed for both the construction
of amino acids and construction of the repair machinery itself, it's fair to
say that this use of the DNA information is tantamount to building the
metabolic machinery f. But from Fig. 7.6 we see that this is exactly the
role of the repair map P_f; hence, we can, roughly speaking, identify P_f with
the operation of *translating* the cell's DNA. It follows that the transcription
operation associated with genetic replication corresponds to the replication
map β_f. Putting these remarks together, we see that the *abstract* diagram
of Fig. 7.6 can be identified with the biological operations of the cell as
indicated in Fig. 7.7.

$$\Omega \underbrace{\xrightarrow{\;f\;}}_{\text{metabolism}} \Gamma \underbrace{\xrightarrow{\;P_f\;}}_{\text{translation}} H(\Omega,\Gamma) \underbrace{\xrightarrow{\;\beta_f\;}}_{\text{transcription}} H(\Gamma, H(\Omega,\Gamma))$$

Figure 7.7. Abstract (M, R)-Systems and Real Cellular Functions

• *Genetic Mutations and "Lamarckism"*—The Central Dogma underlying modern molecular biology is that modifications to the cellular genotype
cannot arise as a result of environmental changes being transmitted through
the repair component, but only by means of direct intervention in the replication map β_f itself. In short, mutations come about either from direct
external perturbation of β_f, as by radiation or cosmic rays, or by means
of internal changes such as copying errors of one sort or another. But in
(M, R)-systems changes in ω **can** lead to changes in β_f. Since we have not
yet addressed the issue of *how* the maps P_f and β_f are constructed, it's
not clear from the diagrams of Figs. 7.6 and 7.7 under what circumstances
these types of *Lamarckian* changes can take place within the framework of
an (M, R)-system. Since it would be of some value to be able to employ the
above abstract framework in other contexts like economics and manufacturing, in which there most assuredly is the possibility for Lamarckian change,
it's highly desirable to know the circumstances under which (M, R)-systems
admit the possibility of such changes. We shall address these matters in
detail in the next section.

The abstract cellular structure outlined here gives rise to a long list of
questions about how the mathematical skeleton of Fig. 7.6 can be fleshed-out
to address matters pertaining to the way in which the genetic and metabolic
components of the cell interrelate in the performance of cellular function.
Two questions of immediate concern are: (1) Under what circumstances
can we make the repair and replication components emerge directly and in
a "natural" manner from the metabolic machinery alone?, and (2) What
type of disturbances in the environmental input ω and/or in the metabolic
map f can the repair component counteract?

The first of these questions has obvious relevance to the problem of the origin of life, since a principal point of debate in such circles centers about what came first, metabolism or replication, the chicken or the egg. Any mathematical light we can shed upon conditions that may favor one or the other, not to mention providing some insight into the possibilities for various types of exobiologies, would be of considerable value. The second question bears upon the "resilience" of the cell to a spectrum of disturbances, both internal and external. Answers to this kind of question will go a long way toward furthering our understanding of the mechanisms by which living systems adapt to change. But in order to address either of these questions, we need to move the abstract framework described above onto a more concrete footing by imposing additional structure on the maps and sets appearing in Fig. 7.6. Although for gaining an overview of the basic concepts of living systems it's fine to remain at the very general levels of abstract sets and relations, when it comes to giving concrete answers to concrete questions we need more structure in order to say anything even halfway interesting and detailed. Our goal in the next section is to provide a starting point for such a program directed toward the study of metabolism-repair systems.

Exercises

1. A sequential machine S is an object consisting of the following sets

$$A = \text{the input set,}$$
$$B = \text{the output set,}$$
$$S = \text{the set of states,}$$

and maps

$$\delta: A \times S \to S,$$
$$\lambda: A \times S \to B.$$

The map δ is called the *state-transition map,* while λ is the *output map.* Note that the sets A and B are finite, but the set of states S may be infinite. (a) Show that sequential machines are a special case of the cellular automata considered in Chapter 3. (b) By matching up the defining sets and maps of both, show that the partial (M, R)-system

$$\Omega \xrightarrow{\; f \;} \Gamma \xrightarrow{\; P_f \;} H(\Omega, \Gamma) \xrightarrow{\; \beta_f \;} H(\Gamma, H(\Omega, \Gamma))$$

can be formally regarded as a sequential machine. (c) What is the biological interpretation of the set of states of the sequential machine in this situation?

2. In the language of sequential machines, the property of complete reachability discussed in Chapter 6 is called *strongly connected*. In general, a sequential machine is **not** strongly connected. But it is possible to embed any given machine within a larger one that is strongly connected. Simply enlarge the input alphabet A and extend the mappings δ and λ appropriately. Show why this kind of embedding becomes a much more delicate matter in the case of (M, R)-systems. (Hint: Consider the fact that the input set and the state set of the (M, R)-system are not independent.)

8. Linear (M, R)-Systems: Repair

The simplest setting in which to address the formalization of (M, R)-systems is when everything in sight is linear. In this case the input and output spaces Ω and Γ are finite-dimensional vector spaces, while the metabolic map f is linear and, to make things as simple as possible, autonomous, i.e., constant. Thus we take our mathematical setting to be exactly that considered in Chapter 6.

More precisely, the cellular input and output spaces are assumed to be sequences of vectors taken from R^m and R^p. Thus,

$$\Omega = \{\omega : \omega = (u_0, u_1, u_2, \ldots, u_N)\}, \qquad u_i \in R^m,$$

$$\Gamma = \{\gamma : \gamma = (y_1, y_2, y_3, \ldots)\}, \qquad y_i \in R^p.$$

Note that to respect causality, we have assumed the cellular outputs begin one time unit after receipt of the inputs. Further, for technical reasons alluded to already in Chapter 6, we assume that only a *finite* number of inputs are applied, but that the outputs can, in principle, continue indefinitely. In view of the linearity of the metabolic map f, we can express the relationship between the cellular inputs and outputs as

$$y_t = \sum_{i=0}^{t-1} A_{t-i} u_i, \qquad t = 1, 2, \ldots, \qquad (I/O)$$

where the coefficient matrices $A_k \in R^{p \times m}$. Under mild technical conditions (See Problem 4) on the pair (ω, γ), the elements $\mathcal{B} = \{A_i\}$ are uniquely determined by the input/output sequence. So, for all intents and purposes, we can identify the metabolic map f with the behavior sequence $\mathcal{B}$, and we shall routinely move back and forth between the two in what follows. Thus,

$$f \longleftrightarrow \mathcal{B}.$$

The relation (I/O) can also be written as

$$y_t = \sum_{i=0}^{t-1} [A_{t-i}^{(1)} \mid A_{t-i}^{(2)} \mid \cdots \mid A_{t-i}^{(m)}] u_i, \qquad t = 1, 2, \ldots,$$

where $A_j^{(r)}$ denotes the rth column of the matrix A_j, while the block row vector $\left[A_j^{(1)} \mid A_j^{(2)} \mid \cdots \mid A_j^{(m)}\right]$ consists of those elements comprising the columns of A_j. We will employ this notation throughout the remainder of this section.

In Chapter 6 we saw that the input/output pair (ω, γ) can be canonically realized by an internal model of finite dimension if and only if the infinite Hankel array

$$\mathcal{H} = \begin{pmatrix} A_1 & A_2 & A_3 & \cdots \\ A_2 & A_3 & A_4 & \cdots \\ A_3 & A_4 & A_5 & \cdots \\ \vdots & \vdots & \vdots & \end{pmatrix}$$

has rank $n < \infty$. In this case, there exist matrices (F, G, H) such that the input/output behavior of the system

$$x_{t+1} = Fx_t + Gu_t, \qquad x_0 = 0, \qquad x_t \in R^n, \qquad t = 0, 1, \dots,$$
$$y_t = Hx_t, \tag{Σ}$$

agrees with that of the system (I/O); i.e., $A_t = HF^{t-1}G$. Furthermore, the system Σ is completely reachable and completely observable, hence, canonical. Under the *assumption* that the system (I/O) possesses a finite-dimensional realization Σ, we saw in Chapter 6 how to compute the elements of this realization $\Sigma = (F, G, H)$ directly from the Hankel array $\mathcal{H}$. These by now almost classical facts from linear system theory form the basis for our attack upon the problem of how to calculate the repair and replication maps P_f and β_f directly from the cellular metabolism f.

Our first task is to see how to calculate the cellular repair map P_f from the metabolic ingredients f, Ω, Γ and $H(\Omega, \Gamma)$. For future reference, let's denote the basal metabolism by

$$f^* \colon \Omega \to \Gamma,$$
$$\omega^* \mapsto \gamma^*.$$

In other words, the cell is operating at its design level if the output produced is γ^* when the environment is ω^*, the corresponding metabolic map then denoted by f^*. From our earlier discussion, we know that the repair map must satisfy the relation

$$P_f(\gamma) = P_f(f(\omega)) = f,$$

for any basal metabolism f and associated output γ. Since we have assumed the repair map is linear, the above relation can be written

$$w_\tau = \sum_{i=0}^{\tau-1} \mathcal{R}_{\tau-i} v_i, \qquad \tau = 1, 2, \dots,$$

where (v_i, w_i) are the input and output to the repair subsystem, with the elements $\{\mathcal{R}_j\}$ being linear maps determined by γ and f. However, when the metabolism is operating correctly, the repair system must accept the input γ^* and produce the output f^*. Consequently, we must have

$$v_\tau = \mathcal{S}(y^*_{\tau+1}), \qquad w_\tau = A^*_\tau,$$

where $\mathcal{S}$ is the "stacking" operator whose action is to stack the columns of an $n \times m$ matrix into a column vector of nm components. Note that here we have used a time parameter different than for the metabolic system, as it will usually be the case that the repair system will operate on a time scale different from that of the metabolism.

It's an easy exercise to see that the coefficients for the repair map must have the form

$$\mathcal{R}_j = [B_{j1} \,|\, B_{j2} \,|\, \cdots \,|\, B_{jp}], \qquad B_{js} \in R^{p \times m}, \qquad j = 1, 2, \ldots .$$

Consequently, in component form we can write the input/output relation for the repair map as

$$w_\tau = \sum_{i=0}^{\tau-1} [\mathcal{R}^{(1)}_{\tau-i} \,|\, \mathcal{R}^{(2)}_{\tau-i} \,|\, \cdots \,|\, \mathcal{R}^{(p)}_{\tau-i}] \mathcal{S}(v_i),$$

where we have written $\mathcal{R}^{(s)}_j \doteq B_{js}$.

Just as the metabolism f was represented by the sequence $\{A_1, A_2, \ldots\}$, we can now see that the repair system P_f can be represented by the sequence

$$P_f \longleftrightarrow \{\mathcal{R}_1, \mathcal{R}_2, \ldots\}.$$

There are two important points to note about the foregoing development:

1) If we write each element A_i as

$$A_i = [A_i^{(1)} \,|\, A_i^{(2)} \,|\, \cdots \,|\, A_i^{(m)}], \qquad A_i^{(j)} \in R^p,$$

the *complexity* of each component of the metabolic map f as measured by the number of its components is $O(pm)$. By the same reasoning, the complexity of each element $\mathcal{R}_j$ of the repair map P_f is $O(p^2m)$. Thus, already we see that the often conjectured complexity increase associated with living systems begins to emerge through purely natural mathematical requirements linking the metabolic and the repair maps.

2) It's easy to see that $\dim \Sigma = n < \infty$ implies that the elements $\{A_1, A_2, \ldots, A_{2n}\}$ are linearly dependent. This condition also implies that the canonical realization of the repair sequence $\{\mathcal{R}_1, \mathcal{R}_2, \ldots\}$ has dimension $n_P \leq n$. Consequently, we can again employ Ho's Algorithm to produce a system $\Sigma_P = (F_P, G_P, H_P)$ realizing the repair map P_f.

Example: The Natural Numbers

To fix the foregoing ideas, suppose we have a cell whose "design" environmental input is

$$\omega^* = \{1, 1, 0, 0, \ldots, 0\}.$$

In this environment, we assume that the cell is designed to produce the set of natural numbers

$$\gamma^* = \{1, 2, 3, 4, \ldots\},$$

i.e., $f^*(\omega^*) = \gamma^*$. In this case it's easy to work out that the metabolic map f^* has the form

$$f^* \approx \{A_1^*, A_2^*, \ldots\} = \{1, 1, 2, 2, 3, 3, \ldots\}.$$

It can be shown that this behavior sequence has a canonical realization of dimension $n^* = 3$, so using Ho's Algorithm as in Chapter 6 we obtain the canonical system matrices

$$F^* = \begin{pmatrix} 0 & 1 & 0 \\ 1 & -1 & 1 \\ 1 & -2 & 2 \end{pmatrix}, \qquad G^* = \begin{pmatrix} 1 \\ 1 \\ 2 \end{pmatrix}, \qquad H^* = (\,1 \quad 0 \quad 0\,).$$

The dynamics for the metabolic subsystem are then

$$x_{t+1}^* = \begin{pmatrix} 0 & 1 & 0 \\ 1 & -1 & 1 \\ 1 & -2 & 2 \end{pmatrix} x_t^* + \begin{pmatrix} 1 \\ 1 \\ 2 \end{pmatrix} u_t^*, \qquad x_0 = 0, \qquad (\Sigma^*)$$

$$y_t^* = (\,1 \quad 0 \quad 0\,)\, x_t^*, \qquad t = 0, 1, 2, \ldots.$$

Turning now to the repair component, we must have $P_{f^*}(\gamma^*) = f^*$, which leads to the 1×1 matrices

$$\mathcal{R}_i^* = \begin{cases} (+1), & i \text{ odd} \\ (-1), & i \text{ even}. \end{cases}$$

Thus, the Hankel array associated with the map P_{f^*} is given by

$$\mathcal{H}_{P_{f^*}} = \begin{pmatrix} 1 & -1 & 1 & -1 & 1 & \cdots \\ -1 & 1 & -1 & 1 & -1 & \cdots \\ 1 & -1 & 1 & -1 & 1 & \cdots \\ \vdots & \vdots & \vdots & \vdots & & \end{pmatrix}.$$

Since we know the repair sequence has a canonical realization of dimension $n_P^* \leq n^* = 3$, experimenting a bit with Ho's Algorithm (or computing the rank of $\mathcal{H}_{P_{f^*}}$), we find $n_P^* = 1$. The resultant canonical repair realization is $\Sigma_P^* = (F_P^*, G_P^*, H_P^*)$, where

$$F_P^* = (-1), \qquad G_P^* = (1), \qquad H_P^* = (1).$$

The repair dynamics are then

$$
\begin{aligned}
z_{\tau+1}^* &= (-1)z_\tau^* + (1)v_\tau^*, \qquad z_0^* = 0, \\
w_\tau^* &= (1)z_\tau^*, \qquad\qquad\quad \tau = 0, 1, 2, \ldots .
\end{aligned}
\qquad (\Sigma_P^*)
$$

From our earlier remarks, we connect this system with the metabolic map f^* via inputs and outputs as $w_\tau^* = A_\tau^*$, $v_\tau^* = y_{\tau+1}^*$.

Remark:

At first glance there appears to be a contradiction here between our earlier claim that the repair system is more "complex" than the metabolism, and the fact that $\dim \Sigma_P^* = 1 < \dim \Sigma^* = 3$—at least if one measures complexity by the dimensionality of the system state space and not by how many numbers are needed to describe the respective metabolic and repair maps. However, our earlier observation used a somewhat different notion of complexity, one involving the objects of our behavioral description, the elements A_i and $\mathcal{R}_i$. Unless $p = 1$, the objects $\{\mathcal{R}_i\}$ always contain more elements than the objects $\{A_i\}$. Therefore, by this measure of complexity the repair system will always be at least as complex as the metabolism. Roughly speaking, it's more difficult to *describe* the behavior of the repair process, but it may be simpler to *realize* its dynamics. In engineering terms, there may be fewer "integrators" but of a more complicated type.

The preceding development has shown how we can always construct the system repair map P_f directly from any prescribed metabolism f. For the moment let's defer consideration of how to construct the associated replication map β_f, turning our attention instead to the second of our two main classes of questions: Under what circumstances will the canonical repair map P_{f^*} restore, or at least stabilize, disturbances in the cellular environment ω^* and/or the metabolic map f^*? We consider each of these possibilities in turn.

• *Case I: Fixed Environment ω^* and a Fixed Genetic Map P_{f^*} with Variable Metabolism f*—In this case our concern is with perturbations to the basal metabolism f^* when the environmental inputs and the genetic repair map are fixed. Assume we have the metabolic disturbance $f^* \to f$. Then we are interested in categorizing those maps f such that either

$$P_{f^*}(f) = f \qquad \text{or} \qquad P_{f^*}(f) = f^*.$$

In the first case, the repair machinery stabilizes the metabolism at the new set-point f, whereas in the latter case the repair machinery restores the original basal metabolism f^*.

To study this situation, it's helpful to introduce the map

$$\Psi_{f^*,\omega^*} : H(\Omega,\Gamma) \to H(\Omega,\Gamma),$$
$$f \mapsto P_{f^*}(f(\omega^*)).$$

The case in which the repair system stabilizes the system at the new metabolism f corresponds to finding the fixed points of the map Ψ_{ω^*,f^*}, i.e., those metabolisms f such that

$$\Psi_{f^*,\omega^*}(f) = f.$$

On the other hand, the situation in which the repair system restores the basal metabolism f^* corresponds to finding those perturbations f such that

$$\Psi_{f^*,\omega^*}(f) = f^*.$$

The map Ψ_{f^*,ω^*} is clearly linear, being a composition of the linear maps f and P_{f^*}. Further, by construction we must have

$$\Psi_{f^*,\omega^*}(f^*) = f^*,$$

so that f^* is a trivial fixed point of Ψ_{f^*,ω^*}. In addition, we have the trivial fixed point $f = 0$. Since each $f \in H(\Omega,\Gamma)$ has the representation $f \approx \{A_1, A_2, \dots\}$, we can formally represent the map Ψ_{ω^*,f^*} by the infinite array

$$\Psi_{f^*,\omega^*} = \begin{pmatrix} \Psi_{11}^* & \Psi_{12}^* & \Psi_{13}^* & \cdots \\ \Psi_{21}^* & \Psi_{22}^* & \Psi_{23}^* & \cdots \\ \vdots & \vdots & \vdots & \end{pmatrix},$$

where each $\Psi_{ij}^* \in R^{p \times p}$. We shall return in a moment to a more detailed consideration of the structure of this matrix. For now it's sufficient just to note that any metabolism $f = \{A_1, A_2, \dots\}$ will be a fixed point of the map Ψ_{f^*,ω^*} if and only if the vector $(A_1, A_2, \dots)'$ is a characteristic vector of the linear map Ψ_{f^*,ω^*} with associated characteristic value 1. In view of the foregoing remarks and observations, we can state the answer to the question of what types of metabolic disturbances will be either stabilized or neutralized by the repair map P_{f^*} by the

METABOLIC REPAIR THEOREM. *i) The cellular metabolic disturbance $f \approx \{A_1, A_2, \cdots\}$ will be stabilized by the repair map P_{f^*} if and only if the vector $(A_1, A_2, \dots)'$ is a characteristic vector of the map Ψ_{f^*,ω^*} with associated characteristic value 1.*

ii) The perturbation f will be "repaired," i.e., restored to the metabolism f^* by the repair system P_{f^*}, if and only if $f = f^* + \ker \Psi_{f^*,\omega^*}$. In other words, when the vector

$$(A_1 - A_1^*, A_2 - A_2^*, \dots)' \in \ker \Psi_{f^*,\omega^*}.$$

The Metabolic Repair Theorem shows clearly that the two properties of stabilization at the new basal metabolism f and restoration of the old basal metabolism f^* are diametrically opposed. If we want to be able to restore the old metabolism f^* in the face of many types of disturbances, we need to have $\ker \Psi_{f^*,\omega^*}$ "large," since in this case there will be many perturbations f that would satisfy condition (ii) of the theorem. On the other hand, if $\ker \Psi_{f^*,\omega^*}$ is big, then there must necessarily be only a "small" number of characteristic vectors of Ψ_{f^*,ω^*} with associated characteristic value 1, i.e., there are only a relatively small number of disturbances f that can be stabilized by P_{f^*} at this new metabolic level. Thus, a cell has the choice of being adaptable to a wide variety of metabolic disturbances or being able to restore the original metabolism in the face of a wide variety of metabolic changes—but not both! We examine this point in greater detail below. But for the moment, let's return to the map Ψ_{f^*,ω^*}.

The above discussion makes evident the central role of the map Ψ_{f^*,ω^*} in determining those metabolic changes that can be either neutralized or stabilized. But in order to say anything more definite about the structure of such types of changes, we need to be able to express the map Ψ_{f^*,ω^*} in terms of its component maps f^* and P_{f^*}.

From the detailed representation of the input/output behavior of the repair system, we have

$$A_\tau = \sum_{j=0}^{i} \sum_{i=0}^{\tau-1} [\mathcal{R}_{\tau-i}^{*(1)} \mid \cdots \mid \mathcal{R}_{\tau-i}^{*(p)}] A_{i-j+1} u_j^*,$$

which is clearly a triangular (in fact, Toeplitz) representation, since A_τ depends only upon the elements $A_1, A_2, \dots, A_\tau$. Under our standard assumptions discussed above, we can always find a solution to this system of equations in the components of the matrices $\{\mathcal{R}_{\tau-i}^*\}$ and the elements $\{u_j^*\}$, $i = 0, 1, 2, \dots, \tau - 1; j = 1, 2, \dots, m$. On the other hand, the requirement that f^* be a fixed point of the map Ψ_{f^*,ω^*} means that there must be a solution to the system

$$\begin{pmatrix} \Psi_{11}^* & \Psi_{12}^* & \Psi_{13}^* & \cdots \\ \Psi_{21}^* & \Psi_{22}^* & \Psi_{23}^* & \cdots \\ \vdots & \vdots & & \end{pmatrix} \begin{pmatrix} A_1^* \\ A_2^* \\ \vdots \end{pmatrix} = \begin{pmatrix} A_1^* \\ A_2^* \\ \vdots \end{pmatrix},$$

for some *triangular* choice of the elements Ψ_{ij}^*. In particular, this means that we must have $\Psi_{ij}^* = 0$, $j > i$ and, hence,

$$A_\tau^* = \Psi_{\tau 1}^* A_1^* + \Psi_{\tau 2}^* A_2^* + \cdots .$$

But we also have the expression for A_τ^* from above involving the elements $\{\mathcal{R}_{\tau-1}^*\}$ and $\{u_j^*\}$. Equating these two expressions, we obtain

$$\Psi_{\tau 1}^* A_1^* + \Psi_{\tau 2}^* A_2^* + \cdots = A_\tau^* = \sum_{j=0}^{i} \sum_{i=0}^{\tau-1} [\mathcal{R}_{\tau-1}^{*(1)} | \cdots | \mathcal{R}_{\tau-1}^{*(p)}] A_{i-j+1}^* u_j^*, \quad (\S)$$

$$\tau = 1, 2, \ldots .$$

The relation ($\S$) enables us to identify the elements of the triangular array $\{\Psi_{ij}\}, j > i$. Unfortunately, the algebra involved in carrying out the identification of these elements is better done by a computer when p and/or m is larger than 1. But we can give a simple expression for Ψ_{f^*, ω^*} in the single-input/single-output case. When $m = p = 1$, we have

$$\Psi_{f^*, \omega^*} = \begin{pmatrix} \mathcal{R}_1^* u_0^* & 0 & 0 & \cdots \\ \mathcal{R}_2^* u_0^* + \mathcal{R}_1^* u_1^* & \mathcal{R}_1^* u_0^* & 0 & \cdots \\ \mathcal{R}_3^* u_0^* + \mathcal{R}_2^* u_1^* + \mathcal{R}_1^* u_2^* & \mathcal{R}_2^* u_0^* + \mathcal{R}_1^* u_1^* & \mathcal{R}_1^* u_0^* & \cdots \\ \vdots & \vdots & \vdots & \end{pmatrix}.$$

Example: The Natural Numbers (cont'd.)

Let's use the above result to examine the repair mechanism for our earlier example involving the natural numbers. In that case we had

$$\omega^* = (1, 1, 0, 0, \ldots) = \{u_0^*, u_1^*, u_2^*, \ldots\},$$

$$f^* \approx \{1, 1, 2, 2, 3, 3, \ldots\} = \{A_1^*, A_2^*, A_3^*, \ldots\},$$

$$P_{f^*} \approx \{1, -1, 1, -1, \ldots\} = \{\mathcal{R}_1^*, \mathcal{R}_2^*, \mathcal{R}_3^*, \ldots\}.$$

Let's suppose the design metabolism f^* is perturbed to the new metabolism

$$f \approx \{1, 2, 2, 2, 3, 3, 4, 4, \ldots\} = \{A_1, A_2, A_3, \ldots\},$$

i.e., there is a change only in the second element. The cellular output under the metabolism f is now

$$\gamma = f(\omega^*) = (1, 3, 4, 4, 5, 6, 7, \ldots).$$

Thus we see that the metabolic change results in a change of output from $\gamma^* = \{$the natural numbers$\}$ to the closely related sequence γ, which differs from γ^* only in the second and third entries. The question is how the cellular repair mechanism P_{f^*} will treat this new metabolism.

To address this issue, we compute the matrix Ψ_{f^*,ω^*} which, using the result given above for the case $m = p = 1$, yields

$$\Psi_{f^*,\omega^*} = \begin{pmatrix} 1 & 0 & 0 & \cdots \\ 0 & 1 & 0 & \cdots \\ 0 & 0 & 1 & \cdots \\ \vdots & \vdots & \vdots & \end{pmatrix} = \text{identity.}$$

(In fact, it can be shown that this relation holds for *all* scalar metabolisms. See Problem 11 for a multiple-input, multiple-output example where Ψ_{f^*,ω^*} is not the identity.)

So, in the scalar case *every* metabolism f is a fixed point of the induced map Ψ_{f^*,ω^*} with associated characteristic value 1; hence, by the Metabolism Repair Theorem, the repair mechanism will *stabilize* the cell at the new metabolism f (see Problem 8). In other words, the cell will "lock-on" to the new metabolism f, and thereafter operate with this metabolism rather than the old metabolism f^*. We see that there can be no restoration with this kind of cellular repair process, only stabilization at whatever kind of metabolic perturbation may present itself. This is the ultimate in adaptability, as such a cell can immediately adjust to whatever change may occur to its original metabolic machinery.

Now let's turn to the case of disturbances in the environment ω^*, leaving the cellular metabolism f^* fixed.

Case II: A Varying Environment ω with Fixed Basal Metabolism f^*— Now we turn to the situation in which there is a change of the external environment $\omega^* \to \omega$. We want to identify all such changes that can be neutralized by the repair mechanism P_{f^*}. Thus, we seek to characterize all environments ω such that

$$P_{f^*}(f^*(\omega^*)) = P_{f^*}(f^*(\omega))(= f^*)$$

implies

$$f^*(\omega^*) = f^*(\omega).$$

In other words, we want to know when the repair map P_{f^*} is one-to-one.

But it's easy to see from the earlier definition of P_{f^*} that the repair map can be represented in Toeplitz form by the linear operator

$$P_{f^*} = \begin{pmatrix} \mathcal{R}_1^* & 0 & 0 & \cdots \\ \mathcal{R}_2^* & \mathcal{R}_1^* & 0 & \cdots \\ \mathcal{R}_3^* & \mathcal{R}_2^* & \mathcal{R}_1^* & \cdots \\ \vdots & \vdots & \vdots & \end{pmatrix}, \qquad \mathcal{R}_i^* \in R^{p \times pm}.$$

This representation implies that P_{f^*} is one-to-one if and only if we have the ker $\mathcal{R}_1^* = \{0\}$. This will be the case if and only if $m = 1$ and rank $\mathcal{R}_1^* = p$. We can summarize this simple observation in the following theorem:

ENVIRONMENTAL CHANGE THEOREM. *If $m = 1$ and rank $\mathcal{R}_1^* = p$, all environments ω such that $f^*(\omega) = f^*(\omega^*)$ are given by $\omega = \omega^* + \ker f^*$.*

On the other hand, if $m > 1$ and/or rank $\mathcal{R}_1^* = r < p$, then any environmental change of the form $\omega = x + \omega^*$ will be repaired by P_{f^*}, where x stands for any solution of the equation $f^*(x) = \hat{\gamma}$, $\hat{\gamma} \in \ker \mathcal{R}_1^*$.

The Metabolic Repair and Environmental Change Theorems provide clear, complete and computable answers to the questions surrounding the types of changes in the metabolism and/or environment that the cellular repair machinery is capable of fixing. Now let's direct our attention to the problem of repairing the repairer.

Exercises

1. Suppose the cellular design environment is $\omega^* = \{1, 0, 0, \ldots, 0\}$ as given in the text, but that now the design output is the set of Pell numbers $\gamma^* = \{1, 2, 5, 12, 29, 70, \ldots\}$, which are generated by the recurrence relation $a_{i+2} = 2a_{i+1} + a_i$, $a_0 = 0$, $a_1 = 1$. (a) Compute the metabolic map f^* associated with this cell, as well as the repair map P_{f^*}. (b) What are the complexities of these two maps? (c) Suppose now the environmental input shifts from ω^* to $\omega = \{1, 1, 0, 0, \ldots, 0\}$. What happens to the map f^* and the repair map P_{f^*}?

2. Consider the case when the design environment is the *vector* sequence

$$\omega^* = \{u_0, u_1, \ldots, u_N\},$$

where each $u_i \in R^2$. The two components are given by

$$u_i^1 = i\text{th natural number},$$
$$u_i^2 = i\text{th Pell number}, \qquad i = 0, 1, 2, \ldots, N.$$

(a) Compute the basal metabolism f^* and repair map P_{f^*} for this system. (b) Suppose now the environment changes so that the second component of the input shifts from being the Pell numbers to the Fibonacci numbers. Is this environmental shift neutralized by the repair system?

3. Prove the Environmental Change Theorem.

9. Linear (M, R)-Systems: Replication

The cellular replication map

$$\beta_f \colon H(\Omega, \Gamma) \to H(\Gamma, H(\Omega, \Gamma)),$$

can be formalized in much the same fashion as discussed above for the repair map P_f. However, since the functional role of replication is quite different from that of repair, a number of important questions arise for replication that are absent in the case of repair, questions involving mutation, adaptation, Lamarckian inheritance, and so forth. We shall consider these matters in due course. But, for now, let's focus upon the steps involved in the *formal* realization of the map β_f.

By assumption, β_f is a linear map whose inputs are metabolic maps of the form $f \approx \{A_1, A_2, \dots\}$. Moreover, β_f produces as its output the repair map $P_f \approx \{\mathcal{R}_1, \mathcal{R}_2, \dots\}$. Therefore, we must have a representation of the action of the replication map β_f as

$$c_\sigma = \sum_{i=0}^{\sigma-1} U_{\sigma-i} e_i, \qquad \sigma = 1, 2, \dots,$$

for an appropriate set of matrices $\{U_j\}$, where the input $e_i \doteq \mathcal{S}(A_i)$ and the output $c_i = \mathcal{R}_i$. Here again the operator $\mathcal{S}(\cdot)$ is the "stacking" operation discussed in the last section. Arguing just as for the repair map, we conclude that the matrices $\{U_j\}$ must have the form

$$U_j = \left[C_{j1} \,|\, C_{j2} \,|\, \cdots \,|\, C_{j,mp} \right], \qquad j = 1, 2, \dots,$$

where each $C_{jr} \in R^{p \times mp}$. In what follows, we shall write $U_j^{(r)} \doteq C_{jr}$. So just as with f and P_f, we have the representation of β_f as

$$\beta_f \longleftrightarrow \{U_1, U_2, \dots\}.$$

Since the inputs for the replication system must correspond to the metabolism f, while the outputs must be the associated repair map P_f, we have the relations

$$e_\sigma = \mathcal{S}(A_{\sigma+1}), \qquad c_\sigma = \mathcal{R}_\sigma.$$

These conditions are expressed in the time-scale σ of the replication system, which is generally different from that of either the metabolic or repair subsystems.

Using the same arguments made for the repair map, it's easy to establish that if f has a finite-dimensional realization, so does β_f and $\dim \beta_f \leq \dim f$.

Example: The Natural Numbers (cont'd.)

Continuing with the example started in the preceding section, we have

$$f \approx \{1, 1, 2, 2, 3, 3, \ldots\},$$

$$P_f \approx \{1, -1, 1, -1, 1, \ldots\},$$

which, after a bit of algebra, gives rise to the replication map

$$\beta_f \approx \{1, -2, 1, 0, 0, \ldots\}.$$

Thus, only the terms U_1, U_2 and U_3 are nonzero. Note the apparent decrease in the complexity of the sequences f, P_f and β_f as we pass from metabolism to repair and on to replication. We will return to this point below.

Applying Ho's Algorithm for realization of the map β_f, we obtain the canonical system

$$q_{\sigma+1} = \begin{pmatrix} 0 & 0 & 0 \\ 1 & 0 & 0 \\ 0 & 1 & 0 \end{pmatrix} q_\sigma + \begin{pmatrix} 1 \\ 0 \\ 0 \end{pmatrix} e_\sigma, \qquad q_0 = 0, \qquad q_\sigma \in R^3,$$

$$c_\sigma = (\,1 \quad -2 \quad 1\,) q_\sigma, \qquad \sigma = 0, 1, 2, \ldots.$$

In regard to replication, there are two immediate questions:

1. When can phenotypic changes $f^* \to f$ give rise to permanent changes in the genotype?

2. If external disturbances modify P_f^*, what kinds of changes in f^* can result?

The first of these questions is none other than asking when Lamarckian inheritance can take place. The second addresses the problem of mutations. For the sake of brevity, we address here only the Lamarckian problem.

Suppose we begin with a cell that is specified by the basal metabolism $f^*(\omega^*) = \gamma^*$ and associated repair system P_{f^*}. The Lamarckian inheritance problem comes down to the question of when a metabolic (i.e., phenotypic) change $f^* \to f$ ultimately becomes "hard-wired" in to the cellular genotype as a consequence of it's being processed by the new phenotype's cellular repair and replication machinery. Conventional wisdom in biology says that such types of inheritance of "acquired characteristics" can never take place; renegade modern biologists say that, under some circumstances, maybe they can. Here we discuss the situation within the context of the (M, R)-systems, finding mathematical evidence to support the heretical claims of the "neo-Lamarckians."

The most compact way in which to mathematically express what's at issue in the Lamarckian problem is by tracing the individual steps in the following diagram:

$$f^* \longrightarrow f \xrightarrow{\omega^*} \gamma \xrightarrow{P_{f^*}} \hat{f} \xrightarrow{\beta_{f^*}} \hat{P}$$

The initial step in the above chain is the phenotypic change, with the remaining steps being the subsequent effects this change has on both the next phenotype ($\hat{f}$) and the next repair map ($\hat{P}$). The following three conditions must **all** hold if the change $f^* \to f$ is to constitute genuine Lamarckian inheritance:

A. The new phenotype f must be stabilized ($\hat{f} = f$).

B. The new genotype P_f must also be stabilized ($\hat{P} = P_f$).

C. The genetic "mutation" $P_{f^*} \to P_f$ must be a real, not a "pseudo," mutation ($\hat{P}(\gamma) = \hat{f}$, with $\hat{P} \neq P_{f^*}$).

Mathematically, these conditions can be stated as:

A. f is a fixed point of the map Ψ_{ω^*, f^*}.

B. P_f is a fixed point of the map χ_{ω^*, f^*}.

C. $\gamma \in \ker (P_f - P_{f^*})$.

Satisfaction of all of these conditions is a tall order, requiring that the kernels of Ψ_{ω^*, f^*} and χ_{ω^*, f^*} be "small," while at the same time demanding that the kernel of the map $P_f - P_{f^*}$ be "large." It's clear that whether or not a particular phenotypic change can be inherited will depend greatly upon the way that the change interfaces with the original cellular "design specs" ω^* and f^*. One case in which we can guarantee that such changes cannot occur is when the cell is a scalar system, i.e., when $m = p = 1$. In this situation, both conditions (**A**) and (**B**) are satisfied, since the maps Ψ_{ω^*, f^*} and χ_{ω^*, f^*} are both the identity. But the last condition (**C**) can never be satisfied, since all three conditions together imply that

$$\beta_{f^*} * f * \gamma = f,$$

which, using the commutativity of the convolution product for scalar sequences, leads to the requirement that

$$\beta_{f^*} * \gamma = \text{ identity.}$$

This condition, in turn, implies that $\gamma = \gamma^*$, contradicting condition (**A**). Thus, the fine line that the cell needs to walk in order to display Lamarckian inheritance cannot be negotiated by such a simple mechanism as a cell with only a single input and a single output.

Exercises

1. Compute the replication maps for Exercises 1 and 2 of the previous section.

2. Suppose you were given the basal metabolism

$$f^* = \{I_2, I_2, \dots\},$$

where I_2 is the 2×2 identity matrix. Characterize all those metabolic shifts f leading to Lamarckian inheritance from f^*.

10. Cellular Networks and Manufacturing Processes

Up to now our discussion has centered upon the behavior of a single cell. But cells don't live in isolation. So for our (M, R)-framework to make contact with real cellular phenomena, it's necessary for us to consider how to put many cells together into a cellular network and to consider the implications of the genetic components of the cells for the overall viability of the cellular cluster. As we shall see, the repair/replication features of our cells dramatically affect the way such a network can behave in the face of various types of outside disturbances.

Consider a collection of N "cells," each of which accepts a variety of inputs and produces a spectrum of outputs. Assume that at least one cell in the network accepts inputs from the external environment and that at least one cell produces outputs that are sent to this environment. Further, suppose that every cell accepts either an environmental input or has inputs that are outputs from at least one other cell in the collection; similarly, assume that each cell produces either an environmental output or has its output utilized as the input to another cell. Such a network might look like that shown in Fig. 7.8, where we have taken $N = 5$. Here we have the cells $M_1 - M_5$, together with the two environmental inputs ω_1 and ω_2, as well as the single environmental output γ_1.

It's reasonable to suppose that each cell in such a network will have a finite lifetime after which it will be removed from the system. When this happens, all cells whose input depends upon the output from the "dead" cell will also be affected, ultimately failing in their metabolic role as well. In Fig. 7.8, for instance, if the cell M_1 fails, then so will M_2, M_3, M_4 and M_5, all of whose inputs ultimately depend upon M_1's output. Any such cell whose failure results in the failure of the entire network is called a *central* component of the network.

Now suppose that we associate a component $\mathcal{R}_i$ with each metabolic component M_i. The function of $\mathcal{R}_i$ is to repair M_i. In other words, when M_i fails, the repair component $\mathcal{R}_i$ acts to build a copy of M_i back into the network. The repair elements $\mathcal{R}_i$ are constituted so that each $\mathcal{R}_i$ must receive

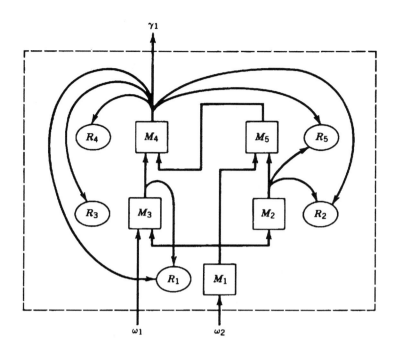

Figure 7.8. A Metabolic Network

at least one environmental output from the network. Moreover, in order to function, each $\mathcal{R}_i$ must receive all of its inputs. So in Fig. 7.8 each repair component $\mathcal{R}_i$ must receive the sole environmental output γ_1. Note also that by the second requirement, any cell M_i whose repair component $\mathcal{R}_i$ receives M_i's output as part of its input cannot be built back into the network. We call such a cell *nonreestablishable*. Thus the cell M_2 is nonreestablishable, but cell M_5 is *reestablishable*.

As already discussed, the repair components $\mathcal{R}_i$ are formed by the repair map P_f and the replication map β_f, which as we've seen, are directly constructed from the cellular metabolism f. These elementary concepts enable us to prove the following important results characterizing the interdependent nature of the components of an (M, R)-network:

NETWORK REPAIR THEOREM. *Every finite (M, R)-network contains at least one nonreestablishable component.*

COROLLARY. *If an (M, R)-network contains exactly one nonreestablishable component, then that component is central.*

This result shows that every (M, R)-network must contain some cells that cannot be built back into the system if they fail. Further, if there are only a small number of such cells, then they are likely to be of prime importance to the overall functioning of the system. This last result has obvious implications for policies devoted to keeping every component of a system alive It may be much better to allow some cells to fail rather than run the risk of incurring a global system failure by trying to prop up weak, noncompetitive components that can't all be saved anyway, as the theorem shows (liberal politicians and other social reformers, please note!). It's worth noting that the Network Repair Theorem follows only from the connective structure of the network and the role of the repair components $\mathcal{R}_i$, making no assumptions about the specific nature of the cells or their metabolic maps. In particular, none of the linearity assumptions invoked earlier are needed to establish the above results on networks. Now let's take a look at how the ideas developed here can be employed to investigate the structure of an industrial manufacturing enterprise.

Example: Industrial Manufacturing Processes

The process of transforming a collection of raw materials like steel, glass, rubber and plastics into a finished product like an automobile provides a perfect setting in which to consider the use of the (M, R)-metaphor outside the confines of cellular biology. Typically, such an enterprise involves a manufacturing process that we can formally represent as a map

$$f: \Omega \to \Gamma,$$

where Ω is a set of available inputs like raw materials, labor equipment, capital, and "know-how," while Γ is a set of finished products. The "metabolism" f is a rule, or recipe, by which the input quantities are transformed and/or assembled into the desired finished goods by the manufacturing process. In this context we usually think of f as consisting of various types of processing elements like men and machines, together with the assembly plan by which the processing elements operate upon the input quantities from Ω. However this picture of manufacturing is incomplete in a variety of ways, not the least of which is that it contains no ready means to account for repair and replication, which are essential ingredients of any truly automated "Factory of the Future." Let's look at how we might characterize some of the major features of such a manufacturing operation in terms of cellular quantities associated with a network of the sort described above.

Since any decent theory of manufacturing must be able to account for operational concerns of day-to-day production and planning—things like costs, capacity utilization and inventories—let's first examine the ways in which various aspects of cellular organization can act as counterparts of such features of manufacturing systems.

• *Direct Labor Costs*—Almost all cellular processes are energy driven. A natural measure of the cost of a process is the number of high-energy phosphate bonds, or molecules of ATP, required to carry out these processes. If desired, this can be expressed in more physical units of kcal/mole. Such units, which play the same role in the accounting of cellular processes that money plays in economics, seem a natural analogue of direct costs.

• *Indirect Labor Costs*—An important part of cellular operation is devoted to actually producing ATP from ingested carbon sources (e.g., sugars or other foods). ATP is needed to drive the cycles that make more ATP, so this seems to be a natural analogue of overhead or other types of indirect costs.

• *Material Costs*—Most of the raw materials required by a cell cannot get through the cellular membranes by themselves. They must be carried across by specific machinery ("permeases"), which must be manufactured by the cell. The cost to the cell of this manufacture and deployment of permease systems appears to be the most natural analogue of the material costs of an industrial manufacturing system.

• *Inventory Costs*—Cells do not generally maintain pools of unutilized materials or inventories. Rather, they seem organized around a "just-in-time" inventory policy.

We can give similar cellular homologs for other manufacturing quantities of concern such as quality costs, reliability in finished product, capacity utilization and flexibility, but the foregoing list is sufficient to get the general idea. The interested reader is invited to consult the papers cited in the Notes and References for a more detailed discussion. Now let's look at how to tailor the (M, R) framework to abstractly represent an industrial manufacturing operation.

We have seen above that the metabolic map $f: \Omega \to \Gamma$ can be thought of as the rule by which the plant operates. Consequently, our earlier development shows that we can construct a repair map P_f and a replication map β_f directly from these metabolic rules. However, there is a small technical point here that cannot be overlooked. Namely, we have considered the output of the manufacturing operation, the set Γ, to be composed of finished products like automobiles or TV sets. But the firm's repair map P_f must accept these items and somehow use them to reconstruct the metabolic map f. As it stands, of course, this idea is nonsense: TV sets or cars can't be directly transformed into hardware and software for manufacturing. To circumvent this difficulty, we simply assume that the output of the manufacturing operation is not the final finished physical product but rather its *monetary equivalent*. Thus, from now on we consider Γ to be just the set $\mathbb{Z}^+$, representing the gross revenues that the firm receives in dollars, say, for production of their particular product mix, be it cars, TVs, wristwatches, or whatever.

Now recall from Chapter 6 that the realization of the map f is abstractly equivalent to finding maps g and h, together with a space X, such that the diagram

$$\Omega \xrightarrow{f} \Gamma$$
$$g \searrow \quad \nearrow h$$
$$X$$

commutes, with g being onto and h one-to-one. In view of our assumptions about the space Γ, we can now attach direct physical meaning to these maps and to the set X. Since g is a map that transforms physical inputs into elements of X, it's natural to regard the X as a set of *physical products*. This interpretation of the elements of X is totally consistent with the role of the map h which is to transform elements of X into monetary revenues. So we can consider h as an abstract embodiment of the firm's *marketing and distribution* procedures. The technical conditions on the maps g and h mean that there is some input $\omega \in \Omega$ that can be processed by the firm's production operation g to achieve any desired product mix in X, and that no two distinct product mixes in X give rise to *exactly* the same gross revenues.

Using precisely the same line of reasoning, we can also consider realization of the repair and replication maps P_f and β_f in diagrammatic terms, and attempt to attach physical meaning to the resulting state spaces and factor maps. Since this would take us too far afield from the major thrust of this book, we leave these considerations to the Discussion Questions and to the material cited in the Notes and References.

Exercises

1. In the text we gave a biological interpretation to the various components of a manufacturing enterprise. Can you give a similar set of biological interpretations for a national economy?

2. Consider a definite industrial sector, such as the automobile industry, and construct a detailed (M, R)-system that models this sector. How would you interface this sector with other sectors like the steel, glass and plastics industries?

3. Prove the Network Repair Theorem. (Hint: Start by choosing a reestablishable component and proceed by induction, using the fact that there are only a finite number of cells in the network.)

4. Consider the (M, R)-network shown below. Determine the central components and reestablishable cells in this network.

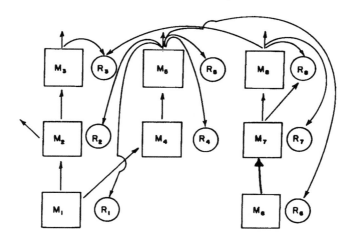

A Metabolism-Repair Network

Discussion Questions

1. Open-loop control is based upon the principle that we calculate the optimal control law $u^*(t)$ assuming that all aspects of the controlled system (dynamics, constraints, external forces, etc.) are known for the entire duration of the process. Feedback control is founded upon the notion that the control law should be calculated using the actual measured state of the system; i.e., the optimal law is based upon how the system is actually behaving. Discuss physical situations in which the assumptions underlying open-loop control might be satisfied. Consider also the possible disadvantages of feedback control laws.

2. One of the growth areas in modern applied control theory is in the area of planning and regulating national economies. Consider the pros and cons of using classical control theory for this task. In particular, discuss the feedback versus open-loop control dichotomy in this setting. How would you go about setting up the dynamics, constraints and optimization criteria for such a process? How do factors like political unrest, international balance of trade, currency exchange rates and tariffs enter into such models?

3. In the Minimum Principle approach to determining an optimal control law, it's necessary to solve a two-point boundary value problem, as well as to minimize the system Hamiltonian in order to find the optimal control. Using dynamic programming, we need to determine the optimal value function and the optimal policy function for all possible initial states and all

time horizons. Compare the computational work involved in each of these procedures, and discuss situations in which one is preferable to the other.

4. In the cosmology of Alfred North Whitehead every step or process in the universe involves both effects from past situations and anticipation of future possibilities. In particular, in Whitehead's view it's admissible to postulate an anticipatory influence on the statistical weights that play a part in quantum physics. This influence could change the probability distribution of the results of atomic reactions, thus accounting for some of the differences between living and nonliving matter. Discuss this idea within the context of the types of anticipatory processes considered in this chapter.

5. One of the main reasons that anticipatory processes have not been studied to any great extent is the fact that any mode of system behavior, *regardless of how it is generated,* can be *simulated* by a purely reactive system. In other words, the reactive paradigm of Newton is *universal.* Discuss whether or not this universality of the reactive paradigm justifies replacing anticipatory processes by their reactive-system simulation. In particular, consider this situation in connection with the Ptolemaic vision of planetary motion as a cascade of epicycles versus Kepler's model based on the heliocentric view.

6. The discussion in the text makes the point that a predictive model M of a system Σ is closed to interactions with the outside world to which the system itself is open. After a period of time, this fact will result in the model M being incapable of dealing with interactions that Σ is experiencing with its environment. This discrepancy will have the following effects: (1) the control will, in general, have other effects on Σ than those that were intended, and (2) the planned modes of interaction between Σ and its controller will be modified by these effects. These observations lead to what we usually term "side effects." Consider the role of side effects in planning operations in, for instance, corporate finance, social welfare or transportation networks. Discuss the degree to which side effects are an inherent aspect of *any* planning process. Examine how Gödel's Incompleteness Theorem, considered in Chapters 6 and 9, bears upon the problem of dealing with side effects.

7. In theoretical biology, the Principle of Function Change describes the phenomenon in which an organ that originally evolved to perform one function changes over the millennia to perform a quite different function, one that was originally only an accidental feature of the original organ. For example, the evolution of the lung from the swim bladders of fish or the development of a seeing eye from an originally photosensitive patch of skin (a proto-eye) used for another purpose. Consider the Principle of Function

Change from the perspective of designing systems that are insulated against side effects. Do you think it's possible to develop such a system? Why?

8. The development in the text emphasized the point that the metabolism, repair and replication subsystems operate on different time scales. This fact introduces time lags into the repair and replication operations, relative to the metabolic timeframe. Consider how these different time scales interact to produce the global behavior of a network. In particular, consider time lags arising from the communication links in the network, i.e., lags associated with the flow of information from one cell to another or to the outside environment. Can you think of examples where such lags are critical to the operation of the network?

9. The Network Theorem showed that every (M, R)-network must contain at least one cell that cannot be repaired if it fails, and that if there is only one such cell then its failure will destroy the entire network. In ecological circles, the term "resilience" is often used to refer to the ability of an ecosystem to absorb unknown, and generally unknowable, disturbances and still maintain, or even enhance, its viability. Discuss the possibility of developing a *theory* of system resilience based upon the notion of reestablishable components in an (M, R)-network. In particular, consider the claim that an ecosystem is more resilient if it contains a large number of nonreestablishable components. Do you think this has anything to do with the idea of species extinction? Or species adaptation?

10. In nature it's generally assumed that changes in the cellular genotype (mutations) can only be directly brought about by environmental influences like radiation or by internal errors like DNA copying mistakes, but never by phenotypic traits, i.e., Lamarckian inheritance is not allowed. Consider the idea of phenotypic variations generating genotypic changes in the context of the manufacturing example discussed in the text. Does the idea of Lamarckian inheritance seem to make sense in this setting? What about if we use an evolutionary paradigm in an economic context? How would the system genotype and phenotype be defined in such a case? What about the possibility of *directed* mutations involving direct feedback from the replication map β_f to the metabolism f? Can you think of any good examples where such mutations might occur?

11. We have emphasized the formula

$$\text{adaptation} = \text{heredity} + \text{variation} + \text{selection},$$

but have omitted any discussion of how to superimpose a selection criterion upon the (M, R)-framework developed in the text. Formally, this problem is a typical one faced in control theory; however, for (M, R)-networks we have a very different situation insofar as the interaction between the controls and system behavior is concerned. Furthermore, the primary goal of living systems is not so much to optimize, but rather to survive. Thus, the selection mechanism must serve two conflicting needs simultaneously: the need to *specialize* in order to exploit a particular ecological niche, and the need to *generalize* in order to remain viable under a variety of environmental disturbances and random mutations. Consider how you would go about developing such a selection criterion for different types of living systems, e.g., a cell, a firm, an economy, a society.

12. Consider a system composed of a collection of individuals. Discuss the contention that by optimizing the adaptability of each individual in the organization, we end up destroying the organization. Can you think of any examples illustrating this claim? (Hint: Consider the human body.) What does this principle have to do with the controversy discussed in Chapter 5 involving the unit of selection (gene, individual, group) in evolutionary processes? Consider the old saying, "What's good for General Motors is good for the USA," in this regard.

13. Natural languages like German, Icelandic and Tagalog possess many of the features of an evolutionary system: a "metabolism" that transforms ideas (semantic content) into a formal string of symbols or sounds (syntactic structure), a "repair" mechanism that acts to protect the language from random disturbances in its communicative role, and a "replication" mechanism that allows various types of "mutations" (new words, expressions, grammatical constructions, etc.) to enter the language. How you would use the formalism of an (M, R)-system to model the evolutionary development of a language? Do you think the same ideas would work for artificial languages like Esperanto or even a computer programming language like Pascal or Fortran?

14. In the Environmental Change Theorem, we have seen that it's possible for the cellular metabolism f to be changed by various types of environmental disturbances $\omega^* \to \omega$. Discuss the situation in which such metabolic changes can be *reversed* by further changes in the environment. In particular, consider this reversibility problem in the context of carcinogenesis, as well as for problems of cellular differentiation and development. Relate this discussion to the notions of reachability considered in Chapters 3 and 6.

15. In classical physics there are two complementary theories of optics: Newton's *particle* theory, which regards light as a collection of the paths of individual photons, and the *wave* theory of Huygens, which regards light as the manifestation of the wavefronts of electromagnetic fields. These two theories are dual to each other in exactly the same manner that points and

lines are dual objects in euclidean geometry. Consider the Minimum Principle and the dynamic programming approaches to optimal control problems in this context, and determine which approach corresponds to a "point" theory and which to a "wave" approach. Can you identify the function describing the "wave front" for a control problem, as well as the function that describes the "particle" trajectory?

16. The Metabolism-Repair paradigm for cellular activity is a purely *functional* theory, i.e., it makes no reference to the material elements out of which a cell is composed. This is a serious difficulty if one wants to make contact with the vast amount of experimental work being carried out in biology and in medical laboratories. Consider how you would go about linking-up these two vastly different views of a cell. Do you see any simple relationship between the functional notion of component and the structure that our methods of observation enable us to see in cells?

17. In the text we have discussed the complexity of the metabolic, repair and replication components of an (M, R)-system in terms of the number of entries that are needed to specify one coefficient in the expansion of the corresponding input/output map. For instance, the metabolic maps are specified in terms of the matrices $\{A_1, A_2, \cdots\}$, where each $A_i \in R^{p \times m}$. So the complexity of the metabolic map f is $O(pm)$. The repair map P_f is given by the matrices $\{\mathcal{R}_1, \mathcal{R}_2, \cdots\}$, where each $\mathcal{R}_i \in R^{p \times pm}$, so the complexity of the map P_f is $O(p^2 m)$. Similarly, the complexity of the replication map β_f was seen to be $O(p^3 m^2)$. These quantities depend only upon the number of metabolic inputs m and outputs p. Discuss and compare this concept of complexity with other possibilities such as the dimension of the corresponding state spaces.

18. We have introduced the notion of a sequential machine in Chapter 3 as consisting of an input set A, an output set B, a state-set X, and two maps: $\lambda: A \times X \to X$, the state-transition map, and $\delta: A \times X \to B$, the output map. How would you reformulate the (M, R) set-up in terms of a sequential machine? In particular, is the state-set X finite? Or even finite-dimensional? Compare the formulation in terms of sequential machines with that developed via the realization theory arguments given in Chapter 6.

19. As noted in Discussion Question 15, dynamic programming and the Minimum Principle regard the optimal trajectory of a control process from dual perspectives: a locus of points (the Minimum Principle) or an envelope of tangents (dynamic programming). This kind of duality works fine as long as the optimal trajectory is a definite, fixed curve. However, when stochastic effects are present, we know that the solution is an *ensemble* of curves, weighted according to a probability density function. In this situation the duality relationship between the two approaches breaks down.

Discuss why this is necessarily the case. Also consider reasons why feedback control seems to be the only general method available for attacking problems involving uncertainty.

20. In our (M, R)-model of a cell, there was no notion of a controller sitting outside the cell broadcasting instructions to the various cellular components. Rather, what we had was a kind of implicit control in which the cellular metabolism, repair and replication components each "knew" what they were supposed to be doing at each moment. Whatever "control" there is in the cell, it arises from the programs hard-wired in to these components. Moreover, there is no explicit criterion function that the cell seems to be trying to minimize or maximize. Consider how you would reformulate this cellular operation in more conventional control-theoretic terms. In particular, what kind of optimality criterion do you think makes sense for this situation?

21. Theoretical biologist Robert Rosen has argued that the concept of a "mechanism" presupposes the identity of two modes of modeling that can be very different from each other. For example, syntactics ↔ semantics or analytic ↔ synthetic. In a mechanism the notion of entailment is represented by the idea of a recursive state transition. But, Rosen argues, this is a very special kind of entailment structure, one in which entailment can in some sense be moved from domain to range. For example, the attempt by mathematicians to formalize all the truths of number theory within a purely syntactic structure, i.e., a abstract symbolic formalism in which the symbols just represent numbers but take no cognizance of the fact that they are numbers and not something else. The consequence of this kind of limited entailment structure is that there's not much causality in a mechanism: almost everything about it is **un**entailed.

At first hearing, this argument seems to be in complete contradiction to what we normally think about a mechanism, which is usually thought of as being something in which the action of one part follows in a direct causal fashion from the actions of other parts. A clock or an automobile are good examples of this kind of causality. Comment on this seeming discrepancy. Discuss also Rosen's further claim that living organisms sit at the other end of the entailment spectrum from mechanisms, in the sense that almost everything about them is entailed by some other aspect of their structure or function.

Problems

1. Consider the scalar control process

$$\min_u J(u) \doteq \min_u \int_0^T (x^2 + u^2)\, dt,$$

where the state x and the control u are related through the linear differential equation

$$\dot{x} = u, \qquad x(0) = c.$$

a) Determine the Hamiltonian $H(x, u, \lambda, t)$ for this system.

b) Using the Minimum Principle, show that the optimal open-loop control law is given by $u^*(t) = c \sinh(t - T)/\cosh T$, with the minimum value of the criterion function being $J(u^*) = c^2 \tanh T$.

c) Prove that as the terminal time $T \to \infty$, we have the asymptotic control law $u_\infty^*(t) = -ce^{-t}$, with the corresponding minimal cost being $J(u_\infty^*) = c^2$.

d) Treat the same problem using dynamic programming, and show that the optimal value function satisfies

$$\frac{\partial I}{\partial T} = \min_v \left\{ c^2 + v^2 + v\frac{\partial I}{\partial c} \right\},$$

$$I(c, 0) = 0.$$

e) Demonstrate that the minimizing v satisfies the equation

$$v^*(c, T) = -\frac{1}{2}\frac{\partial I}{\partial c},$$

and, as a result,

$$\frac{\partial I}{\partial T} = c^2 - \frac{1}{4}\left(\frac{\partial I}{\partial c}\right)^2.$$

f) Using the fact that I is a quadratic in c, i.e., $I(c, T)$ has the structure $I(c, T) = c^2 R(T)$, show that the function R satisfies the scalar Riccati equation

$$\frac{dR}{dT} = 1 - R^2(T), \qquad R(0) = 0.$$

Hence, conclude that

$$I(c, T) = c^2 \tanh T,$$
$$v^*(c, T) = -c \tanh T.$$

g) Show that in the limit as $T \to \infty$, we have the simple optimal control law $v^*(c) = -c$, with the associated optimal cost $I(c) = c^2$.

h) Extend these results for the scalar case to the multidimensional setting of minimizing the integral

$$\frac{1}{2}\int_0^T [(x, Qx) + (u, Ru)]\, dt, \qquad Q \geq 0, \quad R > 0,$$

subject to the linear dynamics

$$\frac{dx}{dt} = Fx + Gu, \qquad x(0) = c.$$

Show that the solution is that given in the text, both by dynamic programming as well as by the Minimum Principle.

2. Consider the multidimensional control process

$$\min_u \int_0^T g(x, u)\, dt,$$

with the dynamics

$$\frac{dx}{dt} = h(x, u), \qquad x(0) = c.$$

a) Show that the Bellman-Hamilton-Jacobi equation for this system is

$$\frac{\partial I}{\partial T} = \min_v \left\{ g(c, v) + \left(h(c, v), \frac{\partial I}{\partial c} \right) \right\}, \qquad I(c, 0) = 0.$$

b) Show that the Hamiltonian for the system is $H(c, v, \lambda, t) = g(c, v) + (\lambda, h(c, v))$. Consequently, derive the Minimum Principle from the B-H-J equation of part (a).

c) Show that

$$\frac{\partial I}{\partial T} = \min_v H,$$

and that

$$\lambda = \frac{\partial I}{\partial c}.$$

(Remark: This means that the co-state vector λ can be interpreted as the marginal rate of change in the optimal cost when we change the initial state c. This Problem also shows that the dynamic programming optimal value function calculation and the minimization of the system Hamiltonian involve essentially the same computation, but via quite different routes.)

3. The following problem illustrates many of the features of the "dual control" problem, in which the control applied at each moment can be used either to regulate the system state or to probe the system to learn more about its structure.

Assume we have a robot that moves about on the grid shown on the next page. The robot is attempting to find its way through this environment to the "home" square located in the upper left-hand corner. There are two shaded squares that may or may not contain barriers the robot is not allowed to pass through. Initially, the robot doesn't know whether or not the barriers are actually present. So part of its control strategy is to learn about the presence or absence of these barriers. The other part, of course, is to move toward the goal square. At the beginning the robot knows there is a barrier at $x_1 = 1$, $x_2 = 2$ with probability 0.4, and a barrier at $x_1 = 2$, $x_2 = 3$ with probability 0.5. The robot can always see one move ahead, i.e., if it's within one move of a barrier location, it can always determine with certainty whether or not a barrier is there. For a price of 0.3 moves the robot can make an observation of all squares that are two moves away, where a move is defined to be either one horizontal or one vertical square away from the robot's current location. In other words, the robot can move or observe diagonally only in two moves. The robot's objective is to get to the home square while minimizing the sum of actual moves and penalties for observation.

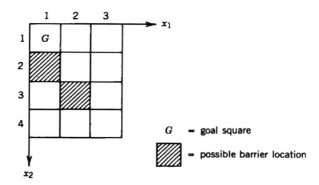

Robot Environment

The Robot's World

a) Set up a dynamic programming formulation of this problem taking the system state to be

$$x = \begin{pmatrix} x_1 \\ x_2 \\ x_3 \\ x_4 \end{pmatrix},$$

where

x_1 = the robot's horizontal coordinate,

x_2 = the robot's vertical coordinate,

x_3 = the robot's state of knowledge about the barrier at $x_1 = 1, x_2 = 2$,

x_4 = the robot's state of knowledge about the barrier at $x_1 = 2, x_2 = 3$.

The components of the state can assume the values

$$x_1 = 1, 2, 3,$$
$$x_2 = 1, 2, 3, 4,$$
$$x_3 = \text{P, A, Q,}$$
$$x_4 = \text{P, A, Q,}$$

where P, A, Q represent the robot's knowing that the barrier is Present, Absent or Unknown, respectively.

For the admissible controls at each moment, take the finite set

$$U = \left\{ \begin{pmatrix} 1 \\ 0 \\ N \end{pmatrix}, \begin{pmatrix} -1 \\ 0 \\ N \end{pmatrix}, \begin{pmatrix} 0 \\ 1 \\ N \end{pmatrix}, \begin{pmatrix} 0 \\ -1 \\ N \end{pmatrix}, \begin{pmatrix} 0 \\ 0 \\ L \end{pmatrix} \right\},$$

where L denotes "make an observation," while N means "there is no observation."

b) Show that the system dynamics are given by

$$x_1(t+1) = x_1(t) + u_3(t),$$
$$x_2(t+1) = x_2(t) - u_2(t),$$
$$x_3(t+1) = \int_3 [x_3(t), u_3(t), m_1(t), w_1(t)],$$
$$x_4(t+1) = \int_4 [x_4(t), u_3(t), m_2(t), w_2(t)],$$

where w_1 and w_2 are quantities taking on the values B or R, with B denoting the barrier is present and R meaning the barrier is absent, while m_1 and m_2 are quantities taking on the values 0 or 1 depending upon whether the presence or absence of the barrier will or will not be determined by the control being used. The quantity $\int_3$ is determined from the table below, while $\int_4$ is found from a similar table.

$x_3(t)$	$u_3(t)$	$m_1(t)$	$w_1(t)$	$x_3(t+1) = \int_3$
P	—	—	—	P
A	—	—	—	A
Q	N	0	—	Q
Q	N	1	B	P
Q	N	1	R	A
Q	L	0	—	Q
Q	L	1	B	P
Q	L	1	R	A

In the above table, the symbol "—" means that the value of $x_3(t+1)$ is the same for all values of the corresponding variable.

c) Work out the optimal control policy, showing that the optimal decision for the robot involves making an observation rather than moving when it is in the states

$$x_1 = 1, \qquad x_2 = 4, \qquad x_3 = Q, \qquad x_4 = P,$$
$$x_1 = 1, \qquad x_2 = 4, \qquad x_3 = Q, \qquad x_4 = Q.$$

d) Solve the same problem under the conditions that the robot cannot make any observations. From this solution show that the "value of information" is 0.5 moves when in the first of the states in part (c), while it is 0.1 moves in the second. That is, the difference in the minimal expected number of moves needed to reach the goal from these two states is lower by this amount when the robot is allowed to observe than when it cannot.

4. Assume we have the input/output relation

$$y_t = \sum_{i=0}^{t-1} A_{t-i} u_i, \qquad t = 1, 2, \ldots,$$

where the input sequence $\omega \in \Omega$ is *finite*, i.e.,

$$\omega = \{u_0, u_1, \ldots, u_N\}, \qquad u_i \in R^m, \qquad m - 1 \le N < \infty.$$

a) Show that the coefficient matrices $\{A_i\}$, $A_i \in R^{p \times m}$, are uniquely determined if and only if the matrix

$$\begin{pmatrix} u_0 & u_1 & u_2 & \cdots & u_N & 0 & 0 & \cdots \\ u_1 & u_2 & u_3 & \cdots & 0 & 0 & 0 & \cdots \\ u_2 & u_3 & u_4 & \cdots & 0 & 0 & 0 & \cdots \\ \vdots & \vdots & \vdots & \cdots & 0 & 0 & \cdots & \\ u_N & 0 & 0 & \cdots & & & & \end{pmatrix}$$

has maximal rank.

b) Prove that the above condition is generic, i.e., is satisfied for an open, dense set of input sequences ω.

c) How would you extend this result to the case of an infinite number of inputs?

(Remark: This exercise shows that the metabolic map f of an (M, R)-system is almost always uniquely determined by the cell's basal metabolism; hence, so are the cellular repair and replication maps P_f and β_f.)

5. In Chapter 6 we considered the input/output relation $f(\omega) = \gamma$ given by the sequences

$$\omega = \{1, 0, 0, \cdots\},$$
$$\gamma = \{1, 1, 2, 3, 5, 8, \cdots\} = \text{ the Fibonacci numbers,}$$

seeing that this "basal metabolism" admitted the canonical realization

$$F = \begin{pmatrix} 0 & 1 \\ 1 & 1 \end{pmatrix}, \qquad G = \begin{pmatrix} 1 \\ 1 \end{pmatrix}, \qquad H = (\,1 \quad 0\,).$$

a) Compute the repair and replication maps P_f and β_f that go along with this metabolism.

b) Suppose the environmental input ω shifts to $\hat{\omega} = \{1, 2, 1, 0, 0, \ldots\}$. Can the repair map neutralize this environmental change?

c) What end effect does the above environment $\hat{\omega}$ have upon the action of the system's replication map β_f?

6. In Chapter 5 we considered replicator dynamics that describe the change in genetic frequency distribution by assuming the selection principle that the relative change in frequency of a genotype is proportional to its advantage, i.e., if x_i represents the fraction of the total population having genotype i, while $f_i(x)$ is the reward for genotype i when competing against the entire population, where Φ is the average reward for the entire population, then the replicator selection mechanism postulates that

$$\frac{\dot{x}_i}{x_i} \propto f_i(x) - \Phi.$$

How could you incorporate this selection principle into the (M, R)-framework of the text?

7. The Metabolic Repair Theorem of the text categorizes all those metabolic changes that can either be neutralized or stabilized by the repair map P_f. In general, if we have a change from the basal metabolism f^* to a new metabolism f, the system could go through a sequence like $f^* \to f \to \hat{f} \to \tilde{f} \to \bar{f} \to \cdots$. Show that this cycle is of period k if and only if the metabolism f is a fixed point of the map $\Psi^k_{f^*, \omega^*}$, thereby generalizing the result given in the text. In particular, this result shows that in the space of metabolisms we have the possibility of three types of motion: a fixed point (equilibrium), a cycle of period $k < \infty$ (periodic motion), and aperiodic motion. Compare this result with the discussion of chaotic motion in Chapter 4 or the classes of cellular automata dynamics presented in Chapter 3.

8. Show that for scalar inputs and outputs (ω^*, γ^*), the map Ψ_{f^*, ω^*} always equals the identity.

9. In Chapter 6 we saw that if we have two serially-connected *independent* linear systems Σ_1 and Σ_2 with transfer matrices W_1 and W_2, respectively, then the transfer matrix of the overall system is just the product $W_2 W_1$. Considering the two subsystems f and P_f in our (M, R) setup, it's clear that the transfer matrix of P_f is not independent of the transfer matrix of f.

a) If the transfer matrix of f is W, what is the transfer matrix of P_f?

b) What is the transfer matrix of the overall system from f through P_f if the two subsystems are connected in series as indicated in the text?

c) Extend the results of parts (a) and (b) to include the system replication map β_f.

10. Consider the (M, R)-network shown below,, which abstractly represents the components of a global industry.

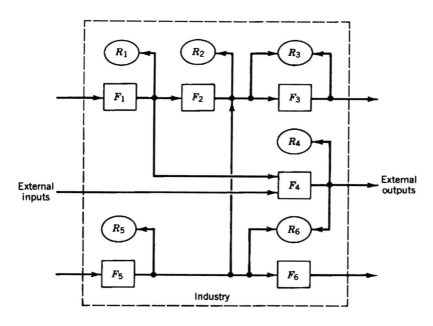

(M, R)-Network for a Global Industry

a) Show that the firm F_6 is reestablishable but all other firms are non-reestablishable.

b) Are there any central components in this network?

c) In this network some of the firms devote their entire output to supporting other firms. Can you think of any real-life situations in which this type of behavior might naturally occur?

11. Suppose we have a cell whose design environment is given by

$$\omega^* = \{u_0^*, u_1^*, 0, 0, \dots\}, \qquad u_i^* \in R^2, \qquad i = 1, 2,$$

where

$$u_0^* = \begin{pmatrix} 1 \\ 0 \end{pmatrix}, \qquad u_1^* = \begin{pmatrix} 0 \\ 1 \end{pmatrix}.$$

Further, assume the design output is

$$\gamma^* = \{y_1^*, y_2^*, y_3^*, \dots\}, \qquad y_i^* \in R^2, \qquad i = 1, 2, \dots,$$

with

$$y_1^* = \begin{pmatrix} 0 \\ 1 \end{pmatrix}, \qquad y_2^* = \begin{pmatrix} 2 \\ 0 \end{pmatrix} \qquad y_3^* = \begin{pmatrix} -2 \\ 2 \end{pmatrix}.$$

The remainder of the (infinite) output sequence is given by the relation

$$y_i^* = y_{i-2}^* - y_{i-1}^*, \qquad i = 4, 5, \dots .$$

a) Show that the behavior sequence defining the basal metabolism is given by

$$f^* = \{A_1^*, A_2^*, \dots\}, \qquad A_i^* \in R^{2 \times 2}, \qquad i = 1, 2, \dots,$$

where

$$A_1^* = \begin{pmatrix} 0 & 1 \\ 1 & 0 \end{pmatrix}, \qquad A_2^* = \begin{pmatrix} 1 & -1 \\ 0 & 1 \end{pmatrix},$$

the remaining elements of f^* being determined by the recurrence relation

$$A_i^* = A_{i-2}^* - A_{i-1}^*, \qquad i = 3, 4, \dots .$$

b) Show that the canonical realization of the cellular metabolic subsystem is of dimension two and is given by

$$x_{t+1}^* = \begin{pmatrix} 0 & 1 \\ 1 & -1 \end{pmatrix} x_t^* + \begin{pmatrix} 1 & 0 \\ 0 & 1 \end{pmatrix} u_t^*, \qquad x_0^* = \begin{pmatrix} 0 \\ 0 \end{pmatrix},$$

$$y_t^* = \begin{pmatrix} 0 & 1 \\ 1 & 0 \end{pmatrix} x_t^*. \qquad\qquad (\Sigma^*)$$

c) Show that the repair system has $m_P^* = 4$ inputs and $p_P^* = 2$ outputs, with the behavior sequence

$$P_{f^*} = \{\mathcal{R}_1^*, \mathcal{R}_2^*, \dots\}$$

being given by the elements

$$\mathcal{R}_1^* = \begin{bmatrix} 0 & 0 & | & 0 & 1 \\ 0 & 0 & | & 1 & 0 \end{bmatrix},$$

$$\mathcal{R}_2^* = \begin{bmatrix} 0 & 0 & | & 1 & -1 \\ 0 & 0 & | & 0 & 1 \end{bmatrix}.$$

The remaining elements are specified by the recurrence relation

$$\mathcal{R}_{i+2}^* = -(\mathcal{R}_i^* + \mathcal{R}_{i+1}^*), \qquad \mathcal{R}_i^* \in R^{2\times 4}, \qquad i = 1, 2, \dots .$$

d) Construct the canonical model for the repair system as the two-dimensional realization given by

$$z_{\tau+1}^* = \begin{pmatrix} 0 & -1 \\ 1 & -1 \end{pmatrix} z_\tau^* + \begin{pmatrix} 0 & 1 \\ 0 & 0 \end{pmatrix} v_\tau^*, \qquad z_0^* = \begin{pmatrix} 0 \\ 0 \end{pmatrix},$$

$$w_\tau^* = \begin{pmatrix} 0 & 1 \\ 1 & 0 \\ 1 & -1 \\ 0 & 1 \end{pmatrix} z_\tau^*, \qquad \tau = 1, 2, \dots . \qquad (\Sigma_P^*)$$

e) Show that the replication subsystem has the inputs and outputs

$$e_i^* = \mathcal{S}(A_{i+1}^*) \quad \text{and} \quad c_{i+1}^* = \mathcal{S}(\mathcal{R}_{i+1}^*), \qquad i = 0, 1, 2, \dots .$$

Thus, the replication system has $m_R^* = 4$ inputs and $p_R^* = 8$ outputs. Compute the behavior sequence for the replication system, showing it to be

$$\beta_{f^*} = \{\mathcal{U}_1^*, \mathcal{U}_2^*, \dots\}, \qquad \mathcal{U}_i^* \in R^{8\times 4},$$

where

$$\mathcal{U}_1^* = \begin{bmatrix} 0 & 0 & 0 & 0 \\ 0 & 0 & 0 & 0 \\ 0 & 0 & 0 & 0 \\ 0 & 0 & 0 & 0 \\ 0 & 0 & 0 & 0 \\ 0 & 1 & 0 & 0 \\ 0 & 1 & 0 & 0 \\ 0 & 0 & 0 & 0 \end{bmatrix}, \qquad \mathcal{U}_2^* = \begin{bmatrix} 0 & 0 & 0 & 0 \\ 0 & 0 & 0 & 0 \\ 0 & 0 & 0 & 0 \\ 0 & 0 & 0 & 0 \\ 0 & 1 & 0 & 0 \\ 0 & 0 & 0 & 0 \\ 0 & -1 & 0 & 0 \\ 0 & -1 & 0 & 0 \end{bmatrix}.$$

Verify that the remaining elements of the behavior sequence are all linear combinations of the above two elements, leading to the conclusion that the canonical replication system has dim $\Sigma_R^* = 2$.

f) Construct the canonical model for the replication system as

$$q_{\sigma+1}^* = \begin{pmatrix} -1 & 1 \\ -2 & 0 \end{pmatrix} q_\sigma^* + \begin{pmatrix} 0 & 0 & 0 & 0 \\ 0 & 1 & 0 & 0 \end{pmatrix} e_\sigma^*, \qquad q_0^* = \begin{pmatrix} 0 \\ 0 \end{pmatrix},$$

$$c_\sigma^* = \begin{pmatrix} 0 & 0 \\ 0 & 0 \\ 0 & 0 \\ 0 & 0 \\ 1 & 0 \\ 0 & 1 \\ -1 & 1 \\ 1 & 0 \end{pmatrix} q_\sigma^*, \qquad \sigma = 0,1,2,\dots . \tag{Σ_R^*}$$

g) Suppose we want to characterize all those metabolic perturbations $f^* \to f$ that can be repaired by the genetic subsystem of the above cell. Show that this means that we must find all those metabolisms f such that $f - f^* \in \ker W$, where W is the triangular Toeplitz operator

$$W = \begin{pmatrix} w_1 & 0 & 0 & \cdots \\ w_2 & w_1 & 0 & \cdots \\ w_3 & w_2 & w_1 & \cdots \\ \vdots & \vdots & \vdots & \end{pmatrix},$$

with the action of the elements $w_i(\cdot)$ given by

$$w_i(Z) = \sum_{j=1}^{i} \mathcal{R}_{i-j+1}^* \, Z \, u_{j-1}^*, \qquad i = 1,2,\dots .$$

h) Using the elements $\omega^* = \{u_0^*, u_1^*, 0, 0, \dots\}$ and $\{\mathcal{R}_i^*\}$ given above, prove that the kernel of the operator W for this cell is given by

$$\ker W = \left\{ Z_i : Z_i = \begin{pmatrix} x & x \\ 0 & 0 \end{pmatrix}, i = 1,2,\dots \right\}.$$

In this expression the elements marked "x" are arbitrary numbers. Use this result to characterize all those metabolic perturbations $f = \{A_1, A_2, \dots\}$ that the repair subsystem will be able to neutralize, i.e., perturbations that can actually be repaired by the cell's genetic component. With the kernel of W given above, show that this means that the elements of a repairable metabolic perturbation must be of the form

$$A_i = \begin{pmatrix} x & x \\ a_i^{*21} & a_i^{*22} \end{pmatrix}, \qquad i = 1,2,\dots .$$

Here again "x" is arbitrary, while the element a_i^{*kj} is just the (k,j)th component of the basal metabolism A_i^*. So for this cell, any metabolic perturbation that affects only the first row of the basal metabolism will be repaired by the cellular genetic machinery.

12. A *category* $\mathcal{A}$ is a family of objects $A, B, C, \ldots$ such that to each ordered pair of objects $(A, B) \in \mathcal{A}$, the set $H(A, B)$ of mappings from $A \to B$ is in $\mathcal{A}$. Further, $\mathcal{A}$ must satisfy the axioms:

Axiom 1: If $f \in H(A, B)$ and $g \in H(B, C)$, then there is a unique map $gf \in H(A, C)$; gf is called the *composite* of f and g.

Axiom 2: If $f \in H(A, B)$, $g \in H(B, C)$ and $h \in H(C, D)$, then we have $h(gf) = (hg)f$.

Axiom 3: To each object A in the category, we associate a unique mapping $i_A \in H(A, A)$ such that: (i) for any object B and any $f \in H(A, B)$, $f \circ i_A = f$, (ii) for any object C and any $g \in H(C, A)$, $i_A \circ g = g$.

A sequence $A, B, C, D, \ldots$ in a category will be called *normal* if no adjacent pair occurs infinitely often in the sequence. The category $\mathcal{A}$ will be called *normal* if for any normal sequence of objects in $\mathcal{A}$, the associated sequence

$$H(A, B), H(B, C), H(C, D), \ldots$$

contains only a finite number of empty sets.

a) Show that if $\mathcal{A}$ is normal and possesses infinitely many objects, then to each object $A \in \mathcal{A}$ there is an object $Z \in \mathcal{A}$ such that $H(Z, A) \neq \emptyset$.

b) Prove that if $\mathcal{A}$ is normal, then any map $\alpha \in H(X, Y)$, $X, Y \in \mathcal{A}$, can be embedded as a metabolic map of an (M, R)-system, i.e., there exists a map $P_\alpha \in \mathcal{A}$ such that P_α can serve as the repair map in the (M, R)-system with α as the metabolic map.

(Remark: Different categories give rise to different classes of (M, R)-systems, i.e., to different abstract biologies. The foregoing results, and their many corollaries and extensions, strongly suggest category theory as a natural setting for comparing different types of (M, R)-systems. This point is pursued in detail in many of the papers cited in the Notes and References.)

13. In Discussion Question 18 we considered the connection between an (M, R)-system and a sequential machine. Let $M = \{A, B, X, \lambda, \delta\}$ and $M' = \{A', B', X', \lambda', \delta'\}$ be two such machines, the notation being as in the earlier discussion. We define a *homomorphism* between M and M' as a triple of mappings $\Psi = \{\psi_1, \psi_2, \psi_3\}$, where

$$\psi_1 : X \to X',$$
$$\psi_2 : A \to A',$$
$$\psi_3 : B \to B',$$

having the property that the diagrams

$$
\begin{array}{ccc}
X \times A & \xrightarrow{\ \lambda\ } & X \\
{\scriptstyle \psi_1 \times \psi_2}\downarrow & & \downarrow{\scriptstyle \psi_1} \\
X' \times A' & \xrightarrow[\ \lambda'\ ]{} & X'
\end{array}
$$

and

$$
\begin{array}{ccc}
X \times A & \xrightarrow{\ \delta\ } & B \\
{\scriptstyle \psi_1 \times \psi_2}\downarrow & & \downarrow{\scriptstyle \psi_3} \\
X' \times A' & \xrightarrow[\ \delta'\ ]{} & B'
\end{array}
$$

commute.

a) Use this definition of homomorphism between machines to prove the following theorem:

THEOREM. *Let $\mathcal{U}$ be a category that is closed under cartesian products, i.e., if $A, B \in \mathcal{U}$, then $A \times B \in \mathcal{U}$, and if f, g are maps in $\mathcal{U}$, then so is the map $f \times g$. Then the set of all sequential machines M whose sets $A, B, X \in \mathcal{U}$ and whose maps λ and δ are also in $\mathcal{U}$ forms a category with the maps $H(M, M')$ in this category being given by the homomorphism definition above.*

b) Let $\mathcal{M}(\mathcal{U})$ denote the above category of machines over $\mathcal{U}$. Prove that if M is a machine induced by an (M, R)-system, with the map ψ_1 being onto and the maps ψ_2, ψ_3 one-to-one, then the machine M' is itself induced by an (M, R)-system.

c) Define a new category $\bar{M}(\mathcal{U})$ in the following manner: The objects of $\bar{M}(\mathcal{U})$ are the same as those of $M(\mathcal{U})$, whereas the set of maps $\bar{H}(M, M')$ consists only of those homomorphisms $\Psi = \{\psi_1, \psi_2, \psi_3\}$ such that ψ_1 is onto and ψ_2, ψ_3 are one-to-one. Show that $\bar{M}(\mathcal{U})$ is a *subcategory* of $M(\mathcal{U})$. Thus, conclude that the totality of all (M, R)-systems over a category $\mathcal{U}$ is a subcategory of the associated category $\bar{M}(\mathcal{U})$.

14. Define a *simple* (M, R)-system on the category $\mathcal{U}$ to consist of the quadruple $\{\Omega, \Gamma, f, P_f\}$, where Ω and Γ are objects of $\mathcal{U}$ such that $H(\Omega, \Gamma)$ is a set of maps on $\mathcal{U}$, $f \in H(\Omega, \Gamma)$ and $P_f \in H(\Gamma, H(\Omega, \Gamma))$.

a) Prove that a sequential machine M on the category $\mathcal{U}$ represents a simple (M, R)-system if and only if M has an output-dependent state function, i.e., if and only if there exists a map $q \in H(B, X)$ such that

$$
\lambda(x, a) = q(\lambda(x, a)).
$$

b) Show that the sequential machine M on the category $\mathcal{U}$ represents an (M, R)-system if and only if the state-space X may be written as a subset of a cartesian product $X_1 \times X_2$ in such a way that there exist maps α, β such that

$$A.\ \pi_1(x) = \pi_2(x') \Rightarrow \delta(x, a) = \delta(x', a),$$

$$B.\ \pi_1(\lambda(x, a)) = \beta(\delta(x, a)),$$

$$C.\ \pi_2(\lambda(x, a)) = \alpha(\delta(x, a), \pi_2(x)),$$

where

$$\pi_1(x_1, x_2) = x_1, \qquad \pi_2(x_1, x_2) = x_2.$$

(Note: Here that we have used the notation $x \doteq (x_1, x_2) \in X_1 \times X_2$.)

15. In his classic novel *Madame Bovary*, Flaubert tells the story of the gravedigger Lestiboudois who cultivates potatoes at the cemetery. For every body buried, Lestiboudois earns an amount M. Moreover, there are harvest revenues $f(S)$ that depend on the amount of land S under cultivation. The decision to bury a dead body diminishes the area S in an irreversible manner. Let $v(t)$ denote the rate of burials at the cemetery.

a) Show that Lestiboudois is faced with the following optimal control problem:

$$\max_v \int_0^\infty e^{rt}[f(S) + Mv(s)]\, dt,$$

subject to

$$\dot{S} = -av, \qquad S(0) = S_0, \qquad 0 \le v \le \bar{v}.$$

Here r is Lestiboudois' discount rate, a the area occupied by a grave, and $\bar{v}$ the maximal rate of funerals that can be carried out per unit time.

b) Assume that $f'(S) > 0$, $f''(S) < 0$. Then there exists a value $\hat{S}$ such that

$$f'(\hat{S}) = \frac{rM}{a}.$$

Using the Minimum Principle, show that the optimal gravedigging policy is to bury bodies at the maximal rate $\bar{v}$ as long as the cultivated area $S(t) > \hat{S}$, and to stop burying altogether as soon as the cultivated area exceeds $\hat{S}$ (assuming that $S_0 > \hat{S}$).

(Remark: A policy of this type is called a "bang-bang" control policy for obvious reasons. The nature of the optimal burial policy is clear: The crossover point for Lestiboudois is reached when the discounted revenue of harvest equals the funeral revenue. But Lestiboudois cannot control the rate of deaths. So as long as the cultivated area is above the value $\hat{S}$, the gravedigger enjoys every funeral; when the cultivated area falls below $\hat{S}$, he mourns each additional death.)

16. As another example of an off-beat control process, consider the problem of a drinker who wants to drink as much wine as possible at a party to maximize his or her enjoyment of the eternal spirits, while evading the penalty of the morning-after hangover.

To formalize this problem mathematically, suppose the drinker derives a utility $U(y, u)$ from consuming an amount of wine u. But in consuming this wine, the drinker reaches a level of drunkenness y, which, of course, generates a disutility if it's too great. Assume that the drunkenness level obeys the linear dynamics

$$\dot{y} = -\alpha y + u, \qquad y(0) = y_0 \geq 0,$$

where $\alpha > 0$ is the rate at which the drinker's body naturally disposes of the alcohol in the blood. Let $S(y)$ denote the disutility of a hangover after the party, with

$$S(0) = 0, \qquad S'(y) > 0, \qquad S''(y) \geq 0.$$

Under these assumptions, the optimal control problem is for the drinker to choose the wine consumption rate $u^*(t)$, as well as the length of time T^* s/he will stay at the party, so as to maximize (over u and T) the quantity

$$\int_0^T U(y, u)\, dt + S(y(T)).$$

Of course, the problem conditions make it clear that the drinker cannot drink a negative amount of wine. So we must impose the condition $u(t) \geq 0$. Furthermore, not even the kind of parties this drinker goes to last forever. Therefore, we impose the condition $T \leq \hat{T} < \infty$, where $\hat{T}$ is the maximal time that the police will let the party continue. Finally, and for the sake of definiteness, assume that the drinker arrives at the party sober, i.e., $y_0 = 0$.

a) Formulate the Hamiltonian function for this problem.

b) Determine the equations for the optimal state and co-state.

c) Formulate conditions on the utility function U and the disutility function S so that the problem has a unique optimal control law $u^*(t)$.

d) Show that unless $S(t)$ is very small and $\hat{T}$ is very large (a minor hangover cost and a long party), the optimal control law has the property that $\dot{u}^*(t) < 0$. In other words, the drinker's optimal policy is to continually decrease his or her wine consumption as the party unfolds.

e) Show, however, that the optimal level of drunkenness continually increases during the course of the party, i.e., $\dot{y}^*(t) > 0$.

17. Consider a plane-parallel atmosphere of finite thickness t, illuminated at the top by parallel rays of radiation of net flux π making an angle $\cos^{-1}\alpha$ with respect to the downward-directed normal to the atmosphere. Let the albedo for single scattering be denoted by λ, $0 < \lambda \leq 1$. Suppose we consider the radiation reflected back out the top of the atmosphere in the direction $\cos^{-1} v$. Denote this reflected radiation by $R(v, \alpha, t)$. If we normalize this function by introducing the new function $\hat{R}(v, \alpha, t) = 4vR(v, \alpha, t)$, then in the theory of radiative transfer it has been shown that $\hat{R}$ satisfies the differential equation

$$\hat{R}_t(v, \alpha, t) = -\left(\frac{1}{v} + \frac{1}{\alpha}\right)\hat{R}(v, \alpha, t) + \lambda\left[1 + \frac{1}{2}\int_0^1 \hat{R}(v', \alpha, t)\,dv'/v'\right] \times$$

$$\left[1 + \frac{1}{2}\int_0^1 \hat{R}(v, \alpha', t)\,d\alpha'/\alpha'\right], \qquad 0 \leq v, \alpha \leq 1.$$

The initial condition, given by considering an atmosphere of zero thickness, is $\hat{R}(v, \alpha, 0) = 0$.

a) Making use of one quadrature rule or another, discretize this equation by replacing the integrals that occur by finite sums. Show that the result is a matrix Riccati equation of the type considered in the text for the multidimensional optimal control problem with quadratic costs and linear dynamics.

b) Now shift the setting and suppose you make observations of a signal in the presence of noise over a time interval $0 \leq t \leq T$. Let the signal process be $z(t)$ and assume the noise is additive. So the observations are

$$y(t) = z(t) + v(t),$$

where $v(t)$ is the unknown noise process. We make the following assumptions on the process $v(t)$:

$$\mathcal{E}v(t) = 0,$$

$$\mathcal{E}v(t)v(\tau) = I\delta(t - \tau).$$

Here $\delta(\cdot)$ is the Dirac delta function. Optimal filtering theory addresses the question of how to make the best estimate of $z(t)$, given the observed process $y(t)$.

A linear estimator of z takes the form

$$\hat{z}(t) = \int_0^t h(t, \tau)y(\tau)\,d\tau,$$

where the kernel $h(t, \tau)$ is what's called the *impulse-response* function. The optimal estimator is then that function $h^*(t, \tau)$ minimizing the quadratic form

$$\mathcal{E}\left[(z(t) - \hat{z}(t))^2\right].$$

It's fairly easy to show that $h^*(t, \tau)$ satisfies the Fredholm integral equation

$$h^*(t, s) = k(t - s) - \int_0^t k(|\tau - s|)h^*(t, \tau)\, d\tau, \qquad 0 \le s \le t.$$

The so-called *Kalman-Bucy filter* solves this estimation problem by assuming that the process y is generated by a finite-dimensional linear system driven by white noise. Thus, in the Kalman-Bucy setup

$$z(t) = M x(t),$$
$$\frac{dx}{dt} = Fx(t) + gu(t), \qquad x(0) = x_0.$$

Here $x(t) \in R^n$ and $u(t) \in R^m$ are vector-values processes. Moreover, we have the additional assumptions on the state and noise process:

$$\mathcal{E}x_0 = 0, \qquad \mathcal{E}x_0 x_0' = \Pi_0, \qquad \mathcal{E}u(t) = 0.$$

$$\mathcal{E}v(t)x_0' = \mathcal{E}u(t)x_0' = 0, \qquad t \ge 0,$$

$$\mathcal{E}u(t)u'(\tau) = Q(t)\delta(t - \tau), \qquad Q(t) \ge 0,$$
$$\mathcal{E}u(t)v'(\tau) = 0.$$

Under these conditions, the Kalman-Bucy solution is

$$\hat{z}(t) = M\hat{x}(t),$$
$$\frac{d\hat{x}}{dt} = F\hat{x}(t) + K(t)\left[y(t) - M\hat{x}(t)\right], \qquad \hat{x}(0) = 0.$$

The quantity $K(t)$ is often called the *Kalman gain matrix*. Show that the gain matrix satisfies the equation $K(t) = P(t)M'$, where $P(t)$ is the solution of the matrix Riccati equation

$$\frac{dP}{dt} = FP(t) + P(t)F' + GQG' - K(t)K'(t), \qquad P(0) = \Pi_0.$$

c) Both the radiative transfer problem of finding the reflected radiation and the filtering problem of finding the optimal linear estimate of the signal involve a matrix Riccati equation for their solution. Using this fact, draw the mathematical parallels between these two very different physical situations. In particular, show that the reflection function $\hat{R}$ corresponds to what filtering theorists call the covariance of the state estimate, i.e.,

$$\hat{R} \approx \Sigma(t) = \mathcal{E}\hat{x}(t)\hat{x}'(t).$$

Notes and References

§1. Classical optimal control theory dates back at least to the well-known "brachistochrone" problem, which involves finding the path that an object follows under gravitational attraction in a minimal-time descent between two points. It's also of interest to note that the theory of feedback control had its origins in the regulator developed by James Clerk Maxwell to control the Watt steam engine. For an account of these and other elements of control theory in the words of their originators, see the volume

Bellman, R., and R. Kalaba, eds., *Mathematical Trends in Control Theory,* Dover, New York, 1964.

Good introductory accounts of optimal control theory are

Kirk, D., *Optimal Control Theory,* Prentice-Hall, Englewood Cliffs, NJ, 1970,

Barnett, S., *Introduction to Mathematical Control Theory,* Oxford University Press, Oxford, 1975.

For an impressionistic summary of the current state of the control theorist's arcane art, see

Modern Optimal Control Theory, E. Roxin, ed., Dekker, New York, 1989.

§2. The Minimum Principle is an outgrowth of techniques and ideas from the calculus of variations. For a discussion of the role of the classical variational calculus in giving rise to today's optimal control theory, see

Gelfand, I., and S. Fomin, *Calculus of Variations,* Prentice-Hall, Englewood Cliffs, NJ, 1963,

Dreyfus, S., *Dynamic Programming and the Calculus of Variations,* Academic Press, New York, 1965.

The classic work describing the Minimum Principle and its uses in optimization theory is

Pontryagin, L., V. Boltyanskii, R. Gamkrelidze, and E. Mischenko, *The Mathematical Theory of Optimal Processes,* Wiley, New York, 1962.

For a good introduction to the ideas underlying the Minimum Principle, along with several detailed examples, see the Kirk book noted under §1.

The determination of the optimal open-loop control law by use of the Minimum Principle involves the solution of a two-point boundary-value problem. This can pose difficult numerical problems if the interval length of the process is large. One way of overcoming some of these difficulties is by use of embedding methods in which the original problem is replaced by a family of problems containing the original one. Relationships are then developed between adjacent members of the family, thereby leading to an *initial-value* procedure for moving from the trivial problem over an interval of length zero to the interval length specified by the original problem. In many cases this approach stabilizes the original unstable situation. For a detailed account of how to carry out such an embedding in a variety of situations, see

Casti, J., and R. Kalaba, *Imbedding Methods in Applied Mathematics,* Addison-Wesley, Reading, MA, 1973,

Lee, E. Stanley, *Quasilinearlization and Invariant Imbedding,* Academic Press, New York, 1968,

Bellman, R. E., and R. E. Kalaba, *Quasilinearization and Nonlinear Boundary-Value Problems,* Elsevier, New York, 1965.

The linear-quadratic control plays a central role in control theory. This is because it has an elegant closed-form solution and because it serves as a good local approximation to more complicated problems. For an extended discussion of this problem, see the works

Anderson, B. and J. Moore, *Linear Optimal Control,* Prentice-Hall, Englewood Cliffs, NJ, 1971,

Casti, J., *Linear Dynamical Systems,* Academic Press, Orlando, FL, 1987.

§3. The theory of dynamic programming was developed and popularized by Richard Bellman in an extensive series of books, papers and articles beginning in the mid-1950s. A representative sampling of this prodigious output is found in

Bellman, R., *Dynamic Programming,* Princeton University Press, Princeton, 1957,

Bellman, R., and S. Dreyfus, *Applied Dynamic Programming,* Princeton University Press, Princeton, 1962,

Bellman, R., and R. Kalaba, *Dynamic Programming and Modern Control Theory,* Academic Press, New York, 1965.

More recent works along the same lines are

Larson, R., and J. Casti, *Principles of Dynamic Programming, Parts I and II,* Dekker, New York, 1978, 1982,

Dreyfus, S., and A. Law, *The Art and Theory of Dynamic Programming,* Academic Press, New York, 1977.

Part II of the Larson and Casti works cited above is notable for its extensive treatment of the computational problems associated with dynamic programming and the many tricks and subterfuges that can be employed to overcome them.

§4. A thorough discussion of the ideas underlying anticipatory control processes is given in

Rosen, R., *Anticipatory Systems,* Pergamon, Oxford, 1985.

Of special note is the opening chapter of this volume, which traces the historical genesis of the ideas developed here.

The idea that what you do today is conditioned by where you are today, what you've done in the past, and what your expectations are for the future is not unknown outside of biology. In fact, that principle underlies what is termed "rational expectations" in economics, a topic that is treated in detail in

Sheffrin, S., *Rational Expectations,* Cambridge University Press, Cambridge, 1983.

The idea of rational expectations seems to have first formally entered the economics literature in the path-breaking article

Muth, J., "Rational Expectations and the Theory of Price Movements," *Econometrica,* 29 (1961), 315–335.

A layman's introduction to rational expectations from the viewpoint of stock prices is given in Chapter Four of

Casti, J., *Searching for Certainty: What Scientists Can Know About the Future,* Morrow, New York, 1991.

For an easy introduction to the ideas underlying adaptive control processes as seen by the control engineering community, we recommend

Bellman, R., *Adaptive Control Processes: A Guided Tour,* Princeton University Press, Princeton, 1961,

Yakowitz, S., *Mathematics of Adaptive Control Processes,* Elsevier, New York, 1969.

The chemical reaction example is found in the Rosen book cited earlier.

§5. For an extensive discussion of the way in which global failures can occur in the seeming absence of any local signs of difficulties, see the Rosen book above as well as the important article

Rosen, R. "Feedforward and Global System Failure: A General Mechanism for Senescence," *J. Theor. Biol.,* 74 (1978), 579–590.

§6. This taxonomy of adaptive systems is considered in greater detail in

Rosen, R., "Biological Systems as Paradigms for Adaptation," in *Adaptive Economic Models,* T. Groves and R. Day, eds., Academic Press, New York, 1974.

§7. The idea of a metabolism-repair system as an abstract metaphor for a living cell was first put forth by Rosen in the late-1950s. After a brief flurry of interest the idea seems to have gone into hibernation, a fate that can be attributed to the absence of the proper system-theoretic framework within which to formulate properly the questions standing in the way of further development of the idea. Part of our exposition in this chapter has been directed toward substantiating this claim. Be that as it may, the original work in the area is summarized, together with references, in

Rosen, R., "Some Relational Cell Models: The Metabolism-Repair Systems," in *Foundations of Mathematical Biology,* Vol. 2, Academic Press, New York, 1972.

A recent account of relational thinking as it pertains to the deepest questions about living things is presented in

Rosen, R., *Life Itself,* Columbia University Press, New York, 1991.

§8–9. Further elaboration of these results can be found in

Casti, J., "Linear Metabolism-Repair Systems," *Int'l. J. Gen. Sys.,* 14 (1988), 143–167,

Casti, J., "The Theory of Metabolism-Repair Systems," *Appl. Math. & Comp.,* 28 (1988), 113–154,

Casti, J., "Newton, Aristotle and the Modeling of Living Systems," in *Newton to Aristotle,* J. Casti and A. Karlqvist, eds., Birkhäuser, New York, 1989, pp. 47–89.

§10. The idea of using (M, R)-networks as an abstract metaphor for a manufacturing enterprise arose in private discussions with Robert Rosen. Some of the details are reported in

Casti, J., "Metaphors for Manufacturing: What Could it be Like to be a Manufacturing System?," *Tech. Forecasting & Soc. Change,* 29 (1986), 241–270,

Casti, J., "(M, R)-Systems as a Framework for Modeling Structural Change in a Global Industry," *J. Soc. Biol. Struc.,* 12 (1989), 17–31.

DQ #4. We have seen that there need be no logical contradiction between the notions of causality and anticipation; i.e., an anticipatory decision procedure need not violate traditional ideas of causality. For further arguments along these lines, see

Burgers, J. M., "Causality and Anticipation," *Science,* 189 (1975), 194–198.

For introductory accounts of Whitehead's philosophy, see

Whitehead, A., *Process and Reality: An Essay in Cosmology,* Cambridge University Press, Cambridge, 1929.

Whitehead, A., *Science and the Modern World,* Macmillan, New York, 1925.

DQ #9. The notion of an ecosystem's resilience has been emphasized in a series of articles by C. S. Holling. A good summary of his views is found in

Holling, C. S., "Resilience and the Stability of Ecological Systems," *Ann. Rev. Ecol. Syst.,* 4 (1973), 1–23,

Holling, C. S., "Resilience in the Unforgiving Society," Report R–24, Institute of Animal Resource Ecology, University of British Columbia, Vancouver, March 1981.

DQ #15. For an account of this duality, mirroring exactly the Minimum Principle versus Dynamic Programming approaches to the solution of optimal control problems, see the Dreyfus book cited under §2 above as well as

Hermann, R., *Differential Geometry and the Calculus of Variations,* 2d ed., Math Sci Press, Brookline, MA, 1977.

DQ #21. For a full account of Rosen's views, see his book *Life Itself,* cited above under §7.

PR #3. The robot example is taken from the Larson and Casti book (Part II) cited under §3. This problem serves as the prototype for all adaptive control problems, in which a control resource must be allocated in order to both control the system and to learn about it. This two-fold purpose of the control gives rise to the term *dual-control process,* which is often used to describe this class of problems. For further details on this important type of control process, see

Feldbaum, A. A., "Dual Control Theory-I," *Automation & Remote Cont.,* 21 (1961), 874–880.

PR #12. These category-theoretic results can be found in the Rosen work cited under §7 above.

PR #13–14. For further information on these matters, see

Arbib, M. A., "Categories of (M, R)-Systems," *Bull. Math. Biophys.,* 28 (1966), 511–517.

These are by no means the only efforts devoted to the use of category theory to formally characterize biological systems. See also

Baianu, I., and M. Marinescu, "Organismic Supercategories-I: Proposals for a General Unitary Theory of Systems," *Bull. Math. Biophysics,* 30 (1968), 625–635.

PR #15-16. For many other examples of the use of control theory in off-beat settings, see the entertaining paper

Feichtinger, G., and A. Mehlmann, "Planning the Unusual: Applications of Control Theory to Nonstandard Problems," *Acta Applic. Math.,* 7 (1986), 79–102.

PR #17. The interconnections between filtering theory and radiative transfer were discovered by noticing that a matrix Riccati equation plays the central role in the mathematical formalization of both theories. For a complete account of these matters, see the articles

Casti, J. and E. Tse, "Optimal Linear Filtering Theory and Radiative Transfer: Comparisons and Interconnections," *J. Math. Anal. Applic.,* 40 (1972), 45–54,

G. Sidhu and J. Casti, "A Rapprochement of the Theories of Radiative Transfer and Linear Stochastic Estimation," *Appl. Math. & Comp.,* 1 (1975), 295–323.

CHAPTER EIGHT

The Geometry of Human Affairs: Connective Structure in Art, Literature and Games of Chance

1. Geometric Paradigms

Without exaggeration, it's fair to say that virtually all the formal mathematical systems used to represent natural systems in the preceding chapters have been variants of the classical dynamical system-based Newtonian paradigm. We have discussed some of the reasons leading to the dominant position of the Newtonian world view, as well as some extensions and generalizations. Here we take an even bolder step away from Newton, introducing a formal system based upon the geometrical ideas of connectivity rather than on the ideas of dynamical change.

Instead of focusing on the twin concepts of particle and dynamical law (i.e., forces) that form the backbone of the Newtonian picture, here we shift attention to the notion of how the various pieces of a natural system fit together, and the manner in which this connective structure serves to characterize the properties of the system. In short, we are concerned with the *global* geometry of the system instead of its *local* dynamics.

In order to capture the geometric essence of any natural system N, we must select an appropriate geometric structure into which the observables of N can be encoded. It turns out to be useful to employ what is termed a *simplicial complex* for this formal mathematical framework. It will soon be made clear that a simplicial complex (or just plain "complex") is a natural generalization of the intuitive idea of a euclidean space, and is formed by "pasting together" a number of pieces of varying dimension. The mathematical apparatus, which has its roots in algebraic topology, gives us a systematic procedure for keeping track of how the pieces fit together to generate the entire object, as well as how each piece contributes to the overall geometrical representation of N.

Since our emphasis now is on the *static* structure of N, we obtain information about the entire system N as opposed to the local knowledge inherent in the dynamical view a là Newton. The construction of the simplicial complex gives us a "picture" of N, and can be carried out with virtually no assumptions about the analytic character of the observables and the state space representing N. This is exactly the sort of generality we need in order to make headway in trying to capture mathematically the "systemness" of systems in the arts and humanities. Ample evidence in support of this claim will appear as we go along.

2. Sets and Relations

The starting point for the construction of a geometrical formalism capturing the connective structure of N is at the most primitive level possible: finite sets and binary relations associating the elements of one set with those of another.

Let X and Y be two sets consisting of finite collections of elements,

$$X = \{x_1, x_2, \ldots, x_m\}, \qquad Y = \{y_1, y_2, \ldots, y_n\}.$$

We make no hypotheses about the nature of the objects in X and Y; they are just abstract elements. Further, let's suppose that we define (or are given) a rule λ enabling us to decide unambiguously whether an element $x_i \in X$ and an element $y_j \in Y$ are *related* according to the criterion specified by λ. Thus, λ is a kind of decision procedure: we feed a pair (x, y), $x \in X$, $y \in Y$ into λ, and out comes either a YES or a NO telling us whether or not x is λ-related to y. This procedure is graphically depicted in Fig. 8.1.

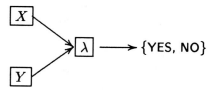

Figure 8.1. A Binary Relation λ between Sets X and Y

Since λ deals with pairs of elements, we call λ a *binary relation* on the sets X and Y.

Example: Flowers and Colors

Let's consider the set

$$X = \{\text{flowers}\} = \{\text{daffodil, rose, carnation, tulip, pansy, orchid}\},$$
$$= \{x_1, x_2, \ldots, x_6\},$$

and the set

$$Y = \{\text{colors}\} = \{\text{red, yellow, green, blue, white}\},$$
$$= \{y_1, y_2, \ldots, y_5\}.$$

A potentially interesting relation λ for gardeners and florists might be

"Flower type x_i is λ-related to color y_j if and only if there exists a strain of flower x_i having color y_j."

Since normally there exist only yellow daffodils, we would expect that $(x_1, y_2) \in \lambda$, i.e., daffodil (element x_1) is related only to the color yellow (element y_2) by the rule λ. A similar argument for roses shows that the pair of elements $(x_2, y_1) \in \lambda, (x_2, y_2) \in \lambda, (x_2, y_5) \in \lambda$. We then continue this process for all the other flowers in X.

As only certain pairs of elements from the cartesian product $X \times Y$ satisfy the condition of λ-relation, we see that, in general, $\lambda \subset X \times Y$. A compact way of expressing the relation λ is by means of an *incidence matrix* Λ. Let's label the rows of Λ by the elements of X and the columns by the elements of Y, agreeing that if $(x_i, y_j) \in \lambda$, then the (i, j) element of Λ equals 1; otherwise it's 0. So for the flower example above, the incidence matrix Λ is

λ	red	yellow	green	blue	white
daffodil	0	1	0	0	0
rose	1	1	0	0	1
carnation	1	0	0	0	1
tulip	1	1	0	0	1
pansy	1	0	0	1	1
orchid	1	0	0	1	1

Exercise

1. Consider a finite set of people P and a finite set of human activities A. For instance, let $P = \{p_1, p_2, \ldots, p_{20}\}$, while $A = \{a_1, a_2, \ldots, a_{10}\}$, where

$a_1 = $ playing golf,

$a_2 = $ gossiping,

$a_3 = $ gardening,

$a_4 = $ employment as a bureaucrat,

$a_5 = $ political activity,

$a_6 = $ employment as a retailer,

$a_7 = $ member of an auto club,

$a_8 = $ interest in foreign languages,

$a_9 = $ interest in young people,

$a_{10} = $ interest in the environment.

(a) Determine a plausible relation $\lambda \subset P \times A$. (b) Construct the incidence matrix Λ that represents this relation.

3. Covers and Hierarchies

Given a *single* set X, one of the most natural ways a binary relation arises is when we have a *cover set* for X. Let Y be a set whose elements y_i are each *subsets* of the elements of X. So, for example, a typical element of Y looks like $y_1 = \{x_2, x_3, x_4\}$ or $y_2 = \{x_1, x_2\}$. We call Y a *cover* for X if

1) Each $x_i \in X$ is contained in at least one element of Y.

2) $\bigcup_j y_j = X$.

In the special case in which each x_i is contained in *exactly* one element of the set Y, we call the cover a *partition* of X.

Example: Flowers and Colors (cont'd.)

Consider the set of flowers X from the last section. Here we have

$$X = \{\text{daffodil, rose, carnation, tulip, pansy, orchid}\}.$$

The set

$$Y_1 = \{\{\text{daffodil, tulip, rose}\}, \{\text{carnation}\}, \{\text{pansy, rose, orchid}\}\}$$
$$= \{y_1^1, y_2^1, y_3^1\}$$

constitutes a cover for X that is not a partition, since the element $\{\text{rose}\}$ appears in two elements of Y. The set

$$Y_2 = \{\{\text{daffodil, carnation}\}, \{\text{rose}\}, \{\text{tulip, orchid}\}, \{\text{pansy}\}\}$$

is a cover that is also a partition, while the set

$$Y_3 = \{\{\text{carnation, rose}\}, \{\text{orchid}\}\}$$

is neither a cover nor a partition since $\bigcup_j y_j \neq X$.

A set cover naturally induces a binary relation λ in the following manner. Let X be a set and Y a cover set for X. Then we define the relation λ by the rule

"$(x_i, y_j) \in \lambda$ if and only if element x_i
belongs to the covering element y_j."

So with the above set X and the cover set Y_1, the incidence matrix Λ_1 is given by

λ_1	y_1^1	y_2^1	y_3^1
x_1	1	0	0
x_2	1	0	1
x_3	0	1	0
x_4	1	0	0
x_5	0	0	1
x_6	0	0	1

Note that the incidence matrix representation provides a simple test for determining whether a cover is a partition: it's a partition if each row of Λ contains a single nonzero entry.

The concept of a cover set allows us to introduce the notion of a hierarchy in a natural manner. Each element of the cover set Y is, in general, an aggregation of elements of X. As a result, we can think of Y as a set existing at a level "higher" than that of X. If we arbitrarily let N denote the level of X, then Y is a level $N+1$ set. Since Y itself is a set, we can now consider a cover set Z for Y, together with an associated binary relation $\mu \subset Y \times Z$. We then regard Z as a level $N+2$ set. This process can go in the other direction too, letting X be a cover set for some set W at level $N-1$. This idea of a hierarchy of sets and relations is shown in Fig. 8.2.

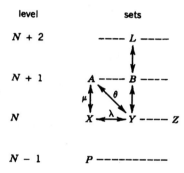

Figure 8.2. A Hierarchy of Sets and Relations

Of special interest in this diagram are the diagonal relations like θ, which enable us to go from one hierarchical level to another in a nontrivial way. It's clear that if the relation λ is given between X and Y at level N, and we induce the relation μ from the cover set A, then we can construct θ via the composition $\theta \circ \lambda = \mu$.

Primitive as the above ideas are, they already contain enough mathematical structure for us to say a few interesting things about matters in the arts and humanities, areas usually thought to be far beyond the boundaries of mathematical investigation. Let's look at a few examples.

Example 1: A Shakespearean Sonnet

Consider the following famous love poem by Shakespeare:

(1)   Shall I compare thee to a summer's day?
       Thou art more lovely and more temperate:
       Rough winds do shake the darling buds of May,
       And summer's lease hath all too short a date:
(5)   Sometime too hot the eye of heaven shines,

And often is his gold complexion dimm'd;
And every fair from fair sometime declines,
By chance, or nature's changing course, untrimm'd:
But thy eternal summer shall not fade,
(10)   Nor lose possession of that fair thou owest;
Nor shall Death brag thou wander'st in his shade,
When in eternal lines to time thou growest.
          So long as men can breathe, or eyes can see,
(14)   So long lives this, and this gives life to thee.

We consider a set at level N consisting of all the concepts referred to by the nouns in the poem. These are Thee (the beloved), May/summer, Sun, Fair (beauty), and Thy summer (the bloom time of the loved one). For a cover set Y^1 at level $(N+1)$, we must relate the N-level elements to general ideas, or properties, that they display. For these properties we could take

y_1^1 : 'being lovely' (lines 2 and 7),

y_2^1 : 'being temperate' (lines 2, 3 and 5),

y_3^1 : 'enduring time' (lines 4, 9 and 12),

y_4^1 : 'growing/diminishing' (lines 6, 7, 9, 11, 12 and 14).

If we let X be the N-level set, the incidence matrix for the natural binary relation λ relating X to its cover set Y^1 becomes

λ_1	y_1^1	y_2^1	y_3^1	y_4^1
Thee	0	1	1	1
May/summer	0	1	1	1
Sun	0	1	0	1
Fair	0	0	1	1
Thy/summer	1	0	1	1

It's important to observe that we can also generate other relations between these sets and the lines of the poem. Such relations would give insight into the structure of the poem at the corresponding hierarchical levels.

Example 2: Laughter and Tears

In his book *Multidimensional Man,* Ron Atkin suggests that the process of evoking laughter or tears from a particular N-level situation corresponds to a movement either up to level $(N+1)$ for laughter or down to level $(N-1)$ for tears. His argument is that in order to be aware of witticisms present at level N, we must be able to contemplate new relationships on the N-level set either by rearranging existing elements or by extending the elements to find new relationships between them—in short, by being aware of level $(N+1)$.

Atkin's claim is that it is this sudden jump to the $(N + 1)$-level set that generates a release of laughter.

In contrast to laughter, which is a widening of our horizons by a movement *up* the conceptual ladder, sorrow and tears represent a movement *downward* that shrinks those horizons. Moving up the hierarchy, we see the possibility for new relationships, a potentially liberating situation. A movement downward contracts or eliminates the potential for new interactions, tending to force us to think we are being imprisoned by the existing order with no way out.

As one of the many illustrations of the laughter/tears hypothesis cited in Atkin's book, let's look at a passage from Joseph Heller's classic work *Catch-22*. The N-level situation involves Yossarian's attempt to get Orr out of flying combat missions by having Doc Daneeka ground him on the basis that he's crazy. Doc states that this is only possible if Orr asks to be grounded. But Doc then adds that as soon as he's asked, he will not be able to ground Orr since the request itself would constitute evidence that he's not crazy!

In Heller's passage, the N-level consists of the *individuals* Yossarian and Orr. At the $(N+1)$-level we have a set consisting of a number of descriptive words like Sane, Missions, Grounded, and Fit for Duty. Finally, we find that at level $(N+2)$ there is the set consisting of the single element Doc, since this is the agent who can decide whether or not a man at level N is a member of the $(N+1)$-level element Fit for Duty.

The analysis of the scene and its humor comes from the fact that Yossarian thinks he is "covered" by the words Insane and Flying Missions at level $(N+1)$, and this would automatically mean that he cannot also be covered by the term Fit for Duty. But Doc reorganizes the cover set at level $(N+1)$ by saying that Yossarian's request is by itself sufficient to demonstrate that he's Sane, therefore covered by Fit for Duty. Here we see Yossarian's frustration at feeling trapped inside the N-level set and having his appeal to the $(N+2)$-level set rejected through a rearrangement of the $(N+1)$-level cover. Thus, if the reader identifies with Yossarian, he is brought to the verge of tears. But if the reader stands outside the book—at, say, level $(N+3)$—then he experiences the urge to laugh at this "Catch-22" situation.

Example 3: The Barber Paradox

In the famous Barber Paradox from elementary logic, there is a village in which the town barber shaves all those men who do not shave themselves. Since the barber is himself a man, it seems to make sense to ask if the barber shaves himself. However, tracing through the logical possibilities, we come to the surprising conclusion that the barber shaves himself if and only if he *doesn't* shave himself. Let's see how we can use the idea of a hierarchy of sets and relations in order to dissolve this paradox.

As our basic sets we take $B = \{\text{barber}\}$, consisting of the single element "barber," while the set $M = \{m_1, m_2, \ldots, m_k\}$ is composed of the men in the village. The obvious relation λ is given by the rule: "$(b, m_i) \in \lambda$ if and only if the barber shaves man m_i." So why can't we determine whether or not the element $(b, b) \in \lambda$?

The point of the paradox is to recognize that the barber is only a barber (i.e., is properly defined) only insofar as he shaves people. Thus, the element "barber" of set B is really a symbol for a subset of men, namely, the set of men who are shaved by the barber; thus, the barber *as a barber* really exists at the level $N + 1$, say. And when we try to put him in among the set M, we fail because we are trying to regard an $(N + 1)$-object as an N-object, which is what leads to the paradox.

Exercises

1. To solve the Barber Paradox, Bertrand Russell introduced what he called a "theory of types" into set theory. The basic idea is that we start with a given set X, which is of Type N, say. Then $\mathcal{P}(X)$, the power set of X, which is the set of all subsets of X, is a set of the next type. Call it Type $N + 1$. The power set of this set, $\mathcal{P}(\mathcal{P}(X))$, is then a set of Type $N + 2$ and so on. (a) Discuss this solution of the Barber Paradox within the context of hierarchical levels considered in the text. (b) Consider the Epimenides Paradox: "This sentence is false." Can you resolve this paradox using the same arguments employed for the Barber Paradox?

2. In the text we discussed the humor in Joseph Heller's famous novel *Catch-22* as a movement **up** a hierarchy of sets and relations. In *One Day in the Life of Ivan Denisovich*, Alexander Solzhenitsyn shows us how sorrow and tears can be represented as the inverse of laughter by moving **down** such a hierarchy. The real problem in the novel comes from the reader trying to identify with the "hero" Shukhov. But this requires moving from our living room at level N, say, into the prison camp with Shukhov, and entering into a daily life that exists at several lower levels. Consider what these lower levels might be, and discuss why the movement down into them from your living room is experienced as being unpleasant.

3. Consider a road traffic network consisting of local roads, main thoroughfares, intracity highways and interstate freeways. How would you represent such a system in terms of a hierarchy of sets and relations?

4. Complexes, Connections and Traffic

The hierarchical analyses based upon the notion of cover sets, interesting as they appear, are really limited in what they can tell us about the overall manner in which the relation λ binds together the elements of the sets X

and Y. For this type of information we need to obtain a multidimensional "snapshot" of λ. The way to do this is to represent λ geometrically by a mathematical gadget termed a *simplicial complex*.

Let us *abstractly* identify the elements of X with the *vertices* of the complex $K_Y(X; \lambda)$ representing λ. What this means is that we label the first vertex of K by the name "x_1", the second vertex has the name "x_2", and so on. Similarly, we name the *simplices* of K by the elements of Y. Thus, we call the first simplex "y_1", the second "y_2", and so forth. Note that there is no intrinsic ordering of these objects since the sets X and Y consist of unordered elements. Since the relation λ associates a *subset* of X with *each* element of Y, we can geometrically represent each $y \in Y$ by connecting its vertices (in X) into an abstract simplex. For example, if λ associates the vertices x_1, x_2, and x_4 with the element $y \in Y$, then we can geometrically envision y as being the 2-simplex shown in Fig. 8.3, consisting of the filled-in triangle whose vertices are x_1, x_2 and x_4.

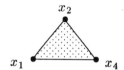

Figure 8.3. The 2-Simplex y

The same procedure can be used to associate an abstract geometrical object (point, line, triangle, tetrahedron, ...) with each $y \in Y$. The geometric dimension of each such simplex y equals one less than the number of vertices making up y.

Example 1: An Abstract Complex

Let the relation λ be given by the incidence matrix

λ	x_1	x_2	x_3	x_4
y_1	1	0	1	1
y_2	0	1	1	0
y_3	1	0	1	0
y_4	1	1	1	1

Here y_1 is the 2-simplex consisting of the vertices x_1, x_3 and x_4; y_2 is the 1-simplex consisting of x_2 and x_3, and so on. Geometrically, Λ can be represented by the complex $K_Y(X; \lambda)$ shown in Fig. 8.4. Here y_4 is the solid tetrahedron $\langle x_1, x_2, x_3, x_4 \rangle$, whereas y_1, y_2 and y_3 are faces of y_4. Thus we see that the simplices of $K_Y(X; \lambda)$ are connected to each other by the sharing of vertices. For instance, y_3 is connected to y_1 by sharing the vertices

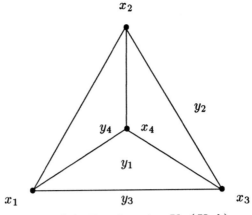

Figure 8.4. The Complex $K_Y(X; \lambda)$

x_1 and x_3 (i.e., the edge x_1–x_3), while y_1 and y_2 are connected by sharing the single vertex x_3 (i.e., the point x_3).

Example 2: Squares and Patterns

As a slightly more concrete illustration of complexes and relations, consider the two squares S_1 and S_2, which are cross-hatched with the line types $\beta, \gamma, \delta, \theta$ and ϵ shown below.

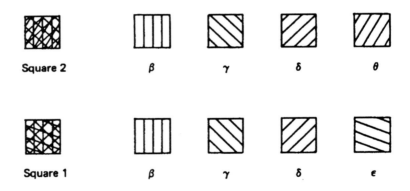

For our vertex set X, we take the set of possible line types: $X = \{\beta, \gamma, \delta, \theta, \epsilon\}$. The simplex set is $Y = \{S_1, S_2\}$. For the relation λ, we use the rule: "S_i is λ-related to vertex x_j if and only if square S_i is cross-hatched with lines of type x_j."

Using the above sets and relation λ, it's an easy matter to verify that square S_1 is a 3-simplex consisting of the vertices β, γ, δ and ϵ, while square S_2 is also a 3-simplex formed by the vertices β, γ, δ and θ. Furthermore, the overlap between the two squares is the 2-simplex (triangle) consisting of the line types β, γ and δ.

Just as there is no intrinsic ordering of the elements of X and Y, allowing us to permute these objects as we like and still obtain the same abstract geometric structure, there is also no intrinsic reason to select X as the vertex set and Y as the simplex set. It's an arbitrary choice. So if we interchange the roles of X and Y, we obtain what's called the *conjugate complex* $K_X(Y; \lambda^*)$. In terms of the incidence matrix Λ, interchanging X and Y corresponds to the operation of matrix transposition. Consequently, the incidence matrix for the conjugate relation λ^* is just Λ'. It should be noted, however, that the geometric structure of the conjugate relation λ^* is generally different from that of λ. We shall exploit this fact as we go along in order to obtain additional information about the connective structure inherent in the relation λ.

Example: A Conjugate Complex

Consider the relation λ given in the preceding example. In this case the incidence matrix Λ' for the conjugate relation λ^* is

λ^*	y_1	y_2	y_3	y_4
x_1	1	0	1	1
x_2	0	1	0	1
x_3	1	1	1	1
x_4	1	0	0	1

The conjugate complex $K_X(Y; \lambda^*)$ is shown geometrically in Fig. 8.5.

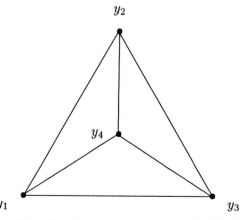

Figure 8.5. The Conjugate Complex $K_X(Y; \lambda^*)$

So far we have said nothing about *dynamics* on a simplicial complex, concentrating our attention solely upon the static connective structure. This structure forms what we could call the *static backcloth,* the playing field upon which the dynamics unfolds. In this sense, the static backcloth plays the role of the state manifold considered in earlier chapters. But what takes the place of a vector field in this simplicial complex setting? Let's look at this question in a bit more detail.

By a *pattern* π defined on a complex K, we mean simply a rule that attaches a number to each simplex $y \in K$, i.e., we have a map $\pi : Y \to R$. A change in the pattern π is often termed *traffic* on the complex K. Since the simplices have a dimensional character, it necessarily follows that the numbers attached to each simplex by the rule π inherit a natural dimension from the simplex on which they "live." This means that π is a graded pattern, i.e., $\pi = \pi_0 \oplus \pi_1 \oplus \pi_2 \oplus \cdots \oplus \pi_D$, where D is the dimension of the complex K. Let's look at a simple example illustrating these ideas.

Example: A Manufacturing Enterprise

Figure 8.6 on the next page shows a simple company with factories in Bristol and Manchester and a head office in London. Part (a) shows how the traffic representing the profits of the firm are generated, while part (b) illustrates the traffic associated with the expenditures for each factory. From this representation, it's clear that some traffic must refer to more than one vertex of a polyhedron. For instance, here we see that the unit cost of material to each factory depends on the quantities required by both factories.

It's clear that as time goes on, the numbers associated with the simplices change, and it is this change that we see as a flow of numbers throughout the structure. However, since the original numbers have a dimensional character, the connective structure of the complex will impose various constraints on the free flow of numbers throughout the system. So, for example, if a number from one 5-simplex must move to another 5-simplex, this movement cannot take place without strain or tension arising somewhere in the system unless there is a chain of 5-connection linking these two simplices. The lack of such a chain presents an obstacle to the free flow of traffic in the complex. This kind of obstacle shows itself in many ways: noise in a communication channel, auto congestion on freeways or tension in a work of art. Whether these obstacles are good or bad cannot be answered without reference to the actual setting that abstract complex represents. But at least we have a way of identifying where these bottlenecks reside, as well as a means of removing them by changing the connective links in the backcloth.

The preceding arguments illustrate the fact that the binary relation $\lambda \subset X \times Y$, the incidence matrix Λ, and the complex $K_Y(X; \lambda)$ are simply different ways of representing the same abstract object. Given any one of them, we can immediately construct the other two. Similarly, given λ, we

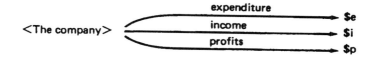

Profits(<The company>) = Income(<The company>) − Expenditure(<The company>)

(a)

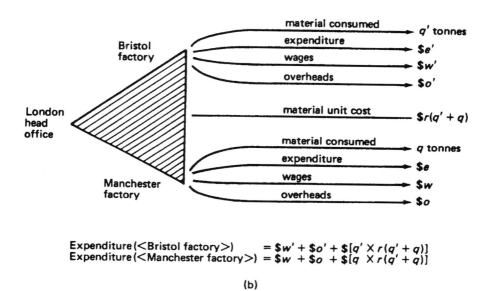

Expenditure(<Bristol factory>)   = \$w′ + \$o′ + \$[q′ × r(q′ + q)]
Expenditure(<Manchester factory>) = \$w + \$o + \$[q × r(q′ + q)]

(b)

Figure 8.6. A Manufacturing Enterprise

can automatically obtain the conjugate (or *dual*) relation λ^* directly. Which of these objects we use in a given situation depends upon considerations of the moment. The matrix Λ is convenient for computations; $K_Y(X; \lambda)$ gives geometrical insight; λ is useful for problem definition. Examples of all three approaches will be seen as we proceed. But for now, let's look at how we can use the complex $K_Y(X; \lambda)$ to define additional objects telling us about both the local and the global structure of λ.

Exercises

1. Two simplices σ_p and σ_r in a complex K are said to be *q-near* if they share a face of dimension q, i.e., if they have $(q + 1)$ vertices in common. Reformulate the concept of q-connection in terms of q-nearness.

2. Discuss the pros and cons of using a change of pattern $\Delta\pi$ to represent a dynamic on K. In particular, consider the constraints imposed on the change $\Delta\pi$ by the geometrical connectivity of the components of K. What is the connection between $\Delta\pi$ and the ordinary idea of a "force" as that term is understood and used in physics?

3. Consider a family consisting of four persons: John, age 48, Vivien, age 47, Alexander, age 21, and Stacie, age 13. Like all families, this one watches a lot of television. The types of shows they watch fall into one of the following six categories: MTV videos, sports, social commentary, news, soaps and feature-length films. Suppose we let the elements of a set Y, consisting of the family members, represent the simplices of a complex, while the set of program types X is the vertex set. Further, suppose we have the relation $\lambda \subset Y \times X$ given by the rule: $(y_i, x_j) \in \lambda$ if and only if family member y_i likes to watch programs of type x_i. (a) Make up a plausible incidence matrix Λ going along with this relation λ. (b) Depict the complex $K_Y(X; \lambda)$ geometrically. (c) Construct the conjugate complex $K_X(Y; \lambda^*)$. (d) Discuss in plain English what these complexes say about the likelihood of the family members ever agreeing on the shows they will watch.

5. Structure Vectors and Eccentricity

The simplices of $K_Y(X; \lambda)$ are connected to each other by sharing vertices. But this does not mean that the simplices are connected *pairwise*. It's perfectly possible for two simplices y_i and y_j to have no vertices in common, yet be connected to each other by an intermediate chain of simplices that serves as a bridge that links y_i and y_j. As a trivial example, consider the three 1-simplices y_1, y_2 and y_3 depicted in Fig. 8.7.

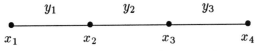

Figure 8.7. A Chain of 1-Connection

Here y_1 consists of the two vertices x_1 and x_2, with y_2 and y_3 defined similarly. The simplices y_1 and y_3 have no vertex in common; they are pairwise disconnected. Nevertheless, there is a chain of simplices consisting of y_1, y_2 and y_3 possessing the following properties: (1) y_1 shares a face with the simplex y_2; (2) y_2 shares a face with y_3. In other words, we can pass from y_1 to y_3 via the intermediate simplex y_2. The geometric dimension of the smallest face in this chain is 0 (the face consisting of either the vertex x_2 or the vertex x_3), so we call such a chain of connection a 0-chain. This idea of passing from one simplex to another via a chain of intermediate simplices forms the basis for the idea of *q-connection* in $K_Y(X; \lambda)$.

Let the dimension of $K_Y(X; \lambda)$ be defined to be D, the dimension of the highest dimensional simplex in $K_Y(X; \lambda)$. For each integer q in the range $0 \le q \le D$, we introduce the relation of q-connection on the simplices of $K_Y(X; \lambda)$ by the

DEFINITION. *The simplices y_i and y_j are q-connected in $K_Y(X; \lambda)$ if there exists a sequence of simplices $\{y_{\alpha_i}\}$, $i = 1, 2, \ldots, r$ in $K_Y(X; \lambda)$ such that*

    i. y_i shares a face of dimension m in y_{α_1}.

    ii. y_j shares a face of dimension n in y_{α_r}.

    iii. y_{α_k} and $y_{\alpha_{k+1}}$ share a face of dimension β_k.

    iv. $q = \min\{m, \beta_1, \beta_2, \ldots, \beta_{r-1}, n\}$.

In short, dimensionally speaking q is the weakest link in the chain connecting the simplices y_i and y_j.

It's easy to verify that the relation of q-connection is an *equivalence relation* on $K_Y(X; \lambda)$ for each $q = 0, 1, 2, \ldots, D$. Consequently, q-connection partitions the complex into equivalence classes whose elements consist of those simplices that are q-connected to each other. Such a relation yields insight into the global geometry of $K_Y(X; \lambda)$. Calling the q-connection equivalence relation λ_q, it then follows that λ_q partitions the complex into Q_q disjoint classes. This fact allows us to define a structure vector Q whose components are the number of classes at each dimensional level $q = 0, 1, \ldots, D$. Thus,

$$Q = (Q_D, Q_{D-1}, \ldots, Q_0).$$

In particular, note that when each element in Q equals 1, there is just a single equivalence class at each dimensional level. This means that the complex is as tightly connected as it possibly can be, with each simplex connected to every other through a chain of q-connection at every dimensional level. It's also clear that if two simplices are q-connected for some value of q, then they are also p-connected for every $p < q$.

Another way of interpreting the structure vector Q is to imagine that you are equipped with a special pair of glasses that enable you to see only in dimensions q and higher. If you put on these special spectacles and look at a picture of $K_Y(X; \lambda)$ (assuming you could draw in all dimension levels up to D), then you would see the complex fall apart into Q_q disjoint pieces.

The simplest way to calculate the connectivity vector Q is to use the incidence matrix Λ. Forming the product $\Lambda\Lambda'$, the element in position (i, j) tells us how many vertices the simplices y_i and y_j have in common. The geometric dimension of this shared face is thus $(\Lambda\Lambda')_{ij} - 1$. With this information available for each pair (y_i, y_j), it's a simple matter to trace out the chain of connection between any two simplices and to determine the smallest dimensional face on this chain.

As a measure of *obstruction* to a free flow of information from one part of the complex to another, we can use the *obstruction vector* $U = Q - (1)$, where (1) is the vector all of whose components equal 1.

Example: Flowers and Colors (cont'd.)

Let's return to the flower example of Section 2. The incidence matrix Λ for that problem is

λ	y_1	y_2	y_3	y_4	y_5
x_1	0	1	0	0	0
x_2	1	1	0	0	1
x_3	1	0	0	0	1
x_4	1	1	0	0	1
x_5	1	0	0	1	1
x_6	1	0	0	1	1

Forming the product $\Lambda\Lambda' - [1]$, where [1] denotes the matrix all of whose entries are 1, we obtain

	x_1	x_2	x_3	x_4	x_5	x_6
x_1	0	0	–	0	–	–
x_2		2	1	2	1	1
x_3			1	1	1	1
x_4				2	1	1
x_5					2	2
x_6						2

where "–" indicates -1. Since the matrix is symmetric, we show only the upper-triangular half. Note that in this example the simplices are the elements of X (the flowers) connected to each other by sharing the vertices in Y (the colors).

In the matrix $\Lambda\Lambda' - [1]$, the diagonal elements are the dimensions of the individual simplices x_i, while the off-diagonal elements are the dimensions of the faces shared by the simplices taken pairwise. An entry "–" means that the two simplices have no vertex in common. Performing the "*q*-analysis," we obtain

$$\begin{aligned}
\text{at } q=2: \quad & Q_2 = 2, \quad \{x_2, x_4\}, \{x_5, x_6\}, \\
q=1: \quad & Q_1 = 1, \quad \{x_2, x_3, x_4, x_5, x_6\}, \\
q=0: \quad & Q_0 = 1, \quad \{\text{all}\}.
\end{aligned}$$

Thus, at level $q = 2$ (sharing three or more colors) there are two disjoint components consisting of the flowers {rose, tulip} and {pansy, orchid}, each of which has three colors in common, but not the *same* three. Dropping to the level $q = 1$ (sharing two colors), all flowers except x_1 (daffodil) have at least two linking colors. At level $q = 0$, all flowers have at least one linking color. But note that this does **not** mean all flowers have at least one color *in common*. That this is not the case can be seen easily from the incidence matrix Λ. What it does mean is that we can pass from one flower to another via a chain of flowers whose adjacent links have at least one common color. The structure vector for $K_X(Y; \lambda)$ in this case is

$$Q = (\overset{2}{2}\ 1\ \overset{0}{1}),$$

with obstruction vector $U = (\overset{2}{1}\ 0\ \overset{0}{0})$, indicating an obstruction only at dimensional level $q = 2$.

We have already seen that the conjugate complex $K_X(Y; \lambda^*)$ comes along free of charge with the complex $K_Y(X; \lambda)$. Thus, we can perform a q-analysis on the conjugate complex too, obtaining a second structure vector Q^*. The two objects Q and Q^* give valuable information about the way in which the simplices in the two complexes relate to each other by vertex sharing, albeit at a distance. This is information about the global geometric structure of the complexes. But what about the local structure? How do individual simplices fit into the overall geometry? Can we define a measure characterizing the way specific simplices are integrated into the local geometry? This query leads to the idea of *eccentricity*.

Consider a single simplex y. Intuitively, we would say that y is not well-integrated into the complex if it does not share many vertices with other simplices in the complex. So if dim $y = n$ (i.e., y is comprised of $n+1$ vertices), and if $m =$ the largest number of vertices that y shares with any other single simplex in the complex, then the number $n - m$ is a measure of how "eccentric" y is as a member of $K_Y(X; \lambda)$. However, it's also reasonable to assume that this difference is more significant at lower-dimensional levels than at higher ones. So we normalize the difference $n - m$ by making it relative to m. This leads to our final measure of eccentricity as

$$\mathrm{ecc}(y) = \frac{n - m}{m + 1},$$

where we have included the 1 in the denominator to avoid possible zero divisors that arise for those y that share *no* vertices with any other simplex in $K_Y(X; \lambda)$.

Example: The Middle East Situation

To illustrate the above ideas, let's consider the Middle East crisis as a relation between the set of participants and the issues that divide them. For the set X we take

$$X = \{\text{issues}\} = \{x_1, x_2, \ldots, x_{10}\},$$

where

$x_1 = $ autonomous Palestinian state on the West Bank and Gaza,

$x_2 = $ return of the West Bank and Gaza to Arab rule,

$x_3 = $ Israeli military outposts along the Jordan River,

$x_4 = $ Israel retains East Jerusalem,

$x_5 = $ free access to all religious centers,

$x_6 = $ return of Sinai to Egypt,

$x_7 = $ dismantle Israeli Sinai settlements,

$x_8 = $ return of Golan Heights to Syria,

$x_9 = $ Israeli military outposts on the Golan Heights,

$x_{10} = $ Arab countries grant citizenship to Palestinians who choose to remain within their borders.

For the set of participants we take

$$Y = \{\text{participants}\} = \{y_1, y_2, \ldots, y_6\},$$

where

$y_1 = $ Israel,

$y_2 = $ Egypt,

$y_3 = $ Palestinians,

$y_4 = $ Jordan,

$y_5 = $ Syria and Iraq,

$y_6 = $ Saudi Arabia.

Let's take the relation $\lambda \subset Y \times X$ specifying the way the participants interact with the issues to be

"$(y_i, x_j) \in \lambda$ if and only if participant y_i is neutral or favorable toward issue (goal) x_j."

A plausible incidence matrix for λ is then

λ	x_1	x_2	x_3	x_4	x_5	x_6	x_7	x_8	x_9	x_{10}
y_1	0	1	1	1	1	1	0	0	1	1
y_2	1	1	1	0	1	1	1	1	1	0
y_3	1	1	0	0	1	1	1	1	1	1
y_4	1	1	0	0	1	1	1	1	1	0
y_5	1	1	0	0	1	1	1	1	0	0
y_6	1	1	1	0	1	1	1	1	1	1

Examination of the complex $K_Y(X;\lambda)$ shows that the most likely negotiating partner for Israel is Saudi Arabia, which is neutral or favorable on all
issues except one (Israel retaining East Jerusalem). However, both Egypt
and the Palestinians are nearly as likely candidates since they are simplices
of dimension only one less than Saudi Arabia. As the Camp David talks
demonstrated some years ago, Egypt is indeed a favored negotiating partner
due also to psychological and other factors not incorporated into the above
relation λ.

Focusing upon goals and issues, we find the high-dimensional objects in
$K_X(Y;\lambda^*)$ being $x_2 =$ return of the West Bank and Gaza to Arab rule, $x_5 =$
free access to religious centers, and $x_6 =$ return of the Sinai to Egypt. These
goals are viewed as neutral or favorable by all six participants. Therefore
they provide a good basis for a negotiated settlement of the conflict. This
observation was also borne out by the Camp David talks, as well as by
subsequent developments.

The structure vector for the complex $K_Y(X;\lambda)$ is

$$Q = (\overset{8}{1}\ 1\ 2\ 1\ 1\ 1\ 1\ \overset{0}{1}),$$

with the obstruction vector being

$$U = (\overset{8}{0}\ 0\ 1\ 0\ 0\ 0\ 0\ \overset{0}{0}).$$

Thus, the only obstruction to a free flow of information is at dimensional
level $q = 6$, where the two components {Israel} and {Egypt, Palestinians,
Jordan, Saudi Arabia} do not share any six-dimensional bridge. In fact,
$q = 6$ is the first-dimensional level at which Israel enters the complex, and
from level $q = 5$ downward all parties are fully connected to one another.
Consequently, as long as the discussions are restricted to six issues or less,
all parties are connected in the same component of the complex and the
basis for a negotiated settlement exists. With more than six issues on the

table, however, the parties become disconnected and an obstruction to a settlement arises.

Now let's look at the dispute from the viewpoint of the issues rather than the disputants, i.e., we examine the conjugate complex $K_X(Y;\lambda^*)$. In this complex the structure vector is

$$Q^* = (\overset{5}{1}\ 1\ 1\ 1\ 1\ \overset{0}{1}),$$

leading to the obstruction vector

$$U^* = (\overset{5}{0}\ 0\ 0\ 0\ 0\ \overset{0}{0}).$$

So with no obstructions at any dimensional level, all issues are tightly connected through sharing parties that are not unfavorably disposed to a given issue. This means that the highest-dimensional single issues in the complex would provide a good starting point for a settlement. In this case, these are the issues involving the return of the West Bank and Gaza to Arab rule, free access to religious centers, and a return of the Sinai to Egypt, as noted earlier.

In terms of *individual* participants or issues, the only parties that show nonzero eccentricity are Israel (ecc=$\frac{1}{6}$) and Saudi Arabia (ecc=$\frac{1}{8}$), with all issues having zero eccentricity. These results are not surprising, since the Middle East situation is basically a conflict between Israel and the Arabs, with Saudi Arabia being, for the most part, the least militant and most flexible Arab state.

While the foregoing example involves only a few states and a handful of issues, the same method can be used to address situations in which there are dozens or even hundreds of participants and factors. Some indications along these lines will be seen later in our treatment of works of art, as well as in some of the case studies cited in the Notes and References.

Exercises

1. Suppose you are given a time series of data $(t_i, y(t_i))$, $i = 1, 2, \ldots, n$. Consider how you might structure this data as a binary relation between two sets. Compute the structure vectors associated with this complex and compare these numbers with the usual correlation coefficients obtained from a least-squares regression analysis of the data.

2. Write a computer program to calculate the structure vectors Q and Q^*, as well as the eccentricities of individual simplices, given the incidence matrix Λ.

3. Consider the problem of land usage in Manhattan. Define the sets

$X = \{$local geographic areas$\}$

$\quad = \{$Upper East Side, Upper West Side, Harlem, Midtown,
      Times Square, Garment District, Chelsea, Greenwich Village,
      Soho, Chinatown, Financial District$\}$

$\quad = \{x_1, x_2, \ldots, x_{11}\},$

$Y = \{$activities$\}$

$\quad = \{$retail trade, cultural amenities, residential, entertainment,
      light manufacturing, heavy industry, financial/business$\}$

$\quad = \{y_1, y_2, \ldots, y_7\}.$

Let the relation $\lambda \subset Y \times X$ be defined by the rule: "$(y_i, x_j) \in \lambda$ if and only if activity y_i takes place in area x_j."

a) Using your knowledge of Manhattan, determine the incidence matrix for λ.

b) Calculate the structure vectors Q and Q^* for the complexes $K_Y(X; \lambda)$ and $K_X(Y; \lambda^*)$.

c) Compute the eccentricities of the simplices for these complexes.

d) Discuss the *interpretations* you can give to these calculations; i.e., what do these numbers tell you about the way goods and services are distributed throughout Manhattan?

4. Two simplices σ and σ' are said to be *q-near* if they share a q-dimensional face, i.e., they have $q + 1$ vertices in common. Reformulate the relation of q-connection in terms of q-nearness.

5. The Middle East example of the text uses a relation λ that involves a party to the dispute being neutral or favorable toward the various issues. Suppose you change the definition of the relation to read "participant y_i is favorable to issue x_j." (a) Using your knowledge of the Middle East, compute the incidence matrix that goes along with this new relation. (b) In what way (if any) does this change affect the connective structure of the complex $K_Y(X; \lambda)$ and its conjugate complex?

6. Complexity and Complexes

The matter of system complexity has been repeatedly addressed in earlier chapters, both from an objective measurement point of view and from a subjective, observer-dependent perspective. Here we take up the issue again, this time focusing upon the static complexity of the complex $K_Y(X; \lambda)$. We develop an objective measure for the complexity of the complex that satisfies three basic axioms distilled from intuitive ideas about the complexity of an object composed of a number of subsystems. We shall relegate to the Problems section the task of relating the notion of complexity developed here with those complexity concepts and measures presented in earlier chapters.

We adopt the following complexity axioms:

Axiom 1. A system consisting of a single simplex has complexity 1.

Axiom 2. A subcomplex (subsystem) has complexity no greater than that of the entire complex (system).

Axiom 3. The new complex formed by combining two complexes has complexity no greater than the sum of the complexities of its component complexes.

Note that Axioms 1-3 implicitly assume that the complex in question is connected at level $q = 0$; i.e., the structure vector Q has $Q_0 = 1$. If not, we compute the complexity function for each of the disconnected components of the complex, using the maximum of these numbers to represent the complexity of the entire complex. This default is equivalent to regarding the complex as the parallel combination of its disconnected components.

A measure that satisfies the foregoing axioms and is readily computable from the structure vector Q is

$$\psi(K) = \frac{2}{(D+1)(D+2)} \sum_{i=0}^{D} (i+1)Q_i,$$

where $D = \dim K_Y(X;\lambda)$ and Q_i = the ith component of the structure vector Q. The normalization factor $2/(D+1)(D+2)$ is introduced to satisfy Axiom 1.

Example: The Middle East Conflict (cont'd.)

Returning to the Middle East conflict, we can easily compute

$$\psi\big(K_Y(X;\lambda)\big) = \frac{52}{45},$$

while the conjugate complex has complexity

$$\psi\big(K_X(Y;\lambda^*)\big) = 1.$$

Consequently, the complex $K_Y(X;\lambda)$, which focuses attention upon the participants, is somewhat "more complicated" than $K_X(Y;\lambda^*)$, which emphasizes the issues.

Having now developed some of the machinery needed to tackle questions of interest in the social sciences, arts and humanities, let's proceed to show how the geometrical structures introduced above can be used in a variety of realistic circumstances.

Exercises

1. In the Middle East conflict example of the text, calculate the complexity measure for the complex given there, and discuss the thesis that a resolution of the dispute will require actions by all parties to reduce the complexity of both the complex and its conjugate.

How would you suggest extending this example to develop a more general conflict resolution scheme?

2. Consider the claim that, in general, the complexity of a complex will be greater if the average eccentricity of its component simplices is large. That is, the complexity of the whole is directly proportional to the heterogeneity of its individual pieces.

7. The Art of M. C. Escher

Nowadays it's almost impossible to walk into the office of a scientist or mathematician without seeing an engraving or two by the well-known Dutch artist M. C. Escher (1898–1971) hanging on the wall. Escher is noted for the remarkable geometrical precision of his work, as well as for its deep connections with mathematical concepts, especially those in group theory. Here we examine one of his more famous works using q-analysis.

A good illustration of the use of simplicial complexes to capture abstract structure is provided by Escher's famous engraving *Sky and Water*, shown in Fig. 8.8. Here we see a collection of what appear to be geese gradually being transformed into fish as the picture is scanned continuously from top to bottom. At the same time, we also see a smooth transition from figure to ground as the shapes constituting the geese become background for the swimming fish. Our goal is to capture some of the structure of these transitions using the machinery of q-analysis.

The first step is to identify relevant sets X and Y and a relation λ that encapsulate some of the connective structure of *Sky and Water*. A bit of reflection soon leads to the realization that the picture is really a statement about the relationship between various geometrical shapes (the birds, fish and their intermediate forms), and features that pertain to the identification of the shapes as being birdlike, fishlike or something in between. In Fig. 8.9 we have identified 39 different shapes that appear in the picture. So we let the elements of the set $Y = \{y_1, y_2, \ldots, y_{39}\}$ be these shapes.

As for the set X, its elements are the following collection of 12 features, each of which plays a prominent role in the picture:

$$X = \{x_1, x_2, \ldots, x_{12}\},$$
$$= \{\text{scales, mouth, gills, fish-tail, fins, fish shape, eye,}$$
$$\text{duck shape, two wings, feathers, beak, legs}\}.$$

Figure 8.8. M. C. Escher, *Sky and Water* (1938)

We make the obvious choice for the relation $\lambda \subset Y \times X : (y_i, x_j) \in \lambda$ if and only if shape y_i displays feature x_j. The incidence matrix Λ for this relation λ is given on page 625.

Using Λ, we form $\Lambda \Lambda' - U$ and obtain the following q-analysis of the connective structure in *Sky and Water:*

$$
\begin{aligned}
q = 6: & \quad Q_6 = 1, & & \{y_1 - y_6\}, \\
q = 5: & \quad Q_5 = 2, & & \{y_1 - y_6, y_8 - y_{10}\}, \{y_{21} - y_{26}, y_{28}, y_{29}\}, \\
q = 4: & \quad Q_4 = 2, & & \{y_1 - y_6, y_8 - y_{10}\}, \{y_{21} - y_{29}\}, \\
q = 3: & \quad Q_3 = 2, & & \{y_1 - y_{13}\}, \{y_{21} - y_{29}\}, \\
q = 2: & \quad Q_2 = 2, & & \{y_1 - y_{13}\}, \{y_{21} - y_{29}, y_{31} - y_{33}\}, \\
q = 1: & \quad Q_1 = 2, & & \{y_1 - y_{13}\}, \{y_{21} - y_{33}\}, \\
q = 0: & \quad Q_0 = 1, & & \{\text{all}\}.
\end{aligned}
$$

The foregoing analysis of the complex $K_Y(X; \lambda)$ focuses attention upon the shapes, showing that the principal shapes in the picture are the "fish" shapes $y_1 - y_6$, followed by the "bird" shapes $y_{21} - y_{26}$. This is a fairly

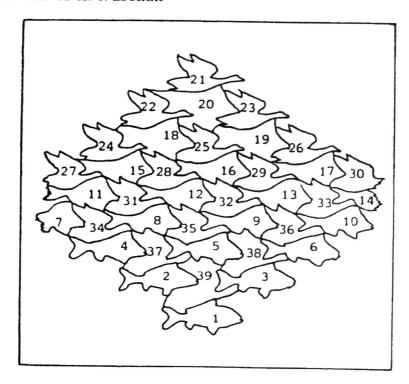

Figure 8.9. Shapes in *Sky and Water*

obvious conclusion; nevertheless, it's satisfying to reach it via our systematic procedures. At the intermediate levels of connectivity $1 \leq q \leq 5$, we see that *Sky and Water* breaks down into two disconnected pieces, essentially fishlike and birdlike shapes, whereas at the extreme levels $q = 6$ and $q = 0$ we have a fully integrated picture.

Looking at individual shapes, all simplices have eccentricity 0. This shows the smooth, almost seamless transition Escher has managed to portray in progressing from the birds to the fish. No shape stands out above any other, our eccentricity measure giving a quantitative confirmation of this observationally evident fact.

Those readers familiar with other works by Escher will recognize that *Sky and Water* is typical of many of his engravings, which feature a smooth passage from one type of figure to another accompanied by a transition from figure to ground. The techniques introduced above, together with the deeper and more refined methods presented in literature cited in the Notes and References, offer the basis for a systematic analysis of many aspects of Escher's style and form. We now turn to a consideration of the structure present in a different artistic style.

λ	x_1	x_2	x_3	x_4	x_5	x_6	x_7	x_8	x_9	x_{10}	x_{11}	x_{12}
y_1	1	1	1	1	1	1	1	0	0	0	0	0
y_2	1	1	1	1	1	1	1	0	0	0	0	0
y_3	1	1	1	1	1	1	1	0	0	0	0	0
y_4	1	1	1	1	1	1	1	0	0	0	0	0
y_5	1	1	1	1	1	1	1	0	0	0	0	0
y_6	1	1	1	1	1	1	1	0	0	0	0	0
y_8	0	1	1	1	1	1	1	0	0	0	0	0
y_9	0	1	1	1	1	1	1	0	0	0	0	0
y_{10}	0	1	1	1	1	1	1	0	0	0	0	0
y_{11}	0	0	0	1	1	1	1	0	0	0	0	0
y_{12}	0	0	0	1	1	1	1	0	0	0	0	0
y_{13}	0	0	0	1	1	1	1	0	0	0	0	0
y_7	0	1	1	0	1	0	1	0	0	0	0	0
y_{21}	0	0	0	0	0	0	1	1	1	1	1	1
y_{22}	0	0	0	0	0	0	1	1	1	1	1	1
y_{23}	0	0	0	0	0	0	1	1	1	1	1	1
y_{24}	0	0	0	0	0	0	1	1	1	1	1	1
y_{25}	0	0	0	0	0	0	1	1	1	1	1	1
y_{26}	0	0	0	0	0	0	1	1	1	1	1	1
y_{28}	0	0	0	0	0	0	1	1	1	1	1	1
y_{29}	0	0	0	0	0	0	1	1	1	1	1	1
y_{31}	0	0	0	0	0	0	1	1	1	0	0	0
y_{32}	0	0	0	0	0	0	1	1	1	0	0	0
y_{33}	0	0	0	0	0	0	1	1	1	0	0	0
y_{27}	0	0	0	0	0	0	1	1	1	1	1	0
y_{30}	0	0	0	0	0	0	0	0	1	1	0	0
y_{34}	0	0	0	0	0	0	0	1	0	0	0	0
y_{35}	0	0	0	0	0	0	0	1	0	0	0	0
y_{36}	0	0	0	0	0	0	0	1	0	0	0	0
y_{37}	0	0	0	0	0	0	0	1	0	0	0	0
y_{38}	0	0	0	0	0	0	0	1	0	0	0	0
y_{39}	0	0	0	0	0	0	0	0	0	0	0	0
y_{14}	0	0	0	1	0	0	0	0	0	0	0	0
y_{15}	0	0	0	0	0	1	0	0	0	0	0	0
y_{16}	0	0	0	0	0	1	0	0	0	0	0	0
y_{17}	0	0	0	0	0	1	0	0	0	0	0	0
y_{18}	0	0	0	0	0	0	0	1	0	0	0	0
y_{19}	0	0	0	0	0	0	0	1	0	0	0	0
y_{20}	0	0	0	0	0	0	0	0	0	0	0	0

Incidence Matrix for *Sky and Water*

Exercises

1. Using the definition of complexity introduced in Section 7, compute the complexity of the complex and its conjugate representing *Sky and Water*. How do you reconcile this measure with the fact that all the simplices in the complex have eccentricity zero?

2. Look in a book of Escher engravings and carry out a similar q-analysis for one of his other works (*Day and Night* or *Metamorphosis II* are good examples to use). Compare the structure vectors, eccentricities and complexities of these cases with *Sky and Water*.

8. Connective Structure in the Work of Piet Mondrian

The Dutch artist Piet Mondrian (1872–1944) was a leading member of the de Stijl abstract art movement centered in The Netherlands. Mondrian's cubist period reached its peak with his works *New York City* and *Broadway Boogie Woogie*, first exhibited in 1943. In this section we use the tools of q-analysis to investigate another one of his paintings of the same style shown in Fig. 8.10. Here we see a black-and-white approximation to Mondrian's famous work *Checkerboard, Bright Colors, 1919*, which was briefly considered earlier in Discussion Question 5 of Chapter 3. This painting consists of a rectangular grid pattern of 256 squares each colored by one of the 8 colors indicated.

At first glance, it appears that the colors are distributed more or less randomly on the canvas. Here we present an analysis by Ron Atkin aimed at testing this hypothesis, in the sense that a random allocation of colors would display no discernible structure or pattern. The task is to see if the q-analysis techniques developed above can be used to tease out any hidden structure in the painting, thereby refuting the randomness hypothesis.

As usual, we begin by identifying relevant sets X and Y and a meaningful relation λ. In this case the choices are easy: the message of the painting (if it has one) clearly resides in the relationship between the squares on the canvas and the colors. Thus, we let

$$X = \{\text{squares on the canvas}\} = \{x_1, x_2, \ldots, x_{256}\},$$

$$Y = \{\text{colors}\} = \{y_1, y_2, \ldots, y_8\}$$
$$= \{\text{white, grey, pale yellow, yellow, light blue,}$$
$$\text{dark blue, mushroom, red}\}.$$

We use the relation $\lambda \subset Y \times X$ defined to be that $(y_i, x_j) \in \lambda$ if and only if color y_i is used to paint square x_j. The 8×256 incidence matrix Λ can be easily (albeit tediously) filled in from Fig. 8.10, once we have agreed on an ordering of the squares. Having the matrix Λ in hand, we could proceed as

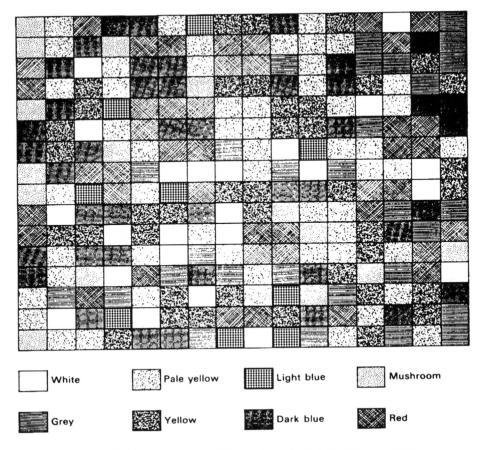

	White		Pale yellow		Light blue		Mushroom
	Grey		Yellow		Dark blue		Red

Figure 8.10. Mondrian's *Checkerboard, Bright Colors, 1919*

before to compute the structure vectors Q, Q^*, as well as the eccentricities of the individual simplices. But let's take another approach.

Suppose we look at the canvas as one single large square X_1 and define a relation $\lambda_1 \subset Y \times X_1$ that just counts the number of times that color y_i appears in X_1; i.e., λ_1 is no longer a binary relation, but takes on integer values. So, for the relation λ_1, the transposed incidence matrix is

λ_1'	y_1	y_2	y_3	y_4	y_5	y_6	y_7	y_8
X_1	19	29	54	40	13	43	15	43

Here we have y_1 (white) appearing 19 times, y_2 (grey) appearing 29 times, and so on. We can now *induce* a binary relation λ by introducing a parameter θ and using the rule: $(y_i, X_1) \in \lambda$ if and only if color y_i appears at least θ times in square X_1. If $\theta = 20$, for example, only the pairs $(y_2, X_1), (y_3, X_1), (y_4, X_1), (y_6, X_1)$ and (y_8, X_1) are in the relation λ.

Setting $\theta = 20$ corresponds to "seeing" only the colors grey (y_2), pale yellow (y_3), yellow (y_4), dark blue (y_6) and red (y_8). So, this relation λ sees X_1 as being the simplex consisting of the vertices $\{y_2, y_3, y_4, y_6, y_8\}$, leading to dim $X_1 = 4$. We call the parameter θ a *slicing parameter*, since it has the effect of "slicing off" the weighted relation λ_1 at various levels of color repetition.

The concept of a slicing parameter is often useful when we want to try to quantify the qualitative 0/1 pattern of the relation λ. Examination of the change in Q and Q^* as we change the slicing parameter is analogous to the kind of parametric variation we saw earlier in Chapters 2 and 4. If we feel that looking at the painting as a single square gives too coarse a view, we might consider decomposing the canvas into subsquares, and then analyze the connectivities introduced by various levels of θ. For instance, suppose we decompose the painting into the 16 subsquares depicted in Fig. 8.11

X_{41}	X_{42}	X_{43}	X_{44}
⋮			⋮
⋮			⋮
X_{11}	X_{12}	X_{13}	X_{14}

Figure 8.11. Decomposition of the Canvas into 16 Subsquares

Defining the weighted relation λ_{16} as above by counting the number of appearances of each color in each subsquare, we obtain the incidence pattern

λ_{16}	X_{11}	X_{12}	X_{13}	X_{14}	X_{21}	X_{22}	X_{23}	X_{24}	$\cdots$	X_{44}
y_1	3	1	1	1	2	3	0	1	$\cdots$	1
y_2	2	3	3	4	0	1	0	3	$\cdots$	7
y_3	5	1	3	3	2	4	5	1	$\cdots$	1
y_4	1	4	2	4	2	4	3	3	$\cdots$	3
y_5	0	2	2	0	1	2	0	0	$\cdots$	0
y_6	2	3	2	2	5	1	2	2	$\cdots$	1
y_7	2	0	1	1	1	0	2	1	$\cdots$	0
y_8	1	2	2	1	3	3	4	5	$\cdots$	3

Slicing the above relation at $\theta = 1$ leads to the structure vector

$$Q = (\overset{15}{1}\ 1\ \cdots\ \overset{0}{1})$$

for the complex $K_Y(X_{16}; \lambda_{16})$, where the colors of highest dimension are y_3 (pale yellow) and y_6 (dark blue), each having dimension 15. Thus, the color simplices require a space of at least 15 dimensions for their representation. Every other color is a subpolyhedron contained in the 16-vertex polyhedron representing y_3 (or y_6).

Turning to the conjugate complex $K_{X_{16}}(Y; \lambda_{16}^*)$, the structure vector is

$$Q^* = (\overset{7}{1} \ 1 \ \cdots \ \overset{0}{1}),$$

with the square X_{13} being of highest dimension (7). Each of the remaining squares is a face of the X_{13} simplex.

If we slice the relation λ_{16} at $\theta = 2$, the structure vector for the complex $K_Y(X_{16}; \lambda_{16})$ becomes

$$Q = (\overset{12}{1} \ 4 \ 3 \ 2 \ 1 \ 2 \ 2 \ 1 \ 2 \ 1 \ 1 \ 1 \ \overset{0}{1}).$$

The dominant colors are still y_3, y_4, y_6 and y_8, all of which are disconnected at dimension $q = 11$. The high number of components at the intermediate dimensional levels means that this view of the painting is much more rigid than at $\theta = 1$, suggesting that the color distribution is not free, artistically speaking.

Finally, looking at the complex $K_{X_{16}}(Y; \lambda_{16}^*)$ when $\theta = 2$ yields the structure vector

$$Q^* = (\overset{5}{1} \ 5 \ 3 \ 1 \ 1 \ \overset{0}{1}),$$

also a significant change from the case $\theta = 1$. Further analysis shows that the squares with nonzero eccentricities tend to be concentrated along the bottom row of the painting.

For more details on this type of analysis, we urge the reader to consult the Atkin's books cited in the Notes and References.

Exercises

1. In the text we considered views of Mondrian's work both as a single square and as 16 squares. Continue this analysis by looking at the painting as 64 squares obtained by subdividing each of the 16 squares into four new squares. Construct the incidence matrix for this situation, taking the slicing parameter to be $\theta = 1$ and $\theta = 2$. Show that in the first case there is nonzero obstruction at all q-levels from 7 to 33. What about the second case?

2. The changes in our view of the painting in going from one slicing level to another can be represented algebraically by a pattern polynomial. For example, consider the painting as consisting of just four squares arranged top to bottom and left to right as $X_{21}, X_{22}, X11$ and X_{12}. (a) Show that in this case the color structure is the same in all squares except X_{22}. So

if we use a slicing level $\theta = 2$, we can describe what we see by the pattern polynomial

$$\pi_2 = y_1 y_2 y_3 y_4 y_5 y_6 y_8 + y_1 y_2 y_3 y_4 y_5 y_6 y_7 y_8,$$

where the y's are the various colors. This pattern corresponds to the square X_{22} along with others, and describes the backcloth at this particular slicing level. (b) Now change the slicing level to $\theta = 10$. Construct the pattern polynomial that defines the structure of the painting at this level. Discuss the assertion that this change of view corresponds to a change of pattern $\Delta\pi$. (c) In terms of the aesthetic experience involved in viewing the picture, what interpretation can you attach to this pattern change?

9. Chance and Surprise

One of the great challenges to both science and philosophy is to provide a rational account of the uncertainty we perceive in the events of daily life. Classical probability theory offers one such approach, but is riddled with many well-known epistemological flaws and paradoxes. The theories of *fuzzy sets, satisficing* and *possibilities* represent recent attempts to rectify some of the deficiencies in the classical methods. Each of these newer schemes has at its heart the basic inequality

$$\text{uncertainty} \neq \text{randomness},$$

expressing the fairly evident fact that the uncertainty we feel over everyday events and situations cannot usually be attributed to the influence of a random mechanism, but appears to stem from an inherent vagueness, or lack of information, either in the linguistic description or other circumstances surrounding the situations we find ourselves in.

Here we want only to indicate the manner in which q-analysis enables us to structure the notions of uncertainty, probability and surprise in a manner providing some insight into the basic difficulties involved, and to give a few suggestions as to how a theory of uncertainty and surprise might be developed.

Consider an experiment in which we toss a fair die four times in succession. Suppose our interest resides in whether or not the face "6" appears. Let's label the elementary events associated with each throw x_1, x_2, x_3 and x_4 and combine them into what is called the set of *elementary events*. Denote this set by X. To be more specific, x_i represents the event "6" occurs on toss i, $i = 1, 2, 3, 4$. Now let's introduce a new set Y consisting of *compound* events associated with the entire experiment of four tosses. In this experiment there are 16 possible outcomes ranging from no occurrences of a "6" to all tosses resulting in "6". Thus we label the elements of Y as

$$Y = \{y_0, y_1, y_2, y_3, y_4, y_{12}, y_{13}, y_{14}, y_{23}, y_{24}, y_{34}, y_{123}, y_{124}, y_{134}, y_{234}, y_{1234}\},$$

where y_0 means no "6" occurred, y_2 means that a "6" occurred only on the second toss, and so forth.

If we take X to be the vertex set for a complex and choose Y as the simplex set, it's easy to develop a relation $\lambda \subset Y \times X$ by the rule: "$(y, x) \in \lambda$ if and only if elementary event x is a component of compound event y." The incidence matrix for λ is easily computed: There is a 1 in the column labeled x_i if and only if the integer i appears as one of the subscript numbers on simplex y, e.g., $(y_{13}, x_1) \in \lambda$ but $(y_{13}, x_2) \notin \lambda$.

Calculating the structure vector Q for the complex $K_Y(X; \lambda)$, we obtain

$$Q = (\overset{3}{1} \ 1 \ 1 \ \overset{0}{1}),$$

indicating that the complex has only a single component at each q-level. In fact, it's easy to see that what we are dealing with here is the *single* simplex y_{1234} and all of its faces. This is exactly the kind of structure for which classical probability theory works well in expressing our sense of the likelihood of things. Let's examine this claim in more detail.

The complex $K_Y(X; \lambda)$ represents what the probabilist would term the *sample space* of the die-tossing experiment. But in contrast to the usual view of events as dimensionless objects, the q-analysis view distinguishes strongly between the compound 0-events (y_1, y_2, y_3 and y_4), the 1-events (y_{12}, y_{13}, etc.), on up to the single 3-event y_{1234}. *Before* the experiment is performed, our sense of the likelihood of the outcome is measured by attaching numbers (probabilities?) to each simplex. *After* the experiment is complete, these numbers have rearranged themselves throughout the complex, all simplices now having value 0 except for that single simplex that corresponds to the actual outcome. By convention, we set that value equal to 1.

So we see that carrying out the experiment corresponds to *traffic* on the complex. But traffic of this sort can move freely about from one simplex to another only if the complex is sufficiently richly connected at all dimensional levels to support such a free flow of dimensionally significant numbers. Basically, this means we must have a structure vector each of whose components is 1—just as above. The case when $K_Y(X; \lambda)$ consists of a single simplex and all of its faces is the simplest example of when this type of situation will occur.

In connection with the die-tossing experiment, the probabilist would attach the following *a priori* estimates to the elements of Y:

$$\mathcal{E}(y_0) = \frac{625}{1296}, \qquad \mathcal{E}(\text{0-simplices}) = \frac{500}{1296}, \qquad \mathcal{E}(\text{1-simplices}) = \frac{150}{1296},$$

$$\mathcal{E}(2 - \text{simplices}) = \frac{20}{1296}, \qquad \mathcal{E}(y_{1234}) = \frac{1}{1296}.$$

These numbers express his sense of likelihood of events, and are formed by weighting the possible outcomes using the binomial distribution, making use of its associated assumptions of independent trials and fixed probabilities for the elementary events x_i. After the experiment is over, these numbers have redistributed themselves so as to coalesce on that one simplex corresponding to the actual outcome. But this rearrangement is possible only if the numbers associated with p-events can freely move about and "reaffiliate" themselves with events at all levels of connectivity p. This can happen only in structures having a single component at each connectivity level p.

The main point about connectivity is that if the numbers assigned to the events in Y are to represent our sense of the likelihood of the outcome of the experiment, then they must do so both before **and** after the experiment. But this requires the kind of free flow of traffic discussed above, a flow that is possible only if the complex representing the events is fully connected at all levels. Thus, we conclude that classical probability theory will, in general, reflect our sense of the likelihood of events only for those structures possessing a single component at each dimensional level.

We note, in passing, that the redistribution of the numbers over the complex when we actually perform the experiment is the discrete analogue of the "collapse" of the Schrödinger wave function in quantum mechanics. In most versions of quantum theory, an object's attributes (position, momentum, spin, etc.) exist as *potentials,* with the various possibilities weighted according to a probability distribution specified by the wave function. Following the actual measurement, a definite value is obtained for any given attribute, the wave function "collapsing" to the single value actually observed. All other possibilities then have probability zero.

Consequently, our complex of events, together with the associated likelihood numbers, is a discrete analogue of the quantum-mechanical wave function with the additional feature that the possible events (simplices) have a dimensional character that must be respected when considering the redistribution of likelihoods following the experiment (observation). The quantum-mechanical implications of these dimensional factors have not been investigated as yet, classical quantum theory having confined itself to the same case as classical probability theory, viz., complexes with $Q = (1 \; 1 \cdots 1)$.

One of the principal uses of probability theory is to provide a numerical measure of our sense of how "surprising" the occurrence of a particular event would be. By the foregoing arguments, it's clear that the concept of surprise is intimately tied up with the connective structure linking events in the space of possible outcomes. In particular, to develop a decent theory of surprises we need a measure of the "reachability" of a q-event σ_q from another base event σ_p^* (the Now!) in the complex. In what follows, we shall adopt the usual convention that an event (a simplex) is denoted by σ with its subscript indicating the dimension of the event represented by σ.

On intuitive grounds, the "surprise value" surp σ_q of the event σ_q should be a number that:

1. Reflects the level of connectivity between σ_q and σ_p^*. In particular, if there is no chain of q-connection between the two, then surp $(\sigma_q) = 0$. That is, we cannot be *surprised* if the event σ_q cannot be experienced from σ_p^* if there is no appropriate dimensional path to move from the base event σ_p^* to σ_q.

2. Is greater if there are a large number of disjoint p-chains from σ_p^* to σ_q, since it is "more surprising" if a large number of p-chains go between the two events than if there are only a small number of such paths.

3. Is smaller if dim $\sigma_p^* = p$ is large, since it is "less surprising" that q-chains exist from σ_p^* to σ_q if σ_p^* has more q-faces.

If we define

$$n_q(\sigma_q, \sigma_p^*) = \text{the number of disjoint } q\text{-chains linking } \sigma_p^* \text{ to } \sigma_q,$$

then a measure of surprise of the event σ_q relative to σ_p^* is given by

$$\text{surp } (\sigma_q \mod \sigma_p^*) = \frac{n_q(\sigma_q, \sigma_p^*)}{p+1},$$

with the conditions

$$\text{surp } (\sigma_q \mod \sigma_p^*) = \begin{cases} 0, & \text{if } \sigma_q \text{ and } \sigma_p^* \text{ are in different} \\ & \text{components of the complex,} \\ 0, & \text{if } p < q, \\ 1/(p+1), & \text{if there are no loops.} \end{cases}$$

The last condition means that we do not distinguish loops beginning and ending at σ_p^* as being different p-chains.

Example: Technological Disasters

An interesting and timely example of the use of surprise theory is given by Ron Atkin, who considers the surprise value of a technological disaster like Three-Mile Island or Chernobyl. Let the vertices X of the complex $K_Y(X; \lambda)$ represent various technological features of the system under study. For a nuclear power plant, the elements of X might be things like the position of control rods, the level of coolants and the pressure in regulators. Let the simplex set Y consist of combinations of such features that we term a "property" or "behavior."

If all vertices are initially in the state OK, then we say that all is well. Assume that during the course of operation of the plant some vertices turn into "anti-vertices"—i.e., their OK activity turns into "not-OK"—and the complex $K_Y(X; \lambda)$ turns into a new complex K^1. As the process of vertices shifting to/from OK$\leftrightarrow$ not-OK unfolds, we have a progression

$$K \longrightarrow K^1 \longrightarrow K^2 \longrightarrow \cdots \longrightarrow K^D,$$

where the event $\sigma = $ DISASTER belongs to K^D. We can now ask the question: Given the event (state) $\sigma_p^* \in K$, what is the value $\text{surp}(\sigma \mod \sigma_p^*)$ for $\sigma \in K^D$? Clearly, we would like to adjust our technology $K \cap_i K^i$ to make this number large.

Exercises

1. The tossing of dice is just one of the standard types of gambling situations from which classical probability theory blossomed forth into a mathematical discipline. Coin-tossing, card-playing and the drawing of lottery numbers are others. (a) Structure these other kinds of "random" events as simplicial complexes. (b) Are the obstruction vectors zero for these complexes? (c) Can you think of ways to alter the experimental situation so as to make the obstruction vectors nonzero for these simple kinds of games?

2. Technological disasters come in many forms, of which nuclear power plant malfunctions are only among the more visible. Think of other kinds of disasters that could result in unpleasant surprises, things like the recent breakdown of the telephone system in New York, the 1977 East Coast power blackout or the failure of a silicon chip in a NORAD computer. Describe the surprise in these situations as a progression of simplicial complexes as in the text. In this same connection, consider also happy surprises like winning a lottery.

10. Simplicial Complexes and Dynamical Systems

With the exception of this chapter, the dominant mathematical paradigm throughout this book has been Newton's legacy: ordinary differential equations. It's appropriate then to conclude our discussion of connectivity with a brief indication of some linkages between the linear dynamical systems considered in Chapter 6 and certain simplicial complexes of the type we have been examining in this chapter.

Consider the single-input linear system

$$x_{t+1} = Fx_t + gu_t, \qquad x_0 = 0, \qquad x_t \in R^n,$$

and its associated reachability matrix

$$\mathcal{C} = \left[g \,|\, Fg \,|\, F^2g \,|\, \cdots \,|\, F^{n-1}g \right].$$

Let us denote the elements of C as $C = \{c_1, c_2, \ldots, c_n\}$. Now we form a new set Y from C by taking as elements all those combinations of subsets of elements from C that are linearly independent. So, for example, if c_1 and c_2 are linearly independent, we write this subset as $c_1 \wedge c_2$ and place this element in the set Y. In general, Y can contain at most $2^n - 1$ elements. To define a relation λ on the product $Y \times C$, we employ the rule that the pair $(y_i, c_j) \in \lambda$ if and only if the vector c_j belongs to the element y_i.

As an example of this set-up, let $\Sigma = (F, g, -)$ be the canonical reachable system with

$$F = \begin{pmatrix} 0 & 1 & 0 \\ 0 & 0 & 1 \\ -\alpha_2 & -\alpha_1 & -\alpha_0 \end{pmatrix}, \qquad g = \begin{pmatrix} 0 \\ 0 \\ 1 \end{pmatrix}.$$

Then

$$C = \begin{pmatrix} 0 & 0 & 1 \\ 0 & 1 & -\alpha_0 \\ 1 & -\alpha_0 & -\alpha_1 + \alpha_0^2 \end{pmatrix} = [c_1 \mid c_2 \mid c_3].$$

The set Y is given by

$$Y = \{c_1, c_2, c_3, c_1 \wedge c_2, c_1 \wedge c_3, c_2 \wedge c_3, c_1 \wedge c_2 \wedge c_3\}.$$

The incidence matrix for the complex $K_Y(X; \lambda)$ is easily seen to be

λ	c_1	c_2	c_3
y_1	1	0	0
y_2	0	1	0
y_3	0	0	1
y_4	1	1	0
y_5	1	0	1
y_6	0	1	1
y_7	1	1	1

The structure vector Q for the complex is

$$Q = (\overset{2}{1}\ 1\ \overset{0}{1}),$$

while for the conjugate complex $K_C(Y; \lambda^*)$ we have the structure vector

$$Q^* = (\overset{3}{3}\ 3\ 1\ \overset{0}{1}).$$

The above results concerning the structure vectors Q and Q^* are generalizable for *all* completely reachable single-input systems as the following result shows.

LINEAR STRUCTURE THEOREM. *The system* $\Sigma = (F, g, -)$ *is completely reachable if and only if the structure vectors for* $K_Y(\mathcal{C}; \lambda)$ *and* $K_{\mathcal{C}}(Y : \lambda^*)$ *have the forms*

$$Q = (\overset{n-1}{1}\ 1\ 1 \cdots \overset{0}{1}),$$

$$Q^* = (2^{n-1} - 1\ \overset{2^{n-1}-1}{\overbrace{\ }}\ 2^{n-1} - 1\ 2^{n-1} - 1 \cdots 2^{n-1} - 1\ \overset{\beta}{1}\ 1\ 1 \cdots \overset{0}{1}),$$

where

$$\beta = \sum_{k=0}^{n-1} \binom{n-2}{k-1}, \qquad \binom{x}{-1} \doteq -1.$$

Example: An Unreachable System

Consider the system Σ defined by

$$F = \text{diag}\,(\lambda_1, \lambda_2, \lambda_3), \qquad g = \begin{pmatrix} 0 \\ 1 \\ 1 \end{pmatrix},$$

which is not completely reachable. Computation of the incidence matrix shows that the element $c_1 \wedge c_2 \wedge c_3$ is not contained in Y, since the three vectors

$$c_1 = \begin{pmatrix} 0 \\ 1 \\ 1 \end{pmatrix}, \qquad c_2 = \begin{pmatrix} 0 \\ \lambda_2 \\ \lambda_3 \end{pmatrix}, \qquad c_3 = \begin{pmatrix} 0 \\ \lambda_2^2 \\ \lambda_3^2 \end{pmatrix},$$

are not linearly independent.

The structure vector for $K_Y(\mathcal{C}; \lambda)$ turns out to be

$$Q = (\overset{1}{3}\ \overset{0}{1}),$$

which is not of the form required for complete reachability by the Linear Structure Theorem. Thus Σ is not reachable.

These results give only the briefest indication of how ideas from algebraic topology can be used to capture the geometric structure of a dynamical system. For extensions to multi-input systems, as well as to nonlinear dynamics, the papers cited in the Notes and References should be consulted (see also Problems 6 and 9).

Exercises

1. Prove the Linear Structure Theorem.

2. Use the Linear Structure Theorem to test the following systems for complete reachability:

$$F = \begin{pmatrix} 1 & 2 & 3 \\ 0 & 1 & 5 \\ 0 & 2 & 2 \end{pmatrix}, \qquad g = \begin{pmatrix} 1 \\ 1 \\ 1 \end{pmatrix}, \tag{a}$$

$$F = \begin{pmatrix} 2 & 3 & -1 & 0 \\ 0 & 0 & -4 & 2 \\ 2 & 3 & -5 & 2 \\ 1 & 1 & 1 & 1 \end{pmatrix}, \qquad g = \begin{pmatrix} 4 \\ 2 \\ 0 \end{pmatrix}. \tag{b}$$

Discussion Questions

1. Suppose you wanted to characterize the difference between the spectrum of goods and services available at a major shopping center and those offered at a small country store. How could you capture this difference using sets and relations?

2. Discuss how you might employ hierarchies of sets and relations to structure a play like *Hamlet* or *Death of a Salesman* as a relationship between the characters and the scenes.

3. We have seen that paintings can be given a simplicial structure that captures some of their abstract geometrical style. Do you think the same concepts can be employed to find a geometrical structure in musical compositions? How would you arrange the sets and relations in this case?

4. The measure of eccentricity discussed in the text suffers from the defect that a simplex will have low eccentricity if there is just one other simplex with which it shares many vertices. This means that we could have two simplices with many vertices in common but sharing very few of their vertices with the remaining simplices in the complex. Nevertheless, our measure of eccentricity would see these two simplices as being "well-integrated" into the complex. Develop and discuss alternative measures of eccentricity that would remove this deficiency.

5. The game of chess can be considered as a relation between the squares of the board and the Black and White pieces. Discuss various binary relations that might be used to define the connective structure of the game. Consider how the resulting complexes and structure vectors might be used in developing computer programs for playing chess. (Hint: There are at least two important relations here, the relation between the Black pieces and the

squares and the relation between the White pieces and the squares. So consider these relations, together with the role of set covers and hierarchies, in your analysis.)

6. Consider the use of q-analysis as a pattern recognition algorithm in the sense of template matching. In other words, suppose that a pattern is displayed on a rectangular grid. Develop a binary relation between the X and Y coordinates of the cells that constitute the displayed pattern. For example, the template for the square below

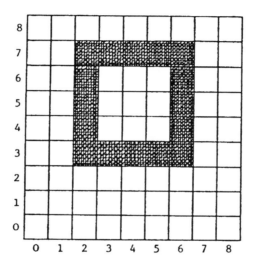

might be represented by the incidence matrix

λ	0	1	2	3	4	5	6	7	8	9
0	0	0	0	0	0	0	0	0	0	0
1	0	0	0	0	0	0	0	0	0	0
2	0	0	0	0	0	0	0	0	0	0
3	0	0	1	1	1	1	1	0	0	0
4	0	0	1	0	0	0	1	0	0	0
5	0	0	1	0	0	0	1	0	0	0
6	0	0	1	0	0	0	1	0	0	0
7	0	0	1	1	1	1	1	0	0	0
8	0	0	0	0	0	0	0	0	0	0
9	0	0	0	0	0	0	0	0	0	0

a) Examine the possibility of distinguishing between a square, a triangle and a circle by means of the structure vectors Q and Q^* associated with their respective simplicial complexes.

b) The geometric nature of a pattern must be invariant under translations, rotations and uniform contractions/dilations. For the foregoing types of geometrical patterns, discuss these conditions in terms of properties of the incidence matrix Λ.

c) Discuss how the above procedure might be developed into a template-matching program.

7. In our consideration of binary relations $\lambda \subset Y \times X$, we have adopted an "all-or-none" position on whether or not an element (y_i, x_j) in contained in the relation λ. Suppose now that we assign a number Λ_{ij} between 0 and 1 as a measure of the degree to which we *believe* that (y_i, x_j) belongs to the relation λ. Consider how to modify the idea of a chain of q-connection in this "fuzzy sets" situation and discuss what it could mean for two simplices to be "fuzzy q-connected."

8. The ecologist C. S. Holling has defined the property of "resilience" of an ecosystem as being the system's "ability to absorb *and benefit* from change and the unexpected." Consider how you might construct a way of measuring the resilience of a simplicial complex. Do you think this measure can be meaningfully defined in a context-free manner?

9. One of the most common methods of solving problems in engineering and physics involving partial differential equations is the so-called *finite-element method.* For example, if we are interested in the temperature distribution of an irregularly-shaped object, the finite-element approach first divides the object into small regular parts like triangles and rectangles (or tetrahedrons and cuboids in higher dimensions). The temperature across an element is then approximated by interpolating the temperatures between values at the vertices. Thus, the problem becomes to find the temperatures at the vertices of the geometric mesh. So it could be said that the finite-element method extends the limits of classical, idealized physics to a physics that can be used to describe and predict the behavior of systems having great geometric complexity.

Consider the finite-element method as a special case of the more abstract geometrical ideas we have discussed in this chapter. In particular, how do you think the transmission of heat throughout the grid is constrained by the geometry of the backcloth and the connective structure of the triangular or tetrahedral elements?

10. Road traffic systems have been one of the most well-developed areas for the application of q-analysis. Consider how you might structure

a complex traffic system in which the road network serves as the backcloth supporting the dynamics of vehicle traffic. (Hint: Think of the roads in the network as being made up of two-dimensional segments having places at which vehicles can enter and leave, called entrance and exit gates. Then consider a link across a segment as being a possible vehicle path from an entrance gate to an exit gate. Represent every such segment by a polyhedron whose vertices are its entrance and exit gates. For more information about this line of attack, see the papers cited in the Notes and References for this Discussion Question.)

11. In geometry we think of Euclidean 3-space as consisting of R^3, together with the standard euclidean metric

$$d(x,y) = \sqrt{(x_1 - y_1)^2 + (x_2 - y_2)^2 + \cdots + (x_n - y_n)^2}.$$

On the other hand, there is *actual space* A^3, which is what we can perceive with our senses. Descartes made the astonishing claim that $E^3 = A^3$, a claim that has dominated physical theories to the present day. Leibniz, on the other hand, took the view that A^3 is the set of relationships between objects, and that we must reject the Newtonian-Galilean claim that A^3 is absolute and that real objects are merely observed in it.

a) Discuss these competing claims within the contexts of both classical physics and its quantum mechanical and relativistic extensions.

b) When we actually make observations, we do so by using some sort of signal. For us humans, this is usually a light signal. Thus, an observer who sets out to identify things s/he wishes to call geometrical points (or points of A^3) does it by relating one such point to another. Moreover, each point is a "place" *seen* with the aid of a signal. So when our observer gathers up his or her measuring sticks and light signals and goes out to observe the geometry of space, he doesn't observe E^3. Rather, s/he obtains a discrete mesh of points whose density is determined by the set of measuring sticks being used, together with the particular pair of spectacles that are worn. Suppose such an observer went out into the desert and started laying out such a mesh. Suddenly, s/he encounters a large boulder. Being a scientist, the observer calls it an object, not a point. Now this object can only be mapped by identifying points on its surface. The light signal enables the observer to notice the coincidence of points at the ends of the measuring rods and points on the boulder's surface. This means that the boulder has the effect of creating a large "hole" in the mesh of points that would have been constructed if the object had not been there. Discuss the claim that *every* such object that we observe in A^3 can be regarded as a hole in some structure that our signal enables us to identify. Think about how you might tie this idea together with the notion that the boulder is a non-bounding cycle in a simplicial complex whose vertices are the points of our mesh (see Problems 2 and 3 for a discussion of chains and cycles in a complex).

12. Consider how you might organize a sets/relations framework to measure the relative strengths of competing teams in sports such as baseball, football or basketball.

13. The classical Newtonian view of time can be represented by the diagram

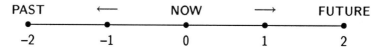

where the numbered vertices represent specific measured moments of time, and the lines correspond to time intervals between the moments. This is a simplicial complex having an infinite set of vertices, with the number attributed to each vertex forming part of a pattern τ_0 on the vertices, i.e., the 0-simplices. The numbers assigned to the edges joining the vertices form another pattern τ_1, referred to as the time interval between successive measured moments. Thus, in Newton's world the time pattern is the *graded* pattern $\tau = \tau_0 \oplus \tau_1$.

The above representation makes it clear why we refer to Newtonian time as a *linear* concept associated with a complex K, which consists of a set of 1-simplices that are 0-connected. When the Newtonian time-axis is used to represent a set of observed real-world events, the idea is to somehow produce a "clock" whose time moments (the vertices) can be put into one-to-one correspondence with the set of events. The pattern τ_0 describes the "NOW" events, while τ_1 describes the interval pattern.

a) Discuss why the pattern of the relativistic time of Einstein has the structure $\tau = \tau_3 \oplus \tau_4$, in which so that time moments are represented by 3-simplices with the time intervals corresponding to 4-simplices. What possible interpretation can you attach to the fact that such relativistic time has also 1- and 2-simplex intervals between appropriate "NOW" moments?

b) Consider the following view of "experienced" time. Let K be a simplicial complex of events. A pattern τ_p represents the "NOW" moments of p-events, and can be thought of as a simple 0/1 function on the p-simplices of K. For this function, the single nonzero value is attached to the particular p-event that is the "NOW" moment. Then the "next" p-event in K is experienced as a change in this pattern, so that the "1" moves to a different p-simplex. This movement involves the connectivity of K, since one p-event cannot follow another unless there exists a $(p + 1)$-interval connecting the two. Thus, the change $\delta\tau_p$ is a *p-force* in the structure of events representing our sense of moving time.

Normally when we speak of time, we refer to the Newtonian pattern $\tau = \tau_0 \oplus \tau_1$. Thus, if an individual experiences time traffic of dimension p, s/he finds it culturally necessary to replace his pattern by the Newtonian τ;

$$\tau_p \oplus \tau_{p+1} \qquad \longrightarrow \qquad \tau_0 \oplus \tau_1$$
$$\text{(experienced time)} \qquad\qquad \text{(Newtonian time)}$$

This picture means that the individual has to force his or her $(p + 1)$-perception of the time interval down into the one-dimensional interval of the Newtonian pattern. The experience of this force is usually expressed by phrases such as "time flies when you're having fun," or "time is hanging heavy on my hands nowadays." These expressions indicate that τ_{p+1} and τ_1 are out of step. Discuss the relevance of this view of multidimensional time in the context of your own experiences.

c) Assume that the experience of a p-event corresponds to the recognition of a p-simplex in K and that there is a 1–1 correspondence between the 1-simplices of K and the Newtonian reference frame. Then the time interval for the gap between one p-event and another will be proportional to the quantity $(p + 1)(p + 2)/2$, the number of edges in the least connection between two p-simplices. For example, if the Newtonian unit interval τ_1 is 1 day, then a 6-event would require $(6 + 1)(6 + 2)/2 = 28$ days to occur, i.e., to "arrive."

Discuss the appropriateness of assuming that a p-event occurs by way of the edges that make up the $(p + 1)$-interval bridging the gap between the current and previous p-events. Consider other possibilities and their implications for the interval between p-events.

14. The North American power-generating network consists of around 6,000 individual generating stations. If the set of generating stations is considered as the N-level in a hierarchy of sets describing the interconnections in the network, try to develop a range of sets and connections from levels N-4 to $N + 4$ describing this system. How many elements do you think there are in the sets at each level?

Suppose there is a major power failure and that it's necessary to restore service immediately. Interpret what it means for a failure to occur in terms of changes in the geometry of the backcloth associated with the complexes at each level of the network.

Discuss how the foregoing considerations relate to the theory of surprises as developed in the text.

15. Jeffrey Johnson has developed an intriguing theory of computer vision based on the principles of q-analysis. The key idea is to define an *n-ary relation* on sets of pixels on a computer screen in the following way. Let $g(p_{ij})$ be the gray-scale level of the pixel located in the ith row and the

jth column of a digital image. Johnson says that pixel p_{ij} is R_0-related to the pixel above it, pixel $p_{i-1,j}$, if $g(p_{ij}) \geq g(p_{i-1,j})$. Similarly, there is a relation R_1 between the gray levels of pixel p_{ij} and the one below it, $p_{i+1,j}$, a relation R_2 for the pixel to the right and a relation R_3 for the pixel to the left. He also defines relations R_4-R_7 for the case when the inequalities go in the other direction. We can then say that p_{ij} and its four neighbors are R_{abcd}-related if and only if $p_{ij} R_a p_{i+1,j}, p_{ij} R_b p_{i,j+1}, p_{ij} R_c p_{i-1,j}, p_{ij} R_d p_{i+1,j}$. So, for instance, if p_{ij} is R_{0123}-related to its four neighbors, this means that p_{ij} is a "locally bright" pixel. On the other hand, if p_{ij} is R_{4127}-related to its neighbors, then the image is getting darker in a northeast direction at pixel p_{ij}.

Using the above scheme for individual pixels at level N, Johnson has shown how to combine them at level $N + 1$ into what he calls *gradient polygons,* which can then be further aggregated to do things like recognize, say, an eye in someone's face. The overall scheme is shown in the diagram below.

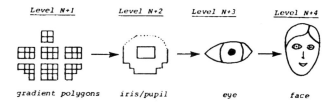

Pixels to Objects via q-Analysis

Consider how you would use this scheme to do things like verifying a signature on a check or identifying a hidden missile silo from a satellite photo. How do you think traffic might be introduced into such a complex in order to attach some sort of dynamics to the patterns?

16. Consider the primary color triangle shown on the next page, whose vertices are Red, Blue and Green (the 0-simplices). The 1-simplices (Violet, Turquoise and Yellow), as well as the 2-simplex (White), are formed by combining the primary colors. So each simplex in this complex is a face of the single simplex $W = \langle R, B, G \rangle$. A person who has normal color vision is capable of seeing three primary colors at once (and their combinations), so will be able to see the entire spectrum of visible colors.

Now consider an experiment with a subject whose color vision is only one-dimensional; i.e., he can see only two colors at a time (or any combination of these two primary colors). For the sake of definiteness, assume the subject cannot distinguish the color red. Let the experiment consist of showing the subject flashcards composed of the seven colors indicated, each

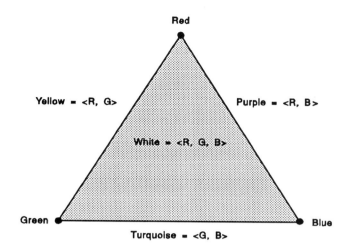

The Color Triangle

appearing randomly with equal likelihood. When a card is displayed, the
subject tells us what color he sees, and we pass on to the next card. With
a subject possessing normal two-dimensional color vision, we would expect
him to identify each color correctly, and our records would show each color
as being named one-seventh of the time. What result would you expect
from the subject who cannot see red? Explain your answer in terms of
connectivity patterns in the color simplex.

Problems

1. Consider the following conjecture: Up to a relabeling of the rows
and columns of the incidence matrix Λ, the associated binary relation λ is
uniquely determined by the structure vectors Q and Q^* obtained from Λ.

a) Prove that this conjecture is false by constructing a counterexample.
That is, display two incidence matrices Λ and $\widehat{\Lambda}$ that differ by more than a
permutation of their rows and columns, but nevertheless generate the same
structure vectors Q and Q^*.

b) The preceding result shows that the elements of Q and Q^* do not
form a complete, independent set of invariants for incidence matrices Λ under
the group of permutations. Consider other invariants that can be adjoined
to Q and Q^* to constitute a complete set.

2. The set of all p-simplices in K form a *chain* C_p under formal addition
and multiplication by scalars from an abelian group J. A typical chain in

C_p can be expressed as the formal sum

$$c_p = m_1\sigma_p^1 + m_2\sigma_p^2 + \cdots + m_r\sigma_p^r, \qquad m_i \in J.$$

We can make C_p into an abelian group by defining $c_p + c_p'$ and αc_p, $\alpha \in J$ in the obvious way. Combining every such group C_p for $0 \leq p \leq n$, we obtain the *chain group*

$$C_\bullet = C_0 \oplus C_1 \oplus \cdots \oplus C_n.$$

With every p-chain c_p we can associate a $(p-1)$-chain ∂c_p, the *boundary* of c_p. The boundary operator ∂ is defined on p-simplices by

$$\partial \sigma_p = \partial \langle x_1 x_2 \cdots x_{p+1} \rangle = \sum_i (-1)^{i+1} \langle x_1 x_2 \cdots \hat{x}_i x_{i+1} \cdots x_n \rangle,$$

where "$\hat{x}_i$" means that the vertex x_i is to be omitted from the term.

   a) Show that $\partial \colon C_p \to C_{p-1}$, $\qquad p = 1, 2, \ldots, n$.

   b) Prove that ∂ is nilpotent; i.e., $\partial^2 = 0$.

   c) Show that ∂ is a homomorphism, i.e.,

$$\partial(c_p + c_p') = \partial(c_p) + \partial(c_p'),$$

$$\partial(\alpha c_p) = \alpha \partial(c_p), \qquad \alpha \in J.$$

   d) Prove that $\partial(C_p) = B_{p-1}$ is a subgroup of C_{p-1}.

   e) Show that $\partial(B_{p-1}) = 0$ in C_{p-2}.

   3. Those p-chains c_p whose boundaries vanish ($\partial c_p = 0$) are called *p-cycles*, denoted Z_p.

   a) Show that the p-cycles form a subgroup of C_p.

   b) Show that B_p, the *bounding cycles*, form a subgroup of Z_p.

   c) Define the *factor group* $Z_p/B_p = H_p$ using the following equivalence relation in Z_p: $z_p \sim z_p'$ if and only if $z_p - z_p' \in B_p$. Show that $H_p = \ker \partial / \mathrm{im}\, \partial$. The groups H_p are called the *Betti* or the *homology groups* of the complex K.

   d) If $H_p \neq 0$, i.e., there are cycles z_p that are not bounding cycles, then H_p as an additive group is isomorphic to a certain number of copies of J, the number being equal to the number of generators (cycles z_p that are linearly independent over J). These numbers, denoted β_p, are called the *Betti numbers* of the complex K. Show that if the complex is arcwise connected, then $\beta_0 = 1$. Moreover, if K possesses k connected components, then $\beta_0 = k$.

   e) Show that the first component of the structure vector Q is such that $Q_0 = \beta_0$, although in general $Q_i \neq \beta_i$, $i > 0$.

4. Consider the Middle East example of the text. Compute the homology groups and the Betti numbers for this complex. What interpretation can you give to the results of your computation?

5.   Consider the elementary complex K, consisting of the three 1-simplices σ_1^1, σ_1^2, and σ_1^3, with vertex set $\{x_1, x_2, x_3\}$.

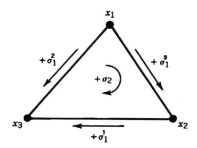

A Simplicial Complex K

a) Show that K has a nontrivial 1-cycle.

b) Show that $H_0 \cong 1$, $H_1 \cong J$ and, hence, that $\beta_0 = 1$, $\beta_1 = 1$.

c) Assume that we modify K to include the 2-simplex $\sigma_2 = \langle x_1 x_2 x_3 \rangle$, i.e., we "fill in" the triangle. Show that for this new complex, $H_1 \cong 0$.

d) Interpret the above result in terms of the number and types of "holes" present in the complex. Show that the Betti number β_i measures the number of $(i+1)$-dimensional holes in K.

6. Consider the single-input linear system

$$x_{t+1} = F x_t + g u_t.$$

a) Show that complete reachability is equivalent to the associated complex having trivial homology, i.e.,

$$H_0 \cong 1, \qquad H_i \cong 0, \qquad i > 0.$$

Thus the system is reachable if and only if it has no "holes."

b) The *Euler characteristic* of a complex is defined in terms of the Betti numbers as

$$\chi(K) = \sum_{i=0}^{n} (-1)^i \beta_i, \qquad n = \dim K.$$

Show that the single-input system is completely reachable if and only if $\chi(K) = 1$.

7. A model predator–prey ecosystem is displayed in the directed graph shown in the diagram below. Here there is an arc from species i to species j if i feeds upon j. Let the set of 15 species be termed $X = \{x_1, x_2, \ldots, x_{15}\}$. Define a predator relation λ_{PRD} by the rule $(x_i, x_j) \in \lambda_{PRD}$ if and only if x_i is a predator of x_j.

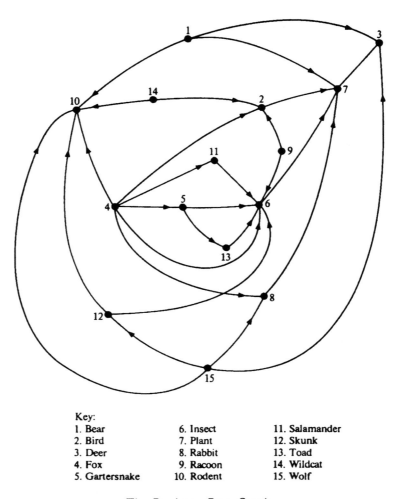

Key:
1. Bear	6. Insect	11. Salamander
2. Bird	7. Plant	12. Skunk
3. Deer	8. Rabbit	13. Toad
4. Fox	9. Racoon	14. Wildcat
5. Gartersnake	10. Rodent	15. Wolf

The Predator–Prey Graph

a) Determine a plausible incidence matrix Λ_{PRD} for this ecosystem.

b) Determine the structure vector Q and the eccentricities for the complex $K_X(X; \lambda_{PRD})$.

c) Define an analogous relation λ_{PRY} for the prey and calculate the same quantities.

d) Compute the complexities of the pair of complexes $K_X(X; \lambda_{PRD})$, $K_X(X; \lambda_{PRY})$ and their conjugates. Interpret the numbers obtained in terms of the food web structure.

e) Show that λ_{PRD} has the nontrivial homology $H_0 \cong 1$, $H_1 \cong J$, while λ_{PRY} has trivial homology. How would you interpret this result?

8. The figure below shows the food web of the insects in the pitcher plant *Nepenthes albomarginata* in West Malaysia. Each line represents a trophic linkage, with the predators being higher in the figure than their prey.

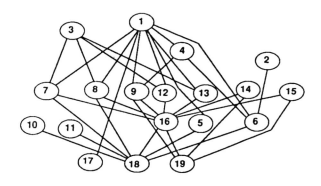

Food Web of *Nepenthes albomarginata*

a) Top predators are defined to be species on which nothing else in the web feeds, while basal species feed on nothing within the web. Identify the top predators and basal species in this web.

b) In such a web, a cycle is where species i feeds on j which eats k which in turn feeds on i. Cannibalism is a cycle in which one species feeds on itself. Are there any cycles in this web? Is there a cannibalistic cycle?

c) Connectance is defined to be the number of realized trophic interactions divided by the number of possible interactions. Usually, it's calculated as twice the number of linkages divided by $S(S-1)$, where S is the number of species. The linkage density is given by the number of links per species. Calculate the connectance and linkage density for this web. Assuming the linkage density remains constant, how does the connectance behave as a function of S?

d) Reformulate this food web in terms of the kind of simplicial complex structure used in Problem 7. Compute the structure vectors and eccentricities and try to relate them to the more conventional measures of connectance and linkage density.

9. Consider the *multi-input* linear system

$$x_{t+1} = Fx_t + Gu_t, \qquad F \in R^{n \times n}, \qquad G \in R^{n \times m}, \qquad m > 1.$$

a) Let the matrices F and G be given by

$$F = \begin{pmatrix} 2 & 4 & 1 & 1 \\ 0 & -1 & 0 & 1 \\ 0 & 0 & -3 & -2 \\ 0 & 0 & 0 & 1 \end{pmatrix}, \qquad G = \begin{pmatrix} 0 & 1 \\ 1 & 1 \\ 0 & 0 \\ 0 & 0 \end{pmatrix}.$$

Show that the associated complex has homology groups

$$H_0 \cong J, \qquad H_1 \cong \underbrace{J \oplus J \oplus \cdots \oplus J}_{21 \text{ copies}}.$$

Using the results of Chapter 6, show that this system is *not* completely reachable.

b) For the reachable system

$$F = \begin{pmatrix} 3 & 0 & 1 \\ 2 & 1 & 0 \\ 1 & 1 & 1 \end{pmatrix}, \qquad G = \begin{pmatrix} 1 & 0 \\ 0 & 1 \\ 0 & 0 \end{pmatrix},$$

show that the associated complex has homology

$$H_0 \cong J, \qquad H_1 \cong 0, \qquad H_2 \cong \underbrace{J \oplus J \oplus \cdots \oplus J}_{10 \text{ copies}}.$$

c) Conclude from the preceding computations that the relationship between trivial homology and reachability for multi-input systems is different than for single-input processes. Can you determine a general result for calculating the homology groups H_i for multi-input systems? (Note: This part of the Problem is quite difficult, requiring more advanced techniques from algebraic topology, e.g., exact sequences and the Mayer-Vietoris sequence.)

10. The relation of q-nearness is defined by the intersection of two polyhedra. But the concept can be extended to the intersections of *many* polyhedra. The set of polyhedra having vertices in common form what we call a *star*, with their common vertices constituting the *hub* of the star. Such a structure can sometimes be used as a pattern-recognition device, as the following problem illustrates.

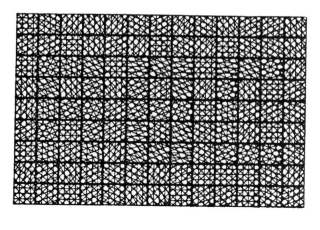

$$\alpha \quad \beta \quad \gamma \quad \delta \quad \epsilon \quad \zeta \quad \eta \quad \theta$$

Squares and Crosshatch Patterns

The figure above shows 117 squares, each related to four of the eight crosshatch line types α-θ indicated.

a) By inspecting the squares, it will be seen that each of their polyhedra has either the face $\langle x, y, z \rangle$ or the face $\langle a, b, c \rangle$—but not both—where a, b, c, x, y, z are some combination of the patterns α-θ. Determine these two disjoint face types.

b) The preceding fact means that the squares, viewed as polyhedra, define two distinct stars. What are they?

c) Display the pattern recognized by the star structure of the squares.

11. Consider a complex $K_Y(X; \lambda)$ and its conjugate. Show that the homological structures of these two complexes are isomorphic, i.e., $H_i \cong H_i^*$, $i = 0, 1, \ldots$.

12. Define the *complementary complex* K^C to K by the following rule: Extend the vertex set to include all the vertices $\bar{x}_i$ that are the *negations* of those x_i in K. Thus, the incidence matrix $\bar{\Lambda}$ of K^C is obtained from Λ by changing the "1's" to "0's," and vice-versa. Show that K and K^C have exactly the same connective structure; i.e., the structure vectors are such that $Q = Q^C$, $Q^* = Q^{*C}$.

13. Let σ_q^1 and σ_q^2 be disjoint events and define

$$\text{surp} \, (\sigma_q^1 \cup \sigma_q^2 \mod L) = \text{surp} \, (\sigma_q^1 \mod L) + \text{surp} \, (\sigma_q^2 \mod L),$$

where L is some base subcomplex of K.

a) Prove that if $K = \bigcup_i \sigma_0^i$ is a space of 0-events only and E a subset of K, then

$$\text{surp } (E \text{ mod } K) = \sum_{\sigma_0 \in E} \text{surp } (\sigma_0^i \text{ mod } K)$$

$$= \sum_{\sigma_0 \in E} \text{surp } (\sigma_0^i \text{ mod } \sigma_0^i)$$

$$= \text{card } E,$$

so that surp $(K \text{ mod } K) = \text{card } K$. Thus, in the space of 0-events

$$\text{Pr } (E \mid K) = \frac{\text{surp } (E \text{ mod } K)}{\text{surp } (K \text{ mod } K)} \ .$$

In other words, in this very special setting we recover the standard "relative frequency" rule from elementary probability theory by using our definition of surprise.

b) How would you use surprise theory to recapture the usual notion of "independent" events? That is, if H and K are disjoint (independent) complexes with A and B being subcomplexes of H and K, respectively, with $\sigma_q^i \in H$, $\sigma_q^j \in K$, how would you *define* the quantity surp (σ_q^i, σ_q^j) mod $(A \times B)$ in order to recapture the usual rule $P(A \cap B) = P(A)P(B)$ used to express the independence of the 0-events A and B in elementary probability theory?

Notes and References

§1–3. The idea of representing binary relations between abstract sets as simplicial complexes, and then using the abstract geometry of the complex to study interesting relationships in the arts and humanities, is due primarily to Ron Atkin. For an introduction to his ideas, including many fascinating worked-out examples, see

Atkin, R. H., *Multidimensional Man*, Penguin, London, 1981.

A recent survey of the state of play in the q-analysis game is found in

Johnson, J., "The Mathematics of Complex Systems," in *The Mathematical Revolution Inspired by Computing*, J. Johnson and M. Loomes, eds., Oxford University Press, Oxford, 1991, pp. 165–186.

See also the article

Johnson, J., "A Theory of Stars in Complex Systems," in *Complexity, Language, and Life: Mathematical Approaches,* J. Casti and A. Karlqvist, eds., Springer, Berlin, 1986, pp. 21–61.

§4. The idea of studying the geometrical properties of a space by looking at the algebraic properties of a suitable representation of the space forms the core of the subject matter of algebraic topology. Good introductions are

Giblin, P., *Graphs, Surfaces and Homology,* Chapman and Hall, London, 1977,

Singer, I. M., and J. A. Thorpe, *Lecture Notes on Elementary Topology and Geometry,* Scott Foresman, Glenview, IL, 1967.

§5. The notion of q-connection as a means for studying the global geometry of a complex is due to Ron Atkin. For an extensive account of how this concept is used in a variety of applied settings, see his pioneering works

Atkin, R., *Mathematical Structure in Human Affairs,* Heinemann, London, 1974,

Atkin, R., *Combinatorial Connectivities in Social Systems,* Birkhäuser, Basel, 1977.

A summary of the basic ideas of q-analysis, along with many examples, is given in

Casti, J., *Connectivity, Complexity and Catastrophe in Large-Scale Systems,* Wiley, Chichester and New York, 1979.

The example on the Middle East crisis is based on discussions with Mel Shakun and is considered in more detail in

Casti, J., "Topological Methods for Social and Behavioral Systems," *Int. J. Gen. Syst.,* 8 (1982), 187–210.

§6. The matter of system complexity for simplicial complex descriptions is considered further in the Casti volume cited for §5. The axiomatic view of complexity in terms of subsystems is developed in great detail from a highly algebraic point of view in

Rhodes, J., "Applications of Automata Theory and Algebra," Lecture Notes, Dept. of Mathematics, University of California, Berkeley, 1971.

For a more readily accessible account of these ideas, together with many examples, see

Gottinger, H., *Coping with Complexity,* Reidel, Dordrecht, 1983.

§7. The analysis of *Sky and Water* given in the text follows the treatment presented in

Johnson, J., *Combinatorial Structure in Digital Pictures,* NERC Remote Sensing Project, Discussion Paper No. 1, Center for Configurational Studies, The Open University, Milton Keynes, UK, January 1985.

Escher's work has by now acquired a certain cult following, at least in academic scientific circles. A good account of the many-sidedness of Escher's work for both science and art is found in the symposium volume

M. C. Escher: Art and Science, H. S. M. Coxeter, M. Emmer, R. Penrose, and M. Teuber, eds., North-Holland, Amsterdam, 1986.

In this same connection, see also

Ernst, B., *The Magic Mirror of M. C. Escher,* Ballantine, New York, 1976,

Gardner, M., "The Art of M. C. Escher," in *The Mathematical Carnival,* Knopf, New York, 1975, pp. 89–103,

The World of M. C. Escher, J. L. Locher, ed., Abrams, New York, 1971,

Hofstadter, D., *Gödel, Escher, Bach: An Eternal Golden Braid,* Basic Books, New York, 1979.

§8. Mondrian's work is examined in far greater detail in the Atkin book cited under §5. Our treatment is but a brief excerpt from this work.

§9. The degree to which the *mathematical* theory of probability accurately reflects our sense of the likelihood of events has been a hotly debated issue in the philosophy of science for many years, with the number of incompatible "meanings" of probability rivaling that of the number of interpretations of quantum mechanics. For an introductory account of the various positions, see the article

Black, M., "Probability," in *Encyclopedia of Philosophy,* P. Edwards, ed., Macmillan, New York, 1967.

For a stimulating view of the whole issue of probability theory and its role in measuring uncertainty, see

Good, I. J., *Good Thinking: The Foundations of Probability and Its Applications,* University of Minnesota Press, Minneapolis, MN, 1983.

A number of alternatives to classical probability theory have been put forward in recent years to account for uncertainty. Prominent among these alternatives is the idea of a *fuzzy set,* in which the usual 0/1 set membership function is replaced by a membership function in which an element's degree of membership in the set can be any number between 0 and 1. This seemingly small change leads to large consequences and dramatic implications for our view of the question of uncertainty versus randomness. For a detailed description of fuzzy sets and the likelihood "calculus" that the theory engenders, see the compendium

Fuzzy Sets and Applications: Selected Papers by L. A. Zadeh, R. Yager, ed., Wiley, New York, 1987.

In this same regard, see the volumes

Shafer, G., *A Mathematical Theory of Evidence,* Princeton University Press, Princeton, 1976,

Dubois, D., and H. Prade, *Possibility Theory,* Plenum Press, New York, 1988,

Kruse, R., and K. D. Meyer, *Statistics With Vague Data,* Reidel, Dordrecht, 1987,

Klir, G., and T. Folger, *Fuzzy Sets, Uncertainty, and Information,* Prentice-Hall, Englewood Cliffs, NJ, 1988,

Kickert, W., *Fuzzy Theories on Decision-Making,* Nijhoff, Leiden, 1978,

Negoita, C., and D. Ralescu, *Applications of Fuzzy Sets to Systems Analysis,* Birkhäuser, Basel, 1975.

The idea of characterizing classical ideas of probability by means of special types of simplicial complexes is developed further in the Atkin book cited under §1–3. The theory of surprises, together with the technological disaster example outlined in the text, follows the development in

Atkin, R. H., "A Theory of Surprises," *Env. & Planning-B,* 8 (1981), 359–365.

§10. A more extensive discussion of these results, enabling us to link-up with the ideas of Chapter 6, can be found in

Casti, J., "Polyhedral Dynamics and the Controllability of Dynamical Systems," *J. Math. Anal. Applic.,* 68 (1979), 334–346.

Extension of the basic idea of the above paper to the much more difficult case of multi-input systems has been made in the work

Ivascu, D. and G. Burstein, "An Exact Homology Sequences Approach to the Controllability of Systems," *J. Math. Anal. Applic.,* 106 (1985), 171–179.

DQ #2. For an example of how to use hierarchies of relations to say something meaningful about Shakespeare's play *A Midsummer Night's Dream,* see the Atkin's book *Multidimensional Man* cited under §1–3.

DQ #5. The game of chess is considered from a *q*-analysis point of view in the Atkin book noted in §5, as well as in

Atkin, R. H., *et al,* "Fred CHAMP, Positional Chess Analyst," *Int. J. Man-Mach. Stud.,* 8 (1976), 517–529,

Atkin, R. H., "Positional Play in Chess by Computer," in *Advances in Computer Chess I,* M. Clarke, ed., Edinburgh University Press, Edinburgh, 1976, pp. 60–73,

Atkin, R. H. and I. H. Witten, "A Multi-Dimensional Approach to Positional Chess," *Int. J. Man-Machine Studies,* 7 (1975), 727–750.

DQ #8. For a more complete account of these views on resilience, stability and surprise in ecosystems, see

Holling, C. S., C. Walters and D. Ludwig, "Surprise in Resource and Environmental Systems," preprint, Institute for Animal Resource Ecology, University of British Columbia, Vancouver, B. C., 1976.

DQ #10. For a detailed account of *q*-analysis in the cause of road traffic flow, see the Atkin book cited under §5, as well as the paper

Johnson, J., "The Dynamics of Large Complex Road Systems," in *Transport Planning and Control,* J. Griffiths, ed., Oxford University Press, Oxford, 1991.

DQ #11. These matters are taken up in far more detail in Chapter 5 of the Atkin book cited under §5.

DQ #13. The role of simplicial complexes in providing an alternate formulation of much of classical and modern physics, including the issue of time, is explored in the works

Atkin, R., "A Homological Foundation for Scale Problems in Physics," *Int. J. Theor. Phys.,* 3 (1970), 449–466,

Atkin, R., "Time as a Pattern on a Multi-Dimensional Structure," *J. Soc. & Biol. Struct.,* 1 (1978), 281–295.

In this same connection, see also the Johnson article in Casti and Karlqvist cited under §1-3.

DQ #14. This discussion of the failure of electrical power networks is carried on to much greater lengths in the article

Gould, P., "Electrical Power Failure: Reflections on Compatible Descriptions of Human and Physical Systems," *Env. & Planning-B*, 8 (1981), 405–417.

DQ #15. Johnson's work on the use of q-analysis for pattern recognition with digital images is reported in

Johnson, J., "Pixel Parts and Picture Wholes," in *From Pixels to Features*, J.-C. Simon, ed., North-Holland, Amsterdam, 1989,

Johnson, J., "Gradient Polygons: Fundamental Primitives in Hierarchical Computer Vision," in *Proc. Symp. in Honor of Prof. J.-C. Simon*, ARCET, University of Paris, 156 Bolve. Péreire, 75017 Paris, October 1990.

PR #1. For an explicit counterexample showing the conjecture is false, see

Earl, C., "A Complex Is Not Uniquely Determined by Its Structure Vectors," *Env. & Planning-B*, 8 (1981), 349–350.

PR #2-3, 5. For more information on these bread-and-butter items from elementary algebraic topology, see the volumes cited earlier under §4.

PR #6 and PR #9. These results appear in the Casti paper already referenced under §10.

PR #7. This food web example has been used in many places as a starting point for a variety of graph-theoretic and q-analysis investigations. The food web itself appears to date back to

Burnett, R., et al., *Zoology: An Introduction to the Animal Kingdom*, Golden Press, New York, 1958.

For more details on the way in which q-analysis sheds light upon the hidden structures in the web, see the Casti volume cited under §5 as well as the paper

Doreian, P., "Analyzing Overlaps in Food Webs," *J. Soc. & Biol. Structures*, 9 (1986), 115–139.

PR #8. This food web and its various connectivity patterns is discussed in the survey article

Pimm, S., J. Lawton and J. Cohen, "Food Web Patterns and Their Consequences," *Nature,* 350 (25 April 1991), 669–674.

PR #10. This problem is discussed in more detail in the Johnson contribution to the Casti and Karlqvist volume cited under §1-3 above.

PR #11. The conclusion of this problem is known as Dowker's Theorem. It first appeared in

Dowker, C. H., "Homology Groups of Relations," *Annals of Math.,* 56 (1952), 84–95.

PR #12. For a discussion of the complementary complex, see Appendix D of the first Atkin book cited in §5, as well as the second volume by Atkin noted under the same section.

PR #13. The surprise paper by Atkin referenced under §9 contains much more information on the relationship between the concept of surprise discussed here and the more familiar notions from elementary probability theory.

CHAPTER NINE

The Mystique of Mechanism: Computation, Complexity and the Limits to Reason

1. Computing the Cosmos

MIT computer scientist Ed Fredkin likes to tell the parable of a computer simulation he calls the "Heaven Machine." It goes like this. One day you read an advertisement from the Heaven Machine Corporation, offering you the opportunity to have an exact copy of your brain states loaded into a gigantic computer simulation. If you accept the company's offer, however, the duplication process destroys your original brain, at which point your life on Earth is unfortunately over. By way of compensation, though, the Heaven Machine Company promises you eternal life within the machine. Further, they claim that this life in the machine is nothing short of "heavenly."

Because you look pretty skeptical about the whole business, the HMC salesman offers to let you talk the whole matter over with your neighbor Joe, who recently signed-up with the company and had his brain duplicated in the machine. So they take you into a room with a huge computer screen, which initially is pretty fuzzy. But after awhile the picture clears up, and you see your neighbor. "Hi there, Joe," you say. "What's happening?" Joe replies: "Life is fantastic. Everything is just heavenly up here. You wouldn't believe some of the conversations I've had lately. There are so many amazing people to talk with—Aristotle, Newton, Buddha. And when I'm not trading ideas with these guys, I play a lot of golf, sun myself on the beach and, in general, relax and enjoy all the things I always wanted to do but never had the time for when I lived next door to you. And the social life! My datebook looks like a dentist's calendar. It's enough to make Warren Beatty look like a fumbling teenager. Believe me when I tell you it's like I died and went to heaven." Hearing all this you think, that's my old pal Joe, all right. No question about that!

So is this a heaven-on-Earth machine or just a nightmarish fantasy? Is it even faintly plausible that by running a computer program—that is, by following a mere set of rules—it would be possible to capture the essence of what it means to be Joe? Or me? Or you? While it's far from apparent on the surface, this query is only a special case of the more general Big Question that we've been leading up to throughout this book: Is every observable real-world phenomenon just the result of following a *recipe?* Put more prosaically, is the universe just one big computer?

At first hearing this question sounds outrageously extravagant, meta-physical even, and certainly beyond the bounds of what we could ever hope to answer. Our principal goal in this chapter is to show that the question not only makes good scientific sense, but that we have a formidable array of tools at our disposal with which to address it. But before entering into some of the details, let's first look at how this "computability" question ties-in with what we've seen in the preceding chapters.

In the book's opening chapter we laid great emphasis on what was there called the "modeling relation." The essence of this relation resided in what we termed "encoding' and "decoding" operations, $\mathcal{E}$ and $\mathcal{D}$, respectively. These are the operations that enable us to translate freely between the real world N of observable quantities and the mathematical world M of formal symbolic systems. Our standing claim has been that success or failure in the system modeling game ultimately comes down to the choices we make for the operations $\mathcal{E}$ and $\mathcal{D}$. Throughout most of this book we have chosen the formal mathematical system M to be a dynamical system, although in the last chapter we departed from this "default option," taking M to be a simplicial complex. A central part of the message of this chapter is to show that in a certain well-defined sense, the appearance of having a choice for M is really an illusion. In fact, we will show here that **any** formal system M is completely equivalent to some dynamical system, which in turn is completely equivalent to a computer program. Moreover, we will finally see that the very definition of what it means to model *scientifically* a natural process N comes down to the choice of a set of rules to plug into the box labeled M. But for the moment we ask the reader to accept these equivalences on faith while we attend to some necessary preliminaries. We'll establish these grandiose claims later as we go along.

With the fact that a dynamical system and a formal system are actually the same abstract object, it should come as no surprise to find dynamical systems playing such a key role in our development. There really is nothing else—at least not from a mathematical point of view. With this fact in hand, it's also easier to understand the modern emphasis on a computer program as constituting what we mean by the *answer* to a mathematical question. Since there is no abstract difference between the output of a computer pro-gram and the result of a long chain of reasoning in a formal logical system, they are really like two sides of the same coin. Now that cheap, powerful computers are readily available, the shift in emphasis from formal pencil-and-paper proofs to computer programs merely reflects the technology of the times. So as we go along in this chapter, the reader should continually bear in mind our two primary goals: (1) To establish the equivalences shown in Fig. 9.1, and (2) to show that there's more to life than following rules. To put it more formally, our aim is to produce a convincing argument that not every observable physical process can be thought of as the output of a

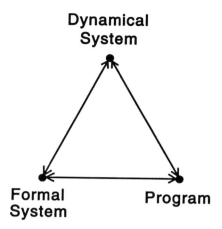

Figure 9.1. Equivalent Mathematical Formalisms

computer program. The road to these results is a long and arduous one, the starting point being a more clear-cut idea of what we mean when we talk about doing a "computation."

Exercises

1. Discuss why it might not be possible, *even in principle,* to duplicate the behavior, personality, likes and dislikes of our friend Joe in a computer program. Assuming that this is indeed the case and that we cannot capture Joe in a set of rules, does that mean that duplication of human cognitive capacities in a machine is also impossible?

2. In 1976 Wolfgang Haken and Kenneth Appel used a computer to solve the Four-Color Map Conjecture, which asserted that every planar map can be colored using no more than four colors. Haken and Appel made crucial use of a computer program in their "solution," a program that tested several thousand special configurations for the absence of certain properties. These properties, in turn, implied the desired result. Many mathematicians objected to this proof, claiming that since the computer did the checking and not a human being, it wasn't a *real* proof—at least not as mathematicians understand the use of that term. Discuss whether or not using a computer to carry out key steps in a proof in any way invalidates our saying that we've "proved" the result.

2. Algorithms and Programs

In 1935 Alan Turing was an undergraduate student at Cambridge, sitting in on a course of lectures on mathematical logic. During the course Turing was introduced to Hilbert's *Entscheidungsproblem*, which involves determining whether or not there exists an effective procedure for deciding if a given proposition follows deductively from the axioms of a formal logical system. The central difficulty Turing had in trying to come to grips with this problem was that there was no clear-cut notion of what was to count as an "effective procedure." Despite the fact that humans had been calculating for thousands of years, in 1935 there was still no good answer to the question: "What is a computation?" Turing set out to overcome this difficulty and solve the *Entscheidungsproblem.* To do so, he had to invent a theoretical gadget that ended up serving as the keystone in the arch of the modern theory of computation.

What Turing had to do was to find a way to replace the intuitive, informal notion of an "effective process" with a formal mathematical structure. What he came up with is what we now call an *algorithm,* an idea he modeled on the steps a human being actually goes through when carrying out a calculation. In essence, Turing saw an algorithm as a rote process or set of rules telling one how to proceed under any given set of circumstances. To get the general idea, let's look at some examples.

Example 1: A Caesar Salad Algorithm

Most general cookbooks have a recipe for the popular Caesar salad, which is a combination of romaine lettuce, olive oil, anchovies, garlic, parmesan cheese, roasted garlic croutons, coddled eggs, lemon juice, salt and pepper. Once these ingredients have been assembled, the recipe specifies in a step-by-step fashion how they are to be processed and combined to form the salad. For example, the first few steps might read:

1. Wash and dry the lettuce.

2. Tear the leaves of lettuce into bite-sized chunks.

3. Rub the inside of a wooden salad bowl with a raw clove of garlic.

4. Place the chunks of lettuce into the bowl and pour one cup of olive oil over them.

$$\vdots$$

These steps constitute what we might call a "Caesar-salad algorithm." At each and every stage, the algorithm specifies an unambiguous action that's to be performed, culminating in a stopping rule telling us when the salad is complete. And, in fact, it's not hard to see that given the raw ingredients and a sufficiently clever engineer, such a recipe could be followed in a purely mechanical fashion by a "Caesar Salad Machine." Now let's leave the kitchen and look at a less down-to-earth and more down-to-mathematics type of example.

Example 2: The Euclidean Algorithm

Consider the well-known euclidean algorithm for finding the largest number dividing two given whole numbers a and b. Assume that a is larger than b and let "rem $\{\frac{x}{y}\}$" denote the remainder after dividing x by y. Then the euclidean algorithm consists of calculating the sequence of integers $\{r_1, r_2, \ldots\}$ by the rules

$$r_1 = \text{rem } \left\{\frac{a}{b}\right\}$$

$$r_2 = \text{rem } \left\{\frac{b}{r_1}\right\}$$

$$r_3 = \text{rem } \left\{\frac{r_1}{r_2}\right\}$$

$$\vdots$$

where the process continues until a remainder zero is obtained, i.e., we obtain a quantity r_n such that $\text{rem}\{\frac{r_{n-1}}{r_n}\} = 0$. The process halts with the number r_n, which indeed turns out to be the largest integer dividing both a and b. To illustrate the process concretely, suppose we let $a = 14$ and $b = 6$. Then following the steps of the euclidean algorithm, we obtain

$$r_1 = \text{rem } \left\{\frac{14}{6}\right\} = 2,$$

$$r_2 = \text{rem } \left\{\frac{6}{2}\right\} = 0.$$

Thus, we conclude that 2 is the greatest common divisor of 14 and 6.

As with the Caesar salad example, what's important for us here is that the steps of the euclidean algorithm are rigidly prescribed and unvarying: One and only one operation is specified at each step, and there is no interpretation of the intermediate results or any skipping of steps—just a boring, basically mechanical repetition of the operations of division and keeping the remainder. This blind following of a set of rules is the distilled essence of what constitutes an algorithm. To reflect the mechanical nature of what's involved in carrying out the steps of an algorithm, Turing invented a hypothetical kind of computer now called a *Turing machine*. He then used the properties of this "machine" to formalize what it means to carry out a computation. Here's how.

A Turing machine consists of two components: (i) an infinitely long tape marked off into squares that can each contain one of a finite set of symbols, and (ii) a scanning head that can move along the tape, reading the squares and writing one of the symbols onto each square, thereby replacing whatever symbol happened to be there before. The behavior of the Turing machine is governed by an algorithm, which is manifested in what we now call a *program*. The program is composed of a finite number of instructions, each of which is selected from the following set of possibilities:

PRINT 0

PRINT 1

GO LEFT ONE SQUARE

GO RIGHT ONE SQUARE

GO TO STEP i IF THE CURRENT SQUARE CONTAINS 0

GO TO STEP i IF THE CURRENT SQUARE CONTAINS 1

STOP

That's it. From just these seven simple instructions we can compose what are called *Turing-Post programs*. These programs tell the machine what kind of computation it should carry out.

While the Turing-Post programming language makes it particularly simple to describe the actions of a Turing machine, it can lead to uncomfortably long programs if there are many branching statements. Therefore, the programming language is often modified so as to eliminate the two branching statements, replacing them with the possibility for the scanning head to exist in one of a finite number of internal *states*. For ease of writing and exposition, we shall adopt this convention in the examples and discussions of this chapter. The overall situation is shown schematically in Fig. 9.2 for a 12-state Turing machine.

The operation of a Turing machine is simplicity itself. We first feed in a tape containing a certain pattern of 0's and 1's (the input data). The machine then begins by setting the scanning head in its first state and placing it at some agreed-upon starting square, usually the first square containing a 1 (reading the tape from left to right). Thereafter, the actions taken by the machine are completely governed by the instructions contained in its program. But rather than continuing to speak in these abstract terms, it's simpler just to run through an example in order to get the gist of how such a gadget operates.

Example: Addition of Two Positive Integers

Consider a Turing machine with 3 internal states A, B and C, and assume that the tape symbols that the head can read/write are "0" and "1". Suppose we want to use this machine to add two whole numbers. For definiteness, we agree to represent an integer n by a string of n consecutive 1's

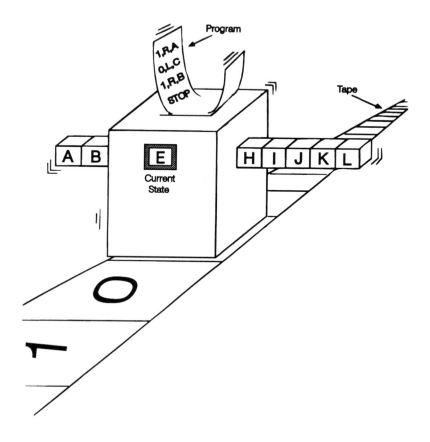

Figure 9.2. A 12-State Turing Machine

on the tape. The program shown in Table 9.1 serves to add any two whole numbers using this 3-state Turing machine. Let's see why.

Table 9.1. A Turing Machine Program for Addition

	Symbol Read	
State	1	0
A	1, R, A	1, R, B
B	1, R, B	0, L, C
C	0, STOP	STOP

The reader should interpret the entries in Table 9.1 in the following way: The first item is the symbol the head should print, the second is the direction the head should move, L(eft) or R(ight), while the final element is the internal state the head should move into. Note that the machine stops as soon as the head goes into state C. Let's see how it works for the specific case of adding 2 and 5.

Since our interest is in using the machine to add 2 and 5, we place two 1's and five 1's on the input tape, separating them by a 0 to indicate that they are two distinct numbers. Thus the machine begins by reading the input tape

...	0	0	0	**1**	1	0	1	1	1	1	1	0	0	0	0	...

Assume the head starts in state A, reading the first nonzero symbol on the left. (Note: Here, and in the other examples of this chapter, the position of the reading head will always be shown in boldface.) Since this symbol is a 1, the program tells the machine to print a 1 on that square and move to the right, retaining its internal state A. The head is still in state A and the current symbol read is again a 1, so the machine repeats the previous step and moves one square further to the right. Now, for a change, the head reads a 0. The program tells the machine to print a 1, move to the right, and switch to state B. I'll leave it to the reader to complete the remaining steps of the program, verifying that when the machine finally halts the tape ends up looking just like the input tape above, except with the 0 separating 2 and 5 having been eliminated, i.e., the tape will have seven 1's in a row, as required.

Before looking at the revolutionary implications of Turing's idea, let's pause here to emphasize the point that Turing machines are definitely not "machines" in the usual sense of being physical devices. Rather they are "paper computers," completely specified by their programs once we have agreed on the number of states and the set of symbols that can be written onto their tapes. So when we use the term "machine" in all that follows, the reader should interpret the word machine to mean "program" or "algorithm," i.e., software, putting all notions of hardware out of sight and out of mind. This abuse of the term "machine" should have been clear as soon as we started talking about an *infinite* storage tape. But it's important to make the distinction as hard and fast as possible: Turing machine = program. Period.

Modern computing devices, even home computers like the one I'm using to write this book, look vastly more complicated and powerful in their calculational power than a Turing machine with its handful of internal states and very circumscribed repertoire of scanning-head actions. Nevertheless, this turns out just not to be the case. And a large part of Turing's genius was to recognize that *any* algorithm, i.e., program, executable on *any* computing machine—idealized or otherwise—can also be carried out on a particular version of his machine, termed a *universal Turing machine* (or *UTM*, for short). So except for the speed of the computation, which definitely *is* hardware dependent, there's no computation that my machine (or anyone else's) can do that can't be done with a UTM.

To specify his UTM, Turing realized that the Turing-Post program for any computer can also be coded by a series of 0's and 1's. Table 9.2 shows one of many ways to do this. Consequently, we may regard the program itself as another kind of input data, and write it onto the input tape along with the data it's to operate upon. Using this crucial fact, Turing constructed a *fixed* program that could simulate the action of any other program P when given P as part of its input, i.e., he created a UTM. The operation of a UTM is simplicity itself.

Table 9.2. A Coding Scheme for the Turing-Post Language

Turing-Post statement	Code
PRINT 0	000
PRINT 1	001
GO RIGHT	010
GO LEFT	011
GO TO STEP i IF THE CURRENT SQUARE CONTAINS 0	101 $\underbrace{00\ldots\ldots}_{i\text{ repetitions}}$ 01
GO TO STEP i IF THE CURRENT SQUARE CONTAINS 1	110 $\underbrace{11\ldots\ldots}_{i\text{ repetitions}}$ 10
STOP	100

Suppose we have a particular Turing machine with program P. Since a Turing machine is completely determined by its program, all we need do is feed the program P into the UTM along with the input data. Thereafter the UTM will simulate the action of P on the data; in short, there will be no recognizable difference between running the program P on the original machine or having the UTM pretend it **is** the Turing machine P. The computer-literate reader will recognize that the fixed program characterizing the UTM is really analogous to interpreter software in a modern digital computer.

With these ideas of input tapes, Turing machine programs and the like under our belts, we are in a position to produce the dictionary filling-in the Dynamical System-Program leg on our modeling triangle shown earlier in Fig. 9.1. That matchup is shown in Table 9.3.

Table 9.3. The Dynamical System-Program Dictionary

Dynamical System	Program
number field	tape symbols
state manifold	all possible tape patterns
state	a tape pattern
constraints	set of admissible tape patterns
initial state	input tape pattern
vector field	Turing-Post program
trajectory	a sequence of tape patterns
attractor	tape pattern when the program halts or goes into an infinite loop

So we see that Turing's machine formalizes the intuitive idea of what it means to carry out a computation. A computation is a program. It's just as simple—and as complicated—as that. The natural question that emerges at this juncture is to ask: Are there limits to what can be computed? Or, to put it more specifically, is every number computable?

Exercises

1. Rewrite your favorite cake recipe as an algorithm.

2. Consider the action of the 6-state Turing machine having three tape symbols 0, 1, and 2, whose program is given in the following table:

State	Symbol Read		
	0	1	2
A	Print YES, STOP	0, R, B	0, R, C
B	0, L, D	1, R, B	2, R, B
C	0, L, E	1, R, C	2, R, C
D	STOP	0, L, F	0, Print NO, STOP
E	STOP	0, Print NO, STOP	0, L, F
F	0, R, A	1, L, F	2, L, F

Assume the input tape is

| ··· | 0 | 0 | 0 | **1** | 2 | 2 | 1 | 0 | 0 | 0 | 0 | 0 | 0 | 0 | 0 | ··· |

(a) This Turing machine is a pattern-recognition device. What kinds of patterns does it recognize? (b) How many steps does it take before the program stops when processing the above input?

3. Suppose you bought a used Turing machine tape at your local computer shop, and that this tape had a finite number of 1's written on it. Of course you want to clean up the tape before using it, changing all the 1's back to 0's. Write a Turing machine program that will do this? Does this "tape-cleaner" program ever halt?

4. Show that there are exactly $(4n+4)^{2n}$ Turing machines with n states. That is, there are this many distinct programs that can be written for an n-state machine.

3. Computation and Reality

The notion of a Turing machine finally put the idea of a computation on a solid scientific footing, enabling us to pass from the vague, intuitive idea of an "effective process" to the precise, mathematically well-defined notion of an algorithm. In fact, Turing's work, along with that of the American logician Alonzo Church, forms the basis for what has come to be called the

– Turing-Church Thesis –

Every effective process is implementable
by running a suitable program on a UTM

The key message of the Turing-Church Thesis is the assertion that any quantity that can be computed can be computed by a suitable Turing machine program. This claim is called a "thesis" and not a "theorem" because it's not really susceptible to proof. Rather, it's in the nature of a definition, or a proposal, suggesting that we agree to equate our informal idea of carrying out a computation with the formal mathematical idea of a program.

To bring this point home more forcefully, it's helpful to draw an analogy between a Turing machine and a typewriter. A typewriter is also a primitive device, allowing us to print sequences of symbols on a piece of paper that is potentially infinite in extent. A typewriter also has only a finite number of 'states' that it can be in: upper and lower case letters, red or black ribbon, different symbols balls, and so on. Yet despite these limitations, any typewriter can be used to type *The Canterbury Tales, Alice in Wonderland,* or any other string of symbols. Of course, it might take a Chaucer or a

Lewis Carroll to "tell" the machine what to do. But it can be done. By way of analogy, it might take a very skilled programmer to 'instruct' the Turing machine on how to solve difficult computational problems. But, says the Turing-Church Thesis, the basic model—the Turing machine—suffices for all problems that can be solved at all by carrying out a computation.

The universal Turing machine also gives us a way of identifying just what kinds of things are actually computable, at least in principle. By definition, a number is computable if and only if it can be obtained as the output of a program P run on a UTM. More formally, we have:

DEFINITION. *A natural number* x *is computable if there exists a Turing machine program* P *such that whenever* P *is started in state A on a blank tape (i.e., one filled with 0's), the program eventually halts after writing a block of* x *1's on the tape.*

This definition is fine for an arbitrary, but fixed, **natural number** x, since any integer, no matter how large, contains only a finite number of digits. The situation for real numbers is a bit more delicate, however, since unless the number is rational, it consists of an infinite number of non-repeating digits. The way around this difficulty is to extend the definition of computability so that a real number r is computable if and only if there is a Turing machine program P that cranks out the digits of r successively, one after the other. Obviously, such a program will usually not halt. Nevertheless, such a definition allows us to conclude that important transcendental numbers like π and e are indeed computable. But the somewhat surprising fact is that almost all numbers are *not* computable. We'll return to this point later. For now, let's have some fun with Turing machines and look at an example of just one such uncomputable quantity.

Example: The Busy Beaver Function

Suppose you're given an n-state Turing machine and an input tape filled entirely with 0's. The challenge is to write a program for this machine having the following properties: (i) The program must eventually halt, and (ii) the program should print as many 1's as possible on the tape before it stops. It's clear, I think, that the number of 1's that can be printed is a function only of n, the number of internal states that the machine's scanning head can enter. Equally clear is the fact that if $n = 1$, the maximum number of 1's that can be printed is only one, a result that follows immediately from the requirement that the program cannot run forever. Such programs that print a maximal number of 1's before halting are called *n-state Busy Beavers*. Table 9.4 gives the program for a 3-state Busy Beaver, while Figure 9.3 shows how this program can print six 1's on the tape before stopping.

Now for our uncomputable function. Define the quantity $BB(n) =$ the number of 1's written by an n-state Busy Beaver program. Thus, the

Table 9.4. A 3-State Busy Beaver

State	Symbol Read	
	0	1
A	1, R, B	1, L, C
B	1, L, A	1, R, B
C	1, L, B	1, STOP

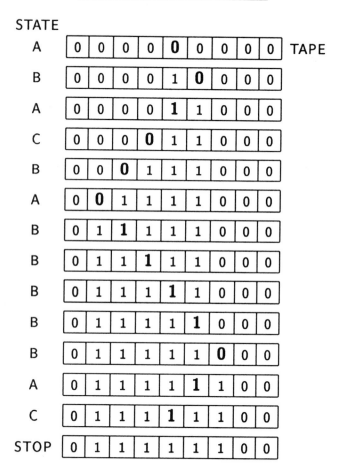

Figure 9.3. The Action of a 3-State Busy Beaver

Busy Beaver function $BB(n)$ is the maximal number of 1's that any halting program can write on the tape of an n-state Turing machine. We have already seen that $BB(1) = 1$ and $BB(3) = 6$, while in the Exercises it's shown that $BB(2) = 4$. From these results for small values of n, you might think that the function $BB(n)$ doesn't have any particularly interesting

properties as n gets larger. But just as you can't judge a book by its cover
(or title), you also can't judge a function from its behavior for just a few
values of its argument. In fact, it can be shown that

$$BB(12) \geq 6 \times 4096^{4096^{4096^{4096^{\cdots4096^{4}}}}}$$

where the number 4096 appears 166 times in the dotted region! So in trying
to calculate the value of $BB(12)$, we reach the point where it becomes impossible to distinguish between the finite and the infinite. It turns out that
for large enough values of n, the quantity $BB(n)$ exceeds that of *any* computable function for that same argument n. In other words, the Busy Beaver
function $BB(n)$ is uncomputable (see Problem 1). So for a concrete example of an uncomputable number, just take a Turing machine with a large
number of states n. Then ask for the value of the Busy Beaver function for
that value of n. The answer is uncomputable.

In our consideration of dynamical systems, we saw that about the most
fundamental question we can ask is how the system behaves in the long-run.
That is, what kind of attractor does the system enter from a given initial
state? Referring to Table 9.3, we see that the computational equivalent
of this question is to ask: Given a particular input tape configuration $\mathcal{I}$
and a particular Turing machine program $\mathcal{P}$, does the program halt when
processing the input data $\mathcal{I}$? Considered in the light of our definition of
computability, we see why this turns out to be the rock-bottom question
upon which all of the theory of computation ultimately rests. So, of course,
we'd like to know if there is any systematic way of answering it. In slightly
more formal terms, our question is known as:

– The Halting Problem –

Is there a general algorithm for determining if a program will halt?

Put more precisely, the Halting Problem asks if there is a single program that
will process any other program $\mathcal{P}$ and a set of input data set $\mathcal{I}$, outputting
a YES if the program $\mathcal{P}$ eventually halts when processing the data set $\mathcal{I}$,
while outputting a NO if it doesn't. Of course, for some programs and some
data sets such an algorithm certainly does exist. For example, most cake
recipes contain an explicit instruction saying, in effect, "Stop now, the cake
is finished." But the Halting Problem asks a much stronger question: Does
there exist a **single** algorithm that will give the correct answer in **all** cases?

To see that the question is far from trivial, suppose we have a program $\mathcal{P}$ that reads a Turing machine tape and stops when it comes to the
first 1. So, in essence, the program says: "Keep reading until you come to
a 1, then stop." In this case, the input data $\mathcal{I}$ consisting entirely of 1's would

result in the program stopping after the first step. On the other hand, if the input data were all 0's, then the program would never stop. Of course, in this situation we have a clear-cut procedure for deciding whether or not this program $\mathcal{P}$ will halt when processing some input tape $\mathcal{I}$: The program $\mathcal{P}$ will stop if and only if $\mathcal{I}$ contains even a single 1; otherwise, $\mathcal{P}$ will run on forever. So here's an example of a halting rule that works for any data set $\mathcal{I}$ processed by this especially primitive program.

Unfortunately, most real computer programs are vastly more complicated than this, and it's far from clear by simple inspection of the program what kinds of quantities will be computed as the program goes about its business. After all, if we knew what the program was going to compute at each step, we wouldn't have to run the program, would we? Moreover, the stopping rule for real programs is almost always an implicit rule of the foregoing sort, saying something like: "If such and such a quantity satisfying this or that condition appears, stop; otherwise, keep computing." The essence of the Halting Problem is to ask if there exists any "effective process" that can be applied to the program and its input data to tell *in advance* whether or not the program's stopping condition will ever be satisfied.

Turing settled the matter once and for all in the negative with his 1936 result that given a program $\mathcal{P}$ and an input data set $\mathcal{I}$, there is no way in general to say if $\mathcal{P}$ will ever finish processing the input $\mathcal{I}$. Here's a short proof of this fundamental result.

Suppose such a Halting Algorithm exists and let $\mathcal{I}$ be the input data. Consider the following UTM program:

1. Check to see if $\mathcal{I}$ is the code for a UTM program $\mathcal{P}$. If not, go back to the start and repeat.

2. If $\mathcal{I}$ is the code for a UTM program $\mathcal{P}$, double the input string to get $\mathcal{I} \cdot \mathcal{I}$.

3. Use the assumed Halting Algorithm for the UTM with input data $\mathcal{I} \cdot \mathcal{I}$. If it stops, go back to the beginning of this step and repeat.

4. Otherwise, halt.

Call the above program $\mathcal{H}$. Since $\mathcal{H}$ is a program, it has its own code H. Thus, we can ask: "Does $\mathcal{H}$ halt for input H?" It surely gets past step 1, since by definition H is the code for the program $\mathcal{H}$. And $\mathcal{H}$ gets past step 3, as well, if and only if the UTM doesn't halt with input $H \cdot H$. Thus we conclude that $\mathcal{H}$ halts with input data H if and only if the UTM does not halt with input data $H \cdot H$. But the UTM simulates a program $\mathcal{P}$ by starting with the input data $\mathcal{P} \cdot \mathcal{I}$, and then behaving just like $\mathcal{P}$ operating on input data $\mathcal{I}$. Therefore, we see that $\mathcal{P}$ halts with input data $\mathcal{I}$ if and only if the UTM halts with input data $\mathcal{P} \cdot \mathcal{I}$. So if we put $\mathcal{P} = \mathcal{H}$ and

$\mathcal{I} = H$, then we find that $\mathcal{H}$ halts with input data H if and only if the UTM halts with input data $H \cdot H$—a direct contradiction to the result obtained a moment ago. Thus we conclude that there is no such Halting Algorithm.

Let's conclude this section with an amusing game-theoretic example, showing that just like with the rest of life, in game theory, too, there is both good and bad news: The good news is that one of the players has a winning strategy; the bad news is that this winning strategy cannot be computed. Hence, the game is unsolvable.

Example: The Turing Machine Game

In this game there are two players, call them A and B. They take turns choosing positive integers as follows:

Step 1: Player A chooses a number n.

Step 2: Knowing n, Player B picks a number m.

Step 3: Knowing m, Player A selects a number k.

Player A wins if there is some n-state Turing machine that halts in exactly $m + k$ steps when started on a blank tape. Otherwise, Player B wins. It's fairly clear that this is a game of finite duration, since once the players have chosen their integers, all we need do is list the $(4n + 4)^{2n}$ Turing machines having n states, then run each of them for exactly $m + k$ steps to determine a winner.

It's a well-known fact from game theory that any game of fixed, finite duration is determined, in the sense that there is a winning strategy for one of the players. In this case, it's Player B. Nevertheless, the Turing machine game is nontrivial to play, since neither player has an *algorithm*, i.e., a computable strategy, for winning the Turing Machine game. Here's a quick proof of this fact, making use of our earlier result for the Busy Beaver function $BB(n)$.

Let $S(n)$ be the maximum number of steps that an n-state Turing machine can perform before halting. Clearly, $BB(n) \leq S(n)$. A strategy for Player B is a formula $m = f(n)$, specifying what number to select as a function of the number n chosen by Player A. By definition, f is a winning strategy for Player B if and only if $S(n) \leq f(n)$. Consequently, a winning strategy for Player B is just to take $f(n)$ to be the function $S(n)$ itself. However, for every computable function f it's eventually the case that $f(n) < BB(n) \leq S(n)$. Thus Player B has no *computable* winning strategy.

Up to now, we have focused our attention on the "programs" vertex of our triangle of formalisms in Figure 9.1. In the next section we turn to the "formal system" side of the house.

Exercises

1. If $n = 2$, show that the maximum number of 1's that a Busy Beaver can print before halting is four, i.e., $BB(2) = 4$.

2. The definition given in the text for the computability of real numbers contains a small technical flaw. Can you find it?

3. Use the fact that the Busy Beaver function is uncomputable to prove that the Halting Problem is undecidable, i.e., has no solution. (Hint: Assume there is an algorithm $\mathcal{A}$ that solves the Halting Problem and derive a contradiction.)

4. Formalization and Hilbert's Program

In a famous epigram on the nature of mathematics, Bertrand Russell claimed that "pure mathematics is the subject in which we do not know what we are talking about, or whether what we are saying is true." This pithy remark summarizes the content of a famous research program proposed earlier this century by David Hilbert for the formalization of mathematics. What Hilbert wanted to do was develop a purely syntactic framework for all of mathematics. He believed that the way to eliminate the possibility of paradoxes and inconsistencies arising in mathematics was to create a purely formal, essentially "meaningless," framework within which to speak about the truth or falsity of mathematical statements. Such a framework is now termed a *formal system,* and it constitutes the jumping-off point for our investigation of the gap between what can be proved and what is actually true in the universe of mathematics. We briefly considered formal systems in Chapter 1. Now we'll look at them in much greater detail, keeping mind our goal of filling-in the legs of the triangle linking formal systems, programs and dynamical systems.

The "meaningless statements" of a formal system are finite sequences of abstract *symbols,* usually called *symbol strings.* A finite number of these strings are taken as the *axioms* of the system. To complete the system, there are a finite number of *transformation rules* telling us how a given string of symbols can be converted into another such string. The general idea is to start from one of the axioms and apply a finite sequence of transformations, thereby converting the axiom into a sequence of new strings, each string in the sequence being either an axiom or a string derived from its predecessors by application of the transformation rules. The terminal string in such a sequence is called a *theorem.* The totality of all theorems constitutes the *provable statements* of the system. But note carefully that these so-called "statements" don't actually say anything; they are just strings of abstract symbols. Let's see how this setup works with a simple example based on Douglas Hofstadter's famous MIU system.

Example: The ★-✠-☁ *System*

Suppose the symbols of our system are the three more or less culturally-free objects ★ (star), ✠ (maltese cross), and ☁ (cloud). Let the two-element string ✠☁ be the sole axiom of the system, and take the transformation rules to be:

Rule I: x☁ $\longrightarrow$ x☁★
Rule II: ✠x $\longrightarrow$ ✠xx
Rule III: ☁☁☁ $\longrightarrow$ ★
Rule IV: ★★ $\longrightarrow$ —

In these rules, "x" denotes an arbitrary finite string of stars, crosses, and clouds, while $\longrightarrow$ means "is replaced by." The interpretation of Rule IV is that anytime two stars appear they can be dropped to form a new string. Now let's see how these rules can be used to derive a theorem.

Starting with the single axiom ✠☁, we can prove that the string ✠★☁ is a theorem by applying the transformation rules as follows:

$\longrightarrow$ ✠☁ $\longrightarrow$ ✠☁☁ $\longrightarrow$ ✠☁☁☁☁ $\longrightarrow$ ✠★☁
(Axiom)   (Rule II)   (Rule II)   (Rule III)

Such a sequence of steps, starting from an axiom and ending at a statement like ✠★☁, is termed a *proof sequence* for the theorem represented by the last string in the sequence. Observe that when applying Rule III at the final step, we could have replaced the last three ☁'s from the preceding string rather than the first three, thereby ending up with the theorem ✠☁★ instead of ✠★☁.

You will probably have also noted that all the intermediate strings obtained in moving from the axiom to the theorem begin with a ✠. It's easy to see from the action of the transformation rules for this system that every string will have this property. This is a *metamathematical* property of the system, since it's a statement **about** the system rather than one made **in** the system itself. As we shall see, the distinction between what the system can say from the inside (its strings) and what we can say about the system from the outside (properties of the strings) is of the utmost importance.

But what does all this abstract nonsense have to do with mathematics? What do maltese crosses, clouds and stars have to do with things like the sum of the first n positive integers, the nature of the attractor set of a dynamical system or the angles of a triangle? Indeed, what do these symbols have to do with *anything?* The answer to this eminently sensible query lies in one word: *interpretation.* Depending on the kind of mathematical structure under consideration, we have to make up a dictionary to translate (i.e., interpret) the abstract symbols and rules of the formal system into the objects of the mathematical structure in question. By this dictionary-construction step, we attach semantic meaning to the abstract, purely syntactic structure of the symbols and strings of the formal system. Thereafter all the theorems of the formal system can be interpreted as true statements about the associated mathematical objects. The following diagram illustrates this crucial distinction between the purely syntactic world of formal systems and the meaningful world of mathematics.

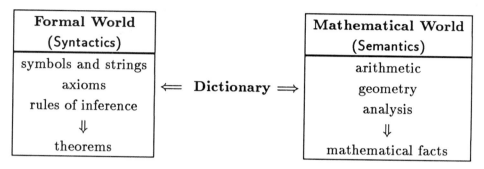

Let's again emphasize that there are two entirely different worlds being mixed here—the purely syntactic world of the formal system and the meaningful world of mathematical objects and their properties. And in each of these worlds there is a notion of truth: theorems in the formal system, correct mathematical statements such as "$2 + 5 = 7$" or "the sum of the angles of a triangle equals 180 degrees" in the realm of mathematics. The connection between the two worlds lies in the interpretation of the elements of the formal system in terms of the objects and operations of the mathematical structure. Once this dictionary has been written and the associated interpretation established, we can then hope along with Hilbert that there will be a perfect, one-to-one correspondence between the facts of the mathematical structure and the theorems of the formal system.

Speaking loosely, what we seek is a formal system in which every real-world truth translates into a theorem of the system, and conversely. We call such a formal system *complete*. Moreover, if the mathematical structure is to avoid contradiction, a real-world statement and its negation should never both translate to theorems of the system. In other words, we should not be able to prove contradictory statements in the formal system. A formal system in which no contradictory statements can be proved is termed *consistent*.

This is a good place to pause briefly to fill-in the remaining legs in the triangle of Fig. 9.1, the legs linking a formal logical system of the type we've been talking about here to both a dynamical system and to a Turing machine program. The relevant dictionaries are shown in Tables 9.5–9.6.

Table 9.5. The Dynamical System-Formal System Dictionary

Dynamical System	Formal System
number field	abstract symbols
state manifold	possible symbol strings
state	a symbol string
constraints	grammar
initial state	starting axiom
vector field	rules of logical inference
trajectory	proof sequence
attractor	theorem

Table 9.6. The Formal System-Program Dictionary

Formal System	Program
abstract symbols	tape symbols
possible symbol strings	possible tape patterns
a symbol string	a tape pattern
grammar	set of admissible tape patterns
starting axiom	input tape pattern
rules of logical inference	Turing-Post instructions
proof sequence	a sequence of tape patterns
theorem	tape pattern when the program halts or goes into an infinite loop

By the time Hilbert proposed his Program, it was already known that the problem of the consistency of mathematics as a whole was reducible to the determination of the consistency of arithmetic. So the problem became to give a "theory of arithmetic," i.e., a formal system that was (i) finitely describable, (ii) complete, (iii) consistent, and (iv) sufficiently powerful to represent all the statements that can be made about the positive integers. By the term "finitely describable," what Hilbert had in mind is not only that the number of axioms and rules of the system be finite in number, but also that every provable statement in the system—i.e., every theorem—should be provable in a finite number of steps. This condition seems reasonable enough, since you don't really have a theory at all unless you can tell other people about it. And you certainly can't tell them about it if there are an infinite number of axioms, rules, and/or steps in a proof sequence.

A central question that arises in connection with any such *formalization* of arithmetic is to ask if there is a procedure by which we can decide the truth or falsity of every arithmetical statement in a finite number of steps. So, for example, if we make the statement: "The sum of two prime numbers greater than 2 is always an even number," we want a finite procedure, essentially a computer program, that halts after a finite number of steps, telling us whether that statement is true or false, i.e., provable or not in some formal system powerful enough to encompass ordinary arithmetic. For example, in the ★-✠-℥-formal system considered earlier, such a decision procedure is given by the far-from-obvious criterion: "A string is a theorem if and only if it begins with a ✠ and the number of ℥'s in the string is not divisible by 3."

The question of the existence of a procedure to decide all statements about arithmetic, i.e., about the natural numbers, is what's called Hilbert's *Entscheidungsproblem* (Decision Problem). Hilbert was convinced that such a formalization of arithmetic was possible, and his manifesto presented at the 1928 International Congress of Mathematicians in Bologna, Italy challenged the world's mathematical community to find/create it. The next section shows how dramatically and definitively wrong even a man as great as Hilbert can be!

Exercises

1. Show how to formulate the game of chess as a formal system. Can you do the same for other board games like Go or Scrabble? What about a game like Monopoly?

2. In the Modeling Relation of Chapter 1, the key elements were the encoding and decoding operations linking the worlds of natural and mathematical systems. The matchups given in Tables 9.3, 9.5–9.6, on the other hand, are dictionaries speaking solely of objects in the realm of mathematics. (a) What are the analogues of the encoding and decoding operations in this setting? (b) Discussion Questions 1–3 of Chapter 1 considered various types of encodings like analogies, metaphors and linkage relations between natural systems and mathematical systems. Consider these same types of encodings, but now within the framework of the dictionaries presented in this section for mathematical objects.

3. Consider the following "statement" made in the ★-✠-℥-system of the text: ✠✠℥★★℥℥★✠ ✠★℥. (a) Is this statement a theorem of the system? (b) If it is, produce a proof sequence demonstrating this fact; if not, determine why it's not a theorem?

5. The Undecidable and the Incomplete

In 1931, less than three years after Hilbert's Bolognese call to arms, Kurt Gödel published the following metamathematical fact, perhaps the most famous, broad-sweeping mathematical (and perhaps philosophical) result of this century:

Gödel's Incompleteness Theorem—Informal Version
Arithmetic is not completely formalizable

Remember that for a given mathematical structure like arithmetic, there are an infinite number of ways we can choose a finite set of axioms and rules of a formal system in an attempt to mirror syntactically the mathematical truths of the structure. What Gödel's result says is that **none** of these choices will work. There does not and cannot exist a formal system satisfying all the requirements of Hilbert's Program. In short, there are no rules for generating **all** the truths about the natural numbers.

Gödel's result is shown graphically in Fig. 9.4 for a given formal system **M**, where the entire square represents all possible arithmetical statements. As we prove an arithmetical statement true using the rules of the system **M**, we color that statement white; if we prove the statement false, we color it black. Gödel's Incompleteness Theorem (which we shall henceforth denote more compactly as simply Gödel's Theorem) says that there will always exist statements like **G** that are eternally doomed to a life in the shadow world of gray. It's impossible to eliminate the gray and color the entire square in black and white. And this result holds for *every* possible formal system **M**, provided only that the system is consistent. So for every consistent formal system **M**, there is at least one statement **G** that can be seen to be true—yet **G** is unprovable in **M**. So as with the rest of life, so it is with arithmetic too: there's no washing away the gray! We call a statement like **G** *undecidable,* since it can be neither proved nor disproved within the framework of that formal system. And if we add that undecidable statement **G** as an axiom, thereby creating a new formal system, the new system will have its own *Gödel sentence* **G**—unprovable, yet true.

By his Theorem Gödel snuffed out once and for all Hilbert's flickering hope of providing a complete and total axiomatization of arithmetic, hence of mathematics. Since Gödel's Theorem represents one of the pinnacles of human intellectual achievement, not to mention forming the basis for a whole host of related developments in mathematics, philosophy, computer science, linguistics, and psychology, let's spend a few pages looking at how one could ever prove such a profound, mind-boggling result.

In arriving at his proof of the incompleteness of arithmetic, Gödel's first crucial insight was to recognize that every formalization of a branch of

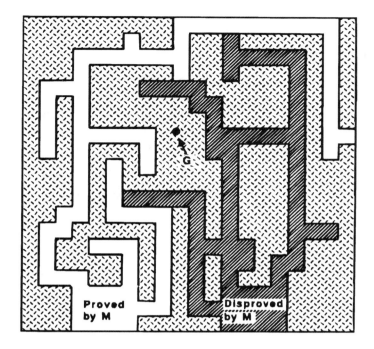

Figure 9.4. Gödel's Theorem in Logic Space

mathematics is itself a mathematical object in its own right. So if we create a formal system intended to capture the truths of arithmetic, that formal system can be studied not just as a set of mindless rules for manipulating symbols, but also as an object possessing mathematical as well as syntactic properties. In particular, since Gödel was interested in the truths about numbers, he showed how it would be possible to represent any formal system purporting to encompass arithmetic within arithmetic itself. In short, Gödel saw a way to mirror—or "code"—all statements about relationships among the natural numbers by using these very same numbers themselves.

To understand this "mirroring" operation a little more clearly, consider the familiar situation at a post office or airline ticket counter where customers are given a number when they enter in order to indicate their position in the queue. Suppose Clint and Brigitte each want to return home for the holidays. To arrange their flights they go to the local office of Swiss Air, where upon entering Clint receives the service number 4, while Brigitte comes in a bit later and gets number 7. By this service-assignment scheme, the real-world fact that Clint will be served before Brigitte is "mirrored" in the purely arithmetical truth that 4 is less than 7. In this way a truth of the real world (that Clint will be served first) has been faithfully translated, or mirrored, in a truth of number theory (that 4 is smaller than 7). Gödel used a tricky variant of this kind of numbering scheme to code all possible statements about arithmetic in the language of arithmetic itself, thereby using arithmetic as both an interpreted mathematical object and as an uninterpreted formal system with which to talk about itself. It's revealing to see how this *Gödel numbering* scheme actually works.

For sake of exposition, let's consider a somewhat streamlined version of the language mathematicians use to describe arithmetic. In this slimmed-down language, there are elementary signs and variables. To follow Gödel's scheme, suppose there are the ten logical signs shown in Table 9.7 along with their Gödel number, an integer between 1 and 10.

Table 9.7. Gödel Numbering of the Elementary Signs

Sign	Gödel Number	Meaning
~	1	not
∨	2	or
⊃	3	if ... then
∃	4	there exists
=	5	equals
0	6	zero
s	7	the immediate successor of
(	8	punctuation
)	9	punctuation
'	10	punctuation

In addition to the elementary signs, the language of arithmetic contains three types of variables: numerical, sentential, and predicate. Numerical variables are symbols like x, y and z for which we can substitute numbers or numerical expressions. Sentential variables, usually denoted $p, q, \ldots$, can be replaced by formulas (sentences). Finally, for predicate variables $P, Q, \ldots$ we can substitute properties such as "prime," "less than," and "odd." Since we have only 10 elementary signs, in Gödel's numbering system numerical variables are coded by prime numbers greater than 10, sentential variables by the squares of prime numbers greater than 10, and predicate variables by the cubes of prime numbers greater than 10, all the prime numbers taken in numerical order.

To see how this numerical coding process works in practice, consider the logical formula $(\exists x)(x = sy)$, which translated into plain English reads: "There exists a number x which is the immediate successor of the number y." Since x and y are numerical variables, the coding rules dictate that we make the assignment $x \rightarrow 11$, $y \rightarrow 13$, since 11 and 13 are the first two prime numbers larger than 10. The other symbols in the formula can be coded using the correspondence in Table 9.7. Carrying out this coding yields the sequence of numbers $(8, 4, 11, 9, 8, 11, 5, 7, 13, 9)$, corresponding to reading the logical expression symbol-by-symbol and substituting the appropriate number according to the coding rule. While this sequence of ten numbers pins the logical formula down uniquely, for a variety of reasons it's more convenient to be able to represent the formula by a single number. Gödel's procedure for doing this is to take the first 10 prime numbers (since there are ten symbols in the formula) and multiply them together, each prime number being raised to a power equal to the Gödel number of the corresponding element in the formula. So, since the first ten prime numbers in order are 2, 3, 5, 7, 11, 13, 17, 19, 23, and 29, the final Gödel number for the above formula is

$$(\exists x)(x = sy) \longrightarrow 2^8 \times 3^4 \times 5^{11} \times 7^9 \times 11^8 \times 13^{11} \times 17^5 \times 19^7 \times 23^{13} \times 29^9$$

Using this kind of numbering scheme, Gödel was able to attach a unique number to each and every statement and sequence of statements that can be made about arithmetic.

By Gödel's coding procedure, every possible proposition about the natural numbers can itself be expressed as a number, thereby opening up the possibility of using number theory to examine its own truths. The overall process can be envisioned by appealing to the metaphor of a locomotive shunting boxcars back and forth in a freight yard. This idea, due to Douglas Hofstadter, is shown in Figure 9.5. In the upper part of the figure we see the boxcars with their uninterpreted numbers painted on the sides of the cars, while looking down from the bird's-eye view we see the interpreted symbols inside each car. The shuffling of the cars according to the rules for manipulating logical symbols and formulas is mirrored by a corresponding transformation of natural numbers—i.e., statements of arithmetic—and vice-versa.

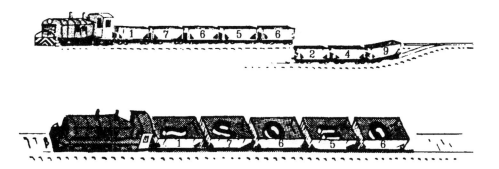

Figure 9.5. Freight Train View of Gödel Numbering and Transformation Rules

Deep insight and profound results necessarily involve seeing the connection between several ideas at once. In the proof of Gödel's Theorem there are two crucial notions that Gödel had to juggle simultaneously. Gödel numbering was the first.

Logical paradoxes of the sort that worried Hilbert are all based on some aspect of self-reference. The granddaddy of all such self-referential conundrums is the so-called Epimenides Paradox, one version of which is

> This sentence is false.

What Gödel wanted to do was find a way to express such paradoxical self-referential statements within the framework of arithmetic. However, a statement like the Epimenides Paradox involves the notion of truth, something that logician Alfred Tarski had already shown could not be captured within the confines of a formal system. Enter Gödel's Big Idea #2.

Instead of dealing with the informal, eternally slippery notion of truth, Gödel somehow had to replace the idea of "truth" with something that is formalizable. This is just exactly the same kind of obstacle that Alan Turing also had to overcome, when he was faced with the difficulty of formalizing the informal idea of a "computation." The idea that Gödel came up with was to translate the real-world concept of truth into the formal system notion of *provability*. Thus, the Epimenides Paradox above translates into the Gödel sentence:

> This statement is not provable.

This sentence, of course, is a self-referential claim about a particular "statement," the statement mentioned in the sentence. But by using his numbering scheme, Gödel was able to mirror this assertion by a corresponding self-referential, metamathematical statement in the language of arithmetic itself. Let's follow through the logical consequences of this mirroring within the framework of any logical system within which we can formulate all possible statements of arithmetic.

If the statement is provable, then it's true; hence, what it says must be true and it is *not* provable. Thus, the statement and its negation are both provable, implying an inconsistency in the underlying formal logical system. On the other hand, if the statement is not provable, then its assertion is true. In this case, the statement is true but unprovable; hence, there is a true statement that does not correspond to any theorem of the formal system. In short, the formal system is incomplete.

Gödel was able to show that for *any* consistent formalization of arithmetic, such a Gödel sentence must exist. Consequently, the formalization must be incomplete. The bottom line then turns out to be that in **every** formal system powerful enough to allow us to express any statement about the natural numbers, there exists a statement that is unprovable using the logical rules of the system. Nevertheless, that statement represents a true assertion about the natural numbers, one that we can see is true by "jootsing," to use Douglas Hofstadter's colorful term for "jumping out of the system." Almost as an aside, Gödel also showed how to construct an arithmetical statement **A**, which translates into the metamathematical claim that "arithmetic is consistent." He then demonstrated that the statement **A** is not provable, implying that the consistency of arithmetic cannot be established by arguments that can be made using the formal system of arithmetic itself. Putting all these notions together, we come to:

Gödel's Theorem—Formal Logic Version

For every consistent formalization of arithmetic, there exist arithmetic truths unprovable within that formal system

---

Since the sequence of steps leading up to Gödel's startling conclusions are both logically tricky and intricately intertwined, the principal landmarks along the road are summarized in Table 9.8.

Exercises

1. Is it possible for two different expressions to have the same Gödel number? If so, give an example. If not, explain why it's not possible.

2. Write out the Gödel number for the following logical expressions: (a) "17 is not the immediate successor of 5." (b) "If 10 is the immediate successor of 9, then 10 does not equal 9."

3. Prove Gödel's Theorem—Algorithmic Version: There is no algorithm $\mathcal{A}$ whose end result consists of all true statements of arithmetic and no false ones. (Hint: Assume there exists such an algorithm and derive a contradiction.)

4. A crucial point that's often overlooked about Gödel's result is that it applies only to formal logical systems that are: (i) strong enough to state every relationship among the natural numbers, and (ii) consistent. Consider what it would mean to drop either of these two conditions. Specifically, produce counterexamples showing that Gödel's Theorem is false if either of these conditions is violated.

5. Gödel formalized the statement "This sentence is not provable." Logician Leon Henkin invented a different kind of sentence: "This sentence is provable." Henkin's sentence is not the negation of Gödel's sentence. Do you see why?

6. A common misunderstanding of Gödel's Theorem is to think that Gödel showed that there are arithmetical truths that we can never prove. What's wrong with this interpretation of Gödel's result?

Table 9.8. The Main Steps in Gödel's Proof

Gödel Numbering: Development of a coding scheme to translate every logical formula in a formalization of arithmetic into a "mirror-image" statement about the natural numbers.

⇓

Epimenides Paradox: Replace the notion of "truth" by that of "provability," thereby translating the Epimenides Paradox into the assertion "This statement is unprovable."

⇓

Gödel Sentence: Show that the sentence "This statement is unprovable" has an arithmetical counterpart, its Gödel sentence **G**, in every conceivable formalization of arithmetic.

⇓

Incompleteness: Prove that the Gödel sentence **G** must be true if the formal system is consistent.

⇓

No Escape Clause: Prove that even if additional axioms are added to form a new system in which **G** is provable, the new system with the additional axioms will have its own unprovable Gödel sentence.

⇓

Consistency: Construct an arithmetical statement asserting that "arithmetic is consistent." Prove that this arithmetical statement is not provable, thus showing that arithmetic *as a formal system* is too weak to prove its own consistency.

6. Complexity, Programs and Numbers

The replacement of the Earth by the Sun as the center of heavenly motions is widely (and rightly) seen as one of the great scientific paradigm shifts of all time. But what is often misunderstood is the reason why this Copernican "revolution" eventually carried the day with the scientific community. The commonly-held view is that Copernicus's heliocentric model vanquished the competition, especially the geocentric view of Ptolemy, because it gave better predictions of the positions of the celestial bodies. In actual fact, the predictions of the Copernican model were a little *worse* than those obtained using the complicated series of epicycles and other curves constituting the Ptolemaic scheme, at least to within the accuracy available using the measuring instruments of the time. No, the real selling point of the Copernican model was that it was much *simpler* than the competition, yet still gave a reasonably good account of the observational evidence.

The Copernican revolution is a good case study in how to wield Occam's Razor to slit the throat of the competition. When in doubt, take the simplest theory that accounts for the facts. The problem is that it's not always easy to agree on what's "simple." The notion of simplicity, like "truth," "beauty," and "effective process" is an intuitive one, calling for a more objective characterization, i.e., formalization, before we can ever hope to agree about the relative complexities of different theories.

In 1964 Ray Solomonoff, a researcher at the Zator Corporation, published a pioneering article in which he presented a scheme to measure the complexity of a scientific theory objectively. He based his idea on the premise that a theory for a particular phenomenon must encapsulate somehow the empirical data available telling about that phenomenon. Thus, Solomonoff proposed to identify a theory with a computer program that reproduces the empirical data. He then argued that the complexity of such a theory could be taken to be the "length" of the shortest such program, measured perhaps by the number of keystrokes needed to type the program or some similarly unambiguous scheme.

By this definition of the complexity of a theory, Solomonoff was anticipating an observation made later by Gregory Chaitin and by the mathematician and philosopher René Thom, who noted that the point of a scientific theory is to reduce the arbitrariness in the data. If a program (read: theory) reproducing the data has the same length as the data itself, then the theory is basically useless since it contributes nothing toward reducing the arbitrariness in the data. Such a putative theory doesn't in any way "compress" the data, and we could just as well account for the observations by writing them out explicitly. But we don't need a theory to do that. So if a set of observations can only be reproduced by programs of the same length as the observations themselves, then we're justified in calling the observations "random" in the sense that they can be neither predicted nor explained using a set of rules shorter than the data itself.

At about the same time Solomonoff was developing these ideas about the complexity of scientific theories, Gregory Chaitin was enrolled in a computer-programming course being given at Columbia University for bright high-school students. At each lecture the professor would assign the class an exercise requiring writing a program to solve it. The students then competed among themselves to see who could write the shortest program solving the assigned problem. While this spirit of competition undoubtedly added some spice to what were otherwise probably pretty dull programming exercises, Chaitin reports that no one in the class could ever begin to think of how to actually prove that the weekly winner's program was really the shortest possible.

Even after the course ended Chaitin continued to ponder this shortest-program puzzle, eventually seeing how to relate it to a different question: How can we measure the complexity of a number? Is there any way that we can objectively claim π is more complex than, say, $\sqrt{2}$ or 759? Chaitin's answer to this question ultimately led him to one of the most surprising and startling mathematical results of recent times.

In 1965, now an undergraduate at the City University of New York, Chaitin arrived independently at the same bright idea as Solomonoff: Define the complexity of a number to be the length of the shortest program for a UTM that will cause the machine to print out the number. This concept of complexity is closely related to Gödel's results, so let's dig a little deeper into the details and show why.

Since the data of any experiment can be encoded as a set of numbers that can also be expressed as a binary string, there is no real difference between Solomonoff's idea about the complexity of a scientific theory and Chaitin's (and, independently, the Russian mathematician Andrei Kolmogorov's) idea regarding the complexity of a number. In a similar fashion, since we saw above that it's possible to code every program for a UTM in the same way that we code its input data, there is also no real difference in talking about complexity for numbers or for programs. In fact, this is the very essence of Chaitin's definition: the complexity of a number equals the complexity of the shortest program that generates the number.

Using these ideas, we can transfer the notion of a random, or incompressible, data string to numbers, coming up with Chaitin's definition of a random number. A number is *random* if there is no program for calculating the number whose length is shorter than the length of the number itself. Expressed another way, we say a number is random if it is maximally complex. Here, of course, we take the length of a number (program) to be the number of bits in the number's (program's) binary expression.

But do random numbers really exist? The surprising fact is that almost all numbers are random! To see why, let's compute the fraction of numbers of length n having complexity less than, say, $n - 5$. There are at most $1 + 2 + \ldots + 2^{n-5} = 2^{n-4} - 1$ programs of length $n - 5$ or less. Consequently, there are at most this many numbers of length n having complexity less than or equal to $n - 5$. But there are a total of 2^n numbers of length n. Thus, the proportion of these numbers having complexity no greater than $n - 5$ is at most $(2^{n-4} - 1)/2^n \leq \frac{1}{16}$. So we see that less than one number in sixteen can be described by a program that's at least 5 bits shorter than the number. Similarly, less than one number in five hundred has a length 10 or more greater than its shortest program, i.e., its complexity. Using this kind of argument and letting $n \to \infty$, it's fairly easy to prove that the set of real numbers with complexity less than their length forms a vanishingly small subset of the set of all real numbers. In short, almost every real number is of maximal complexity, i.e., random. Now let's get back to the problem of shortest programs.

The starting point for Chaitin's remarkable results is the seemingly innocent query: "What is the smallest number that cannot be expressed in words?" This statement seems to pick out a definite number. Let's call it $\mathcal{U}$ for "unnameable." But thinking about things for a moment, we see that there appears to be something fishy about this labeling. On the one hand, we seem to have just described the number $\mathcal{U}$ in words. But $\mathcal{U}$ is supposed to be the first natural number that **cannot** be described in words! This paradox, first suggested to Bertrand Russell by a certain Mr. Berry, a Cambridge University librarian, plays the same role in Chaitin's thinking about the complexity of numbers and programs as the Epimenides Paradox played in Gödel's thinking about the limitations of formal systems.

Recall that to bypass the issue of formalizing truth, Gödel had to substitute a related notion, provability, and talk about a statement being unprovable within a given formal system. Similarly, the Berry Paradox contains its own unformalizable notion, the concept of denotation between the terms in its statement and numbers. Part of Chaitin's insight was to see that the way around this obstacle was to shift attention to the phrase, "the smallest number not computable by a program of complexity n." This phrase *can* be formalized, specifying a certain computer program for searching out such a number. This formalization leads to

CHAITIN'S THEOREM. *No program of complexity* n *can ever produce a number having complexity greater than* n.

COROLLARY. *A program of complexity* n *can never halt by outputting the number specified by Chaitin's phrase.*

More generally, Chaitin's Theorem shows that even though there clearly exist numbers of all levels of complexity, it's impossible to prove this fact. That is, given any computer program, there always exist numbers having complexity greater than that program can recognize, i.e., generate. In the words of physicist Joseph Ford, "A ten-pound theory can no more generate a twenty-pound theorem than a one-hundred pound pregnant woman can birth a two-hundred-pound child." Speaking somewhat informally, Chaitin's Theorem says that no program can calculate a number more complex than itself. Here's an outline of the proof of this very fundamental result linking computing, complexity, and information.

Suppose we have a binary string that we suspect of having complexity greater than some fixed value N and we want to prove it. Assume such a proof exists. Then we can use a program of length $\log N + K$ to search for this proof, since it takes only $\log N$ symbols to represent a number having magnitude N. Here the quantity K is of fixed size, representing the overhead in the program for things like reading in the number N, communicating with the printer and so on. Consequently, with this program we can search through all proofs of length 1, length 2, and so on until we come to the one that proves that the complexity of a specific number is greater than N. When such a proof is found, the program of length $\log N + K$ will have generated a string of complexity greater than N. But there will always be some number N that is much larger than $\log N + K$, since K is fixed. So, on the one hand we have computed a string of complexity N with a program having a length much shorter than N. On the other hand, we have proved that the string has complexity greater than N. But, by definition, such a string can only be computed using a program of length greater than N—a contradiction. Thus, we conclude sadly that there is no such proof.

The implication of this result is that for sufficiently large numbers N (those bigger than $\log N + K$), it cannot be proved that a particular string has complexity greater than N. Or, equivalently, there exists a number N such that no number whose binary string is longer than N can be proved to be random. Nevertheless, we know that almost every number is maximally complex, hence random. We just can't prove that any *particular* number is random. The difficulty is that in a random sequence each digit carries positive information since it can't be predicted from its predecessors. Consequently, an infinite random sequence contains more information than all our finite human systems of logic put together. Hence, verifying the randomness of such a sequence lies beyond the powers of constructive proof. Looking at the problem in another way, in order to write down an "arbitrarily long" binary string we need to give a general rule for the entries of the string. But then this rule is shorter than suitably large sections of the string. So the string can't be random, after all! As one might suspect by now, this result is deeply intertwined with the decision problems considered earlier.

To make this connection, consider a formal system whose axioms can be expressed in a binary string of length N. Chaitin's Theorem then says that there is a program of size N that does not halt—but we can't prove that fact within this axiomatic system. On the other hand, this formal system does allow us to determine exactly which programs of size less than N halt and which do not. So Chaitin's result gives us another solution of the Halting Problem, since it says that there always exist programs for which we cannot determine in advance whether or not they will stop. And, of course, Chaitin's Theorem offers another perspective on Gödel:

> ### Gödel's Theorem—Complexity Version
> *There exist numbers having complexity greater than any theory of mathematics can prove*

So if we have some theory of mathematics, i.e., a formal system, there always exists a number t such that our theory cannot prove that there are numbers having complexity greater than t. Nevertheless, by "jootsing," we can see that such strings must exist. To generate one, simply toss a coin a bit more than t times, writing down a "1" when a head turns up and a "0" for a tail. The next section shows how Chaitin's work led him to make the startling claim that, "Arithmetic is random."

Exercises

1. Show that a sequence that's random by the Chaitin-Kolmogorov definition must have an approximately equal number of 0's and 1's.

2. Write out the shortest program you can think of that will produce the sequence $01001000100001\cdots1$ up to n repetitions. Use this program to obtain an upper bound for the randomness of this sequence.

3. Discuss whether the concepts of algorithmic information theory help us to define mathematically the notion of biological complexity?

4. Are the numbers π and e random? What about $\sqrt{2}$? What is the connection between a number being random and a number being computable?

5. Fill-in the details showing how Gödel's Theorem is a corollary of Chaitin's Theorem.

7. The Randomness of Arithmetic

If asked to name the Top Ten Theorems of All Time, just about every mathematician would reserve a place somewhere on the list for the Pythagorean Theorem, stating that if a and b are the lengths of the short sides of a right triangle, c being the length of the hypotenuse, then $a^2 + b^2 = c^2$. This is an example of a polynomial equation with integer coefficients in three variables a, b and c. Of special mathematical interest are the so-called *Diophantine equations,* which are polynomial equations having integer coefficients for which we seek only integer solutions. Thus the term "Diophantine" refers more to the character of the set of solutions we're looking for than it does to the equation itself.

The number of solutions of a given polynomial equation may vary from finite to infinite, depending upon whether or not we think of it as a Diophantine equation. For example, the Pythagorean equation has an infinite number of both real and integer solutions. On the other hand, the equation $a^2 + b^2 = 4$ has only the four integer solutions $a = \pm 2$, $b = 0$ or $a = 0$, $b = \pm 2$, but an infinite number of real solutions (e.g., take a to be any real number between -2 and $+2$, with $b = \sqrt{4 - a^2}$). So, regarded as a Diophantine equation, this equation has a finite solution set. But thought of as a general polynomial equation, it has an infinite number of solutions. Our concern with Diophantine equations here comes from the surprising connection between the nature of the set of solutions to Diophantine equations and the Halting Problem for a UTM.

In a famous lecture to the 1900 International Mathematical Congress in Paris, Hilbert outlined a set of problems for the coming century. The tenth problem on his list involved Diophantine equations. What Hilbert was looking for was a general algorithm enabling us to decide whether or not any given Diophantine equation has a solution. Note carefully that Hilbert did not ask for a procedure to decide whether the solution set is infinite; only for an algorithm to determine if there is *any* solution.

It turns out that there exists an algorithm for listing the set of solutions to any Diophantine equation. So, in principle all we have to do in order to decide if the solution set is empty is to run this program, stopping the listing procedure if no solution turns up. The difficulty, of course, is that it may take a very long time (like forever!) to decide whether or not a solution will appear. For instance, the first integer solution of the simple-looking Diophantine equation $x^2 - 991y^2 - 1 = 0$ is $x = 379516400906811930638014896080$, $y = 12055735790331359447442538767$. How long would you be willing to punch keys on a calculator waiting for that pair to pop up? This example shows that to solve Hilbert's Tenth Problem we can't rely upon a brute force search for the first solution. A solution might not exist, or it might be so large that we'll get tired of looking. In either case, a direct search gives no guarantee of ever coming up with the correct answer about a particular equation's solvability. We need to do something a bit more clever.

In our discussion of Turing machines we introduced the idea of a computable number as being one whose digits can be successively calculated by some UTM. This idea can be extended to sets of integers by saying that a set is computable if, given any integer in the set as input, the program prints a 1 and halts. But if the given number is not in the set, the program prints a 0 and stops. It turns out that this notion is a bit too strong for many purposes, and it's convenient to introduce a weaker version. We say that a set of integers is *listable* if there is a program that, given any integer as input, prints a 1 and stops if the integer is in the set. But if the integer is not in the set, the program may print a 0 and halt or it may not stop at all. So the difference between a set being computable and its being listable is that if the set is listable, the program may or may not halt. But the program always stops when the set is computable. Obviously, computable sets are listable—but not conversely. This distinction forms the basis for an attack mounted on Hilbert's Tenth Problem by the well-known mathematical logician Martin Davis. Let's look at his strategy.

Davis's idea was to prove that for every listable set of integers S, there is a corresponding polynomial $P_S(k, y_1, \ldots, y_n)$ with integer coefficients, such that a positive integer k^* belongs to the set S if and only if the solution set of the Diophantine equation $P_S(k^*, y_1, \ldots, y_n) = 0$ is not empty, i.e., the equation has at least one solution in integers. In short, the solvability or unsolvability of the equation $P_S(k, y_1, \ldots, y_n) = 0$ serves as a decision procedure for membership in the listable set S. Here we subscript the polynomial with a small S to indicate that there may be a different polynomial for each listable set S. Davis showed that Hilbert's Tenth Problem can be resolved negatively if such a polynomial can be found for every listable set of integers. But what's the connection between a particular Diophantine equation having a solution and the existence of a general algorithm for the solvability of such equations? The connection is worth examining in detail, since it begins to reveal for us the interrelationship between the Tenth Problem and the Halting Problem.

The logical chain of reasoning underwriting Davis's approach to the Tenth Problem is composed of the following links:

A. Suppose there were a Diophantine decision algorithm of the type that Hilbert was looking for, and let S be some listable, but not computable, set of integers.

B. Then by the assumed existence of the algorithm there is a Turing machine program (call it $\mathcal{D}$ for Diophantine) which, given the integer k^* as input, halts with output 1 if the Diophantine equation $P_S(k^*, y_1, \ldots, y_n) = 0$ has a solution, and halts with output 0 if there is no solution.

C. But the relationship between S and P_S implies that the existence of such a program $\mathcal{D}$ means that S is computable, since $\mathcal{D}$ definitely stops with a 0 or a 1 as output.

D. But this contradicts the assumption that the set S is not computable. Hence, no such program $\mathcal{D}$ can exist. That is, there is no algorithm of the sort sought by Hilbert and the Tenth Problem is settled in the negative.

Unfortunately, Davis was unable to prove the existence of such a polynomial P_S for every listable set S. However, later work by Davis, Julia Robinson, and Hilary Putnam showed that if there were even one Diophantine equation whose solutions grew at an exponentially-increasing rate in just the right way, then Davis's polynomial P_S would have to exist. In 1970 Yuri Matyasevich, a 22 year-old mathematician at the Steklov Institute of

Mathematics in St. Petersburg, found just such a Diophantine equation. Amusingly, Matyasevich made crucial use of the famous Fibonacci sequence of numbers in constructing the long-sought equation, a sequence originally introduced by Leonardo of Pisa in 1202 to explain the explosive growth of a rabbit population in the wild. Evidently, the well-known procreation habits of rabbits gives rise to just the kind of rapid growth Matyasevich needed to resolve negatively yet another of Hilbert's conjectures.

An interesting corollary of Matyasevich's proof is that for any listable set of natural numbers S, there exists a polynomial $P_S(y_1,\ldots,y_n)$ with integer coefficients such that as the variables $y_1,\ldots,y_n$ range over the nonnegative integers, the positive values of the polynomial are exactly the set S. To illustrate this curious result, such a polynomial involving 26 variables whose positive values are the set of prime numbers is given in Discussion Question 27.

By now the reader should be highly sensitized to the connection between negative solutions to decision problems and Gödel's Theorem. So before continuing our pursuit of the connection between Diophantine equations, the Halting Problem and arithmetic, let's pause to again give Gödel his due.

> **Gödel's Theorem—Diophantine Equation Version**
> *There exists a Diophantine equation having no solution—but no theory of mathematics can prove its unsolvability*

Suppose we have a UTM and consider the set of all possible programs that can be run on this machine. As we already know, every such program can be labeled by a unique string of 0's and 1's, so it's possible to "name" each program by its own unique positive integer. Consequently, it makes sense to consider listing the programs one after the other and to talk about the kth program on the list, where k can be any positive integer. Now consider the question: If we pick a program from the list at random, what is the likelihood that it will halt when run on the UTM? It turns out that this question is intimately tied up with the solvability of Diophantine equations, leading ultimately to Chaitin's remarkable result about the randomness of arithmetic.

The work of Davis, Robinson, and Putnam on Hilbert's Tenth Problem showed that the solvability of decision problems can be expressed as assertions about the solvability of certain Diophantine equations. In particular, there is a Diophantine equation $P(k,y_1,\ldots,y_n) = 0$ such that this equation has a solution if and only if the kth computer program halts when run on a UTM.

The key step in Chaitin's route to ultimate randomness was not to consider whether a Diophantine equation has *some* solution, but to look at the sharper question of whether or not the equation has a finite or an infinite number of solutions. The reason for asking this more detailed question is that the answers to the original query are not logically independent for different values of k. So if we know whether a solution exists or not for a particular value of k, this information can be used to infer the answer for some other values of k. But if we ask whether or not there are an infinite number of solutions, the answers are logically independent for each value of k; knowledge of the size of the solution set for one value of k gives no information at all about the answer to the same question for another value.

Following this reformulation of the basic question, Chaitin's next step was a real *tour de force*. He proceeded to construct explicitly a Diophantine equation involving an integer parameter k, together with over 17,000 additional variables. Let's call this Diophantine equation

$$\chi(k, y_1, y_2, \dots, y_{17,000+}) = 0,$$

using the Greek symbol χ ("chi") in Chaitin's honor. From this equation we can form a binary number consisting of an infinite string of 0's and 1's in the following manner: Let k run through the values $k = 1, 2, \dots$. We set the kth entry in the binary string to 1 if the equation $\chi = 0$ has an infinite number of solutions for that value of k, and set it to 0 if the equation has a finite number of solutions (including no solution). Chaitin labeled the number created in this way by the last letter in the Greek alphabet Ω ("Omega"). And for good reason, too. The properties of Ω show that it's about as good an approximation to "The End" as the human mind will ever make.

Chaitin proved that the quantity Ω is an uncomputable number. Furthermore, he showed that any formal system described in N bits (read: program of length/complexity N) can yield at most a finite number N of the binary digits of Ω. Consequently, Ω is random, since there is no program shorter than Ω itself for producing all of its digits. Moreover, by the method of construction, the digits of Ω are independent, both statistically and logically. Finally, if we put a decimal point in front of Ω, it represents some decimal number between 0 and 1. And, in fact, when viewed this way Ω can be interpreted as the probability that a randomly-selected Turing machine will halt. Indeed, Chaitin's equation was constructed precisely so that Ω would turn out to be this halting probability.

So while Turing considered the question of whether a given program would halt, Chaitin's extension produces the probability that a randomly chosen program will stop. As an aside, it's worth noting that the two extremes Ω equals zero or one cannot occur, since the first case would mean that no program ever halts, while the second would say that every program will halt. The trivial, but admissible, program "STOP" deals with the first case, while I'll leave it as an Exercise to construct an equally primitive program to deal with the second.

But the real bombshell that Chaitin dropped on the world of mathematics is to prove that the structure and properties of Ω show that arithmetic is fundamentally random. To see why, take the integer k to be finite, but "sufficiently large." For example, let k be greater than the Busy Beaver function value $BB(12)$ considered earlier. For values of k larger than this, there is no way to determine whether the kth entry of Ω is 0 or 1. And there are an infinite number of such undecidable entries. Yet each such entry corresponds to a simple, definite arithmetical fact: For that value of k, either Chaitin's Diophantine equation $\chi = 0$ has a finite or an infinite number of solutions. But as far as human or computer reasoning goes, which of the two possibilities is actually the case may as well be decided by flipping a coin. It is completely undecidable; hence, effectively random.

So Chaitin's work shows that there are an infinite number of arithmetic questions with definite answers that cannot be found using any kind formal axiomatic procedures. The answers to these questions are uncomputable and are not reducible to other mathematical facts. Extending Einstein's famous aphorism about God, dice, and the universe, Chaitin describes the situation by saying that, "God not only plays dice in quantum mechanics, but even with the whole numbers." It's fitting to conclude this section with another tribute to Gödel:

Gödel's Theorem—Dice-Throwing Version

There exists an uncomputable number Ω whose digits correspond to an infinite number of effectively random arithmetic facts

---

The work of Turing, Gödel and Chaitin has given us a wealth of information about what can and can't be done by following a set of rules. But this kind of computability is a pretty black-and-white affair, as can be seen by considering computation of the computable number π. If I ask you for the zillionth digit of π, our definition of computability ensures that there is a program that will eventually produce this quantity. But don't hold your breath waiting for it to pop up on your terminal screen. The zillionth digit of π is only *theoretically* computable, and it's not this kind of computability that the practical man or woman of affairs is interested in when s/he speaks about something being "computable." From a practical point of view, it's the computer time and memory needed to calculate a given quantity that is the binding constraint, not whether the quantity is theoretically computable or not. So for the next few sections our focus will be upon practicalities rather than hypotheticalities, as we examine the difference between the words "easy" and "hard" in the world of computing.

Exercises

1. The negative solution of Hilbert's Tenth Problem shows that there is no algorithm that will decide whether or not a polynomial equation has a solution in integers. (a) Does the same result hold when we seek the solution in complex numbers? (b) Or in the real numbers?

2. Construct a primitive program that never halts, thereby showing that Chaitin's number $\Omega \neq 1$.

8. P and NP: The "Easy" and the "Impossible"

A famous problem of recreational mathematics is the so-called *Tower of Hanoi,* shown in Fig. 9.6 for the case of $N = 3$ rings. In this problem, there are three pegs A, B and C. On the first peg A there are piled N rings whose radii are decreasing. The other two pegs are initially empty. The task is to transfer the rings from A to B, perhaps using peg C in the process. The rules stipulate that the rings are to be moved one at a time, and that a larger ring can never be placed upon a smaller one.

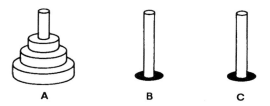

Figure 9.6. The Tower of Hanoi Game

For the case of three rings, it's not too hard to see that the sequence of seven moves

$$A \to B, \quad A \to C, \quad B \to C, \quad A \to B, \quad C \to A, \quad C \to B, \quad A \to B$$

achieves the desired transfer of rings. And, in fact, it can be shown that there is a general algorithm, i.e., a program, solving the game for any number of rings N. This program shows that the minimal number of transfers required is $2^N - 1$. Amusingly, the original version of this puzzle, dating back to ancient Tibet, involves $N = 64$ rings. So it's not hard to see why the Tibetan monks who originated the game claim that the world will end when all 64 rings are correctly piled on peg B. To carry out the required $2^{64} - 1$ steps, even performing one ring transfer every ten seconds, would take well over five *trillion* years to finish the job! Thus, the number of steps needed for the solution of the Tower of Hanoi problem grows exponentially with the number of rings N. This is an example of a "hard" computational

problem—one in which the number of computational steps needed to obtain
a solution increases exponentially with the "size" of the problem.

By way of contrast to the Tower of Hanoi, let's look at an "easy" problem. Consider the problem of solving the set of linear algebraic equations

$$Ax = b, \qquad x, b \in R^N, \qquad A \in R^{N \times N}.$$

Here we measure the size of the problem by the parameter N, the dimension
of the space in which we seek the solution vector x. The Gaussian elimination procedure shows that the solution x can be found in $O(N^3)$ numerical
operations. Thus, the number of computations required to solve a set of
N linear algebraic equations grows only as a polynomial function of the size
of the problem. So if we double the problem size N, the amount of work
increases only by a factor of eight. Such a problem is said to be *polynomially computable* or to "belong to P" (for polynomial). This is what we
mean when we speak of a problem being "easy." Happily, there are a lot
of problems that belong to P: Finding shortest paths or maximal flows in
graphs, sorting names, multiplying matrices and so on.

Note that if a problem belongs to class P it does not necessarily mean
that there is a practical algorithm for computing its solution. It may be that
the coefficients of the polynomial bound are very large or that the degree of
the polynomial is very high. But it's at least a step in the right direction.
On the other hand, if a problem is not in P then we're going to have to
exploit some special structure in the problem or bring some other factors to
bear (like blind luck!) if we hope to be able to solve it for sufficiently large
values of the problem size N.

There is another very important class of problems known as "NP". This
terminology stands for "nondeterministic polynomial time," which does **not**
mean that there is something fuzzy, random or indeterminate about the
problem. Rather, a problem x is in NP if it's possible to verify a proposed
solution of x in a number of steps that grows polynomially in the problem
size. So if you happen to stumble onto what you think is a solution for
a problem in NP, you can verify or refute that it is indeed a solution in
polynomial time. Clearly, $P \subset NP$.

A famous example of an NP problem is the Traveling Salesman Problem, in which a salesman has to plan the shortest route of no greater than,
say, 5,000 miles that takes him, for example, to 60 cities. Furthermore, he
is constrained to pass through each city once and only once. One line of
attack is simply to try all possible routes, looking for the shortest one. Curiously enough, although this brute-force approach is cumbersome and far
from clever, no one has ever found a significantly better way that will work
for an arbitrary number of cities and that will always give the correct answer
to the problem. Unfortunately, if there are $N = 60$ cities, this approach requires more than 10^{79} steps. However, it's easy to see that if I give you a

proposed route, it will take you only 60, i.e., $O(N)$, additions to see if the distance to be traveled is indeed less than 5,000 miles.

There are many other problems in NP for which it's unknown whether or not they are also in P. Examples include: (i) Whether N linear constraints on N variables have a solution in integers, (ii) whether a graph with N vertices contains a clique (a complete subgraph) of at least $N/2$ vertices, and (iii) whether a graph with N vertices has a Hamiltonian circuit (a path that visits each vertex exactly once). While as yet we have no ironclad proof showing that $P \neq NP$, most computer scientists would be shocked into a state of total disbelief if that turned out not to be the case. One of the main reasons is that a very large number of NP problems have all been shown to be equivalent in the sense that if one of them turns out to be in P, then they are all in P. This is a consequence of what's called *Cook's Theorem*, which we'll take up in a moment.

As hard as NP problems are, they are only the second step in a hierarchy of hard problems. The next step up the ladder is the set of problems having the general form: "For all ... does there exists ... ? An example of this kind of problem is: "For all ways of removing half the edges of a graph, is there a Hamiltonian circuit that remains?" The fourth step of the hierarchy involves problems of the form, "For all ... does there exist ... for all ... does there exist ... ? The hierarchy goes on in this way, topping out with the set of problems that mathematicians call *polynomial space*.

A good way to envision this polynomial hierarchy is to think of it as consisting of games with alternating moves. Problems belonging to P are like solitaire. There is only one player and the outcome is fixed. The third level in the hierarchy is like a game in which you and your opponent each have one move. Polynomial space itself can also be seen as a game, but a game with no fixed number of moves. Arguing by analogy with board games, polynomial space is analogous to games like chess and Go that have an open-ended number of moves and for which the outcome is not determined in advance, i.e., neither player has a guaranteed winning strategy.

The question that plagues computer scientists is whether this polynomial hierarchy is real or illusory. If the polynomial hierarchy is real, the best that can be hoped for is to find approximate solutions to the hard problems. But if the classes somehow collapse onto each other, then all the problems in the hierarchy are really solvable quickly; in polynomial time, in fact. But if there is at least some true distinction between classes, where does it stop? And, finally, is polynomial space different from the rest of the hierarchy?

The first step in trying to answer these questions is to ask whether $P = NP$. One promising approach that logicians dreamt up involves what's called an "oracle." To illustrate, suppose there were some all-knowing entity that could answer one particular type of question for you. For instance, the oracle might have a list of every type of route involving N cities that is less than 5,000 miles in length. Then if you wanted to know if your proposed 60-city tour will work, all you need do is consult the oracle to find out if your tour is on its list. Suppose we have a problem in NP such that if we use it as an "oracle" for any other NP problem, that NP problem will become solvable in polynomial time. Such a set of NP problems that can be used as oracles in this way is termed *NP-complete.* It turns out that a problem is NP-complete if it is: (1) in NP, and (2) equivalent to a generic NP problem called *satisfiability,* which we'll take up in just a moment. The importance of this class of problems is that if even one NP-complete problem can be shown to actually be in P, we would necessarily have $NP \subset P$. But this then implies that $NP = P$. So it looked for awhile as if the NP-completeness problem could be settled by finding the right kind of oracles.

But the line of attack on the NP-completeness question using oracles had a rather surprising outcome. Work by different research groups showed that there do indeed exist oracles that make NP and P coalesce. On the other hand, there are other oracles that separate P and NP completely! Despite these shocking results, computer scientist still felt that if the polynomial hierarchy really is a hierarchy, then there ought to be an oracle that can separate all the classes of problems. This led to the recent work in which the polynomial hierarchy is expressed in terms of Boolean circuits. Since this work takes us to the edge of the frontier in the computational complexity business, let's take a couple of pages to describe it.

A *Boolean circuit* is a directed graph without cycles that has one sink (the output node) and several sources. We shall label the sources nodes with the variables x_i, $i = 1, 2, \ldots, n$, while the internal nodes are labeled with $\vee$ ("or"), $\wedge$ ("and") or $\neg$ ("negation"). By letting x_i take the value of the ith coordinate of a Boolean-valued function $f(x_1, x_2, \ldots, x_n)$, we can use the circuit to compute the values of f. Note that there is no loss of generality here in considering only functions taking on the values 0 or 1, since more general functions can always be represented as a collection of such Boolean functions. A typical example of this kind of Boolean circuit is shown in Fig. 9.7.

A key point of contact between the polynomial hierarchy and Boolean circuits is the so-called *satisfiability problem (SAT).* Suppose we are given a Boolean logical expression involving the variables $\{x_i\}$ and their negations. For example, we might have the expression

$$(x_1 + \bar{x}_3 + x_4)(\bar{x}_1 + \bar{x}_2 + \bar{x}_4)(\bar{x}_2 + x_3)(\bar{x}_1 + x_2 + x_4).$$

The problem is to find an assignment of truth values (0 or 1) to the Boolean variables x_1, x_2, x_3 and x_4 so that this expression is true (has the value 1). This means that each of the clauses enclosed within the parentheses must themselves be true. After a bit of fumbling around, it can be seen that

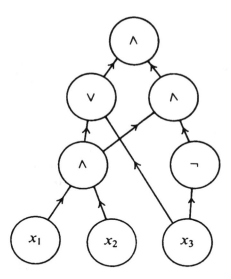

Figure 9.7. A Boolean Circuit

the choice $x_1 = 0$, $x_2 = 1$, $x_3 = 1$, $x_4 = 1$ "satisfies" the expression. It's clear that a brute-force assault on an expression involving n variables could involve trying all 2^n possible combination of truth values for the variables. The number of computations with this "algorithm" clearly grows at an exponential rate with the size of the problem. The satisfiability problem then is to ask if there exists a polynomial-time algorithm that will decide the truth or falsity of every such Boolean expression.

The satisfiability problem also enters in a central way into the question of whether or not $P = NP$ via what has come to be called Cook's Theorem, perhaps the most basic result in computational complexity theory. This surprising result was discovered in 1970 by Stephen Cook as part of his doctoral work in computer science at Berkeley. A very rough statement of the main conclusion is as follows:

COOK'S THEOREM. *There exists a generic polynomial-time transformation taking any problem in NP to a particular instance of the satisfiability problem.*

To illustrate, consider again the traveling-salesman problem (TSP). Cook's transformation converts the TSP to an instance of SAT in such a way that: (i) the transformation is in P, and (ii) the TSP has a solution in P if and only if the corresponding SAT does. The importance of Cook's Theorem is to show that SAT is as hard as any other problem in NP. So if we can find a satisfiability algorithm in P, then we can use Cook's transformation to solve any other NP problem in polynomial time. In other words, $P = NP$. Several thousand NP-complete problems are now known, all of which would be solvable in polynomial time if there were a polynomial algorithm for SAT, and conversely. But since an army of workers have labored mightily without producing a polynomial algorithm for a single one of these problems, the empirical evidence is by now overwhelming great that $P \neq NP$.

The connection between SAT and a Boolean circuit is fairly clear now, I think. If we consider the source nodes of the circuit as being associated with the variables of our Boolean expression, we are asking if there is any input that will cause the circuit to output the value 1. Moreover, by letting the value of a node in the circuit correspond to the content of a particular square on the Turing machine tape, it's not hard to prove that any function computable by a Turing machine in time $T(n)$ can also be computed by a circuit (or family of circuits) of size $O(T(n) \log T(n))$. Here, of course, n is the number of variables in the function f. This fact allows us to obtain lower bounds on the running time of a Turing machine program if we can establish lower bounds on the size of the circuits. By this kind of transference of the setting, we can replace analysis of the behavior of a dynamic object (the Turing machine) by the properties of a static one (the circuit). The idea is that if we can prove that the circuit has to be really big for NP but not so big for P, then we will have proved that $P \neq NP$. So with this equivalence between Turing machines, SAT and Boolean circuits, let's look at some of the mathematical details surrounding these circuits.

The key parameters of a circuit are its size (the number of nodes) and its depth (the length of the longest path from the input to the output). So, for example, the circuit in Fig. 9.7 has size 8 and depth 3.

Unfortunately, difficulties in proving lower bounds on the size of general circuits has led to the study of restricted models. To give some feel for the kinds of results available, here we'll consider just the restriction to circuits of small depth. In other words, suppose the circuits can grow as wide as we like, but that their depth is limited. Other kinds of restrictions are treated in the literature cited in the Notes and References.

In looking at circuits of small depth, it's convenient to consider what's called the *parity function,*

$$f(x_1, x_2, \ldots, x_n) = \sum_{i=1}^{n} x_i \mod 2.$$

The basic result is the

PARITY FUNCTION THEOREM. *The size of a circuit of depth k computing the parity function of n inputs is no smaller than $2^{c_k n^{1/(k-1)}}$ for sufficiently large n, where $c_k = 10^{-k/(k-1)}$.*

Thus, we see that if the depth is restricted, the width of the circuit blows up as $n \to \infty$. In other words, the circuit is bigger than any polynomial.

Powerful and interesting as this result is, even in this restricted circumstance it's still not enough to show that the polynomial hierarchy is distinct. But in 1985 Andrew Yao established the fact that the width of the Boolean circuit for the parity function grows exponentially with the size of the problem. As a result, there must exist an oracle for which the polynomial hierarchy is distinct. According to workers in the field, this result opens up for the first time the possibility of proving that the polynomial hierarchy may be distinct without having to invoke the controversial aid of an oracle. Readers interested in more details of these fascinating matters are invited to consult the original papers cited in the Notes and References.

Computability dealt with the black-or-white issue of whether a function (or number) can be calculated—even in principle. By introducing the polynomial hierarchy, we divide the class of computable functions into those that are *practically* computable and those that aren't. Useful as this division is, it still deals with the matter of worst-case analysis of general functions. But in the everyday world of numerical analysis and computation, we don't compute "general functions." Nor, by definition, do we typically face the worst case. So from a **really** practical point of view, what we're concerned with is the question of how hard it is to compute the solution to the average, or typical, case. This leads us into a quite different corner of computational complexity jungle.

Exercises

1. Order the following functions from fastest to slowest with respect to their rate of growth: 2^n, $n \log n$, n^2, $n!$, n^n, $5n$.

2. The number of permutations of n letters that have no fixed points is given by the function

$$f(n) = \sum_{j=0}^{n} (-1)^j \frac{n!}{j!}.$$

(a) Is $f(n)$ in P? (b) For a fixed value of n, can we calculate the quantity $f(n)$ in polynomial time?

3. Suppose you are given a polynomial $p(z)$ of degree n. Von Schubert's method for factoring p proceeds as follows: First, calculate the numbers $p(1), p(2), \ldots, p(n)$. Next, factor these numbers. Let $d(1), d(2), \ldots, d(n)$ be a particular sequence of divisors of $p(1), \ldots, p(n)$. Then the $d(i)$ define a potential factor of p, one that can be found by interpolation. (a) Prove that this algorithm requires at least 2^n steps to show that a polynomial of degree n is irreducible. Thus, von Schubert's method is not in class P. (b) Is it in class NP?

4. Linear programming problems involve a set of linear constraints like $x \geq 0$, $Ax \leq b$, where $x \in R^n$, $b \in R^m$ and $A \in R^{m \times n}$. (a) Show that to check whether a given $x \in R^n$ satisfies these constraints is a problem in class NP. (b) Is it NP-complete?

5. Display an explicit solution in fifteen moves for the Tower of Hanoi game with $N = 4$ rings.

9. Numerical Analysis, Complexity and Information

Reduced to its basics, most of life comes down to pursuit of the elusive relationship between problems and methods for their solution. In particular, in science the preferred solution method is the algorithm. Here we want to consider both sides of this coin from the perspective of computational complexity theory. Our interest lies in questions like: "Are there some problems that are just intrinsically more complex than others, irrespective of the choice of algorithm?" and "Are there algorithms that are inherently superior to others for a given class of problems?" These are the kinds of matters we'll consider for the next several pages. But to say anything really worth saying on these cosmic issues, it's imperative to restrict the setting somewhat. So in this section we shall confine our attention to the world of the numerical analyst. Let's take a quick look at some of the mountaintops and valleys making up that territory.

The world of the numerical analyst is a world of specifics and practicalities. Here's an example of the kind of problem a resident of this world must regularly confront to earn his or her daily bread. Consider the problem of finding the roots of the equation $f(z) - \omega = 0$, where z and ω are complex numbers. Define the map $G_\omega \colon S \to S$, where S is the Riemann sphere $S = \mathbb{C} \cup \infty$, by the rule

$$G_\omega = z + \frac{\omega - f(z)}{f'(z)}.$$

Thus, the map G_0 is just Newton's method for finding the roots of the polynomial $f(z)$.

Now consider the numerical analyst's problem: Given a quantity ϵ between 0 and 1, as well as a complex polynomial $f(z) = \sum_{i=0}^{n} a_i z^i$, find $z \in \mathbb{C}$ such that $|f(z)| < \epsilon$. To make things simple, assume that we have made a preliminary transformation so that $a_n = 1$ and that coefficients of f all satisfy the bound $|a_i| \leq 1$. So we're not concerned here with finding the exact root of f, but rather with finding an *approximate* root accurate to within an error level ϵ.

Suppose we use the map G_ω to define the following iterative algorithm (due to Stephen Smale):

0. Set $K = 98$.

1. Randomly choose a $z_0 \in \mathbb{C}$ such that $|z_0| = 3$.

2. Define N to be the smallest integer such that:

    a. $N > K(n \log 3 + |\log \epsilon|)$,

    b. $z_i = G_\omega(z_{i-1})$, $\omega_i = M^i f(z_0)$, $i = 1, 2, \ldots, n$, where $M = 1 - (1/K)$.

3. If $|f(z_N)| < \epsilon$, stop; otherwise, go to step 1.

The original question now reduces to asking: On the average, how many steps does it take to find an ϵ-root of the equation using the above algorithm? Note that now we speak of the *average* case, since we could have remarkably bad (or good) luck in our choice of starting point z_0. The answer to this question was given by Smale in the following

ROOT-FINDING THEOREM. *On the average, six steps of the above algorithm suffices to find an ϵ-root for any $0 < \epsilon < 1$ and any polynomial f (normalized as above).*

It's interesting to note that the case $f(z) = z^n$ shows that we can't hope to do better with Newton's method than having the number of steps being linear in the degree of f. Now let's look at another familiar example of this sort from elementary calculus.

The most common methods discussed in introductory calculus courses for computing the integral of a continuous function f are the Riemann method, the trapezoidal rule and Simpson's rule. Specifically, if $f(x)$ is a continuous function on the interval $0 \le x \le 1$, these three rules define real-valued mappings R_h, T_h and S_h on the continuous functions on $[0, 1]$ given by:

Riemann integral:
$$R_h(f) = h \sum_{i=1}^{n} f(ih),$$

trapezoidal rule:
$$T_h(f) = h \left[-(f(1) + f(0))/2 + \sum_{i=1}^{n} f(ih) \right],$$

Simpson's rule:
$$S_h(f) = \frac{h}{3} \left[f(1) + f(0) + 2 \sum_{i=1}^{2n-1} f(ih) \right.$$
$$\left. + 2 \sum_{i=1}^{n} f((2i-1)h) \right]$$

The quantity h in these formulas is the discretization size, so that $h = 1/n$ in the first two methods, while $h = 1/2n$ for Simpson's rule.

Here we'll be concerned with the average cost of computing an approximation to f using these algorithms. For notational simplicity, let's define the mapping J from $C[0, 1] \to \mathbb{R}$ by

$$J(f) = \int_0^1 f(x)\, dx.$$

Our interest will be in the average properties of the error functions

$$\epsilon_R(h,\, f) = |J(f) - R_h(f)|,$$
$$\epsilon_T(h,\, f) = |J(f) - T_h(f)|,$$
$$\epsilon_S(h,\, f) = |J(f) - S_h(f)|.$$

Of course, one crucial property of these quantities is that they all tend to zero as the discretization step size $h \to 0$. This follows from elementary results in numerical analysis. But since our interest is in averaging these functions, we need a probability measure on a space functions. It turns out that the right general setting in which to do this is a Hilbert space of functions, using what's termed the Gaussian measure on this space of functions. It would take us a bit too far afield here to spell out exactly how this averaging process goes, so we refer the reader to the literature cited in the Notes and References for the gory details.

In the context of numerical quadrature, the right Hilbert spaces of functions are the so-called Sobolev spaces $\mathcal{H}^1$ and $\mathcal{H}^2$, which are defined as:

$$\mathcal{H}^1 = \left\{ f \in C^0[0,\,1] : \int |f'|^2 < \infty \right\},$$
$$\mathcal{H}^2 = \left\{ f \in C^1[0,\,1] : \int |f''|^2 < \infty \right\}.$$

Using the Gaussian measure mentioned above, we can now average the error functions above, obtaining

$$\epsilon_R^k(h) = \mathop{\mathcal{E}}_{f \in \mathcal{H}^k} \epsilon_R(h,\, f),$$

for the Riemann method, with similar definitions for the average errors $\epsilon_T^k(h,\, f)$ and $\epsilon_S^k(h,\, f)$ of the other methods.

Stephen Smale has proved the following result linking these average error functions:

NUMERICAL QUADRATURE THEOREM. *(i) For $\mathcal{H}^1$ functions, the average error for the Riemann method is*

$$\epsilon_R^1(h) = h\sqrt{\frac{2}{3\pi}}, \qquad h = 1/n.$$

Moreover, we have

$$\epsilon_T^1(h) = \frac{1}{2}\epsilon_R^1(h), \qquad h = 1/n.$$

(ii) For $\mathcal{H}^2$ functions, we have

$$\epsilon_R^2(h) = h\sqrt{\frac{1}{3\pi}}\left(1 + \frac{h}{2} + \frac{h^2}{10}\right)^{1/2},$$

$$\epsilon_S^2(h) = \frac{h^2}{3}\sqrt{\frac{2}{15\pi}}.$$

Here $h = 1/n$ for Riemann approximation, while $h = 1/2n$ for Simpson's rule. Part (i) shows that to obtain the same accuracy, the Riemann method is twice as expensive as the trapezoidal rule. Part (ii) tells us that, on the average, for $\mathcal{H}^2$ functions Simpson's rule requires far fewer calculations than Riemann integration for the same level of accuracy.

The foregoing results emphasize the level of difficulty one may expect to encounter in computing a good approximation to the solution of the mythical "average" problem using different types of algorithms. On intuitive grounds, it's certainly reasonable to suppose that this difficulty is somehow associated with the complexity of the solution process. So, for example, on the basis of the Numerical Quadrature Theorem we would probably conclude that Riemann approximation is twice as "complex" as the trapezoidal rule. But what if our interest is not on comparing classes of algorithms? What if our interest is in problems instead of methods? That is, suppose we ask if there are some classes of problem that are just plain more complex than others, regardless of what particular algorithm we use for the solution. Is there anything of mathematical and computational interest that can be said about this question? It turns out that there's a lot to be said. But in order to say it we're going to have to shift our emphasis, paying less attention to the properties of methods and more attention to the information content of the problem itself.

Let's go back to the problem of approximating the definite integral

$$J(f) = \int_0^1 f(t)\, dt.$$

To make things interesting, assume the function f has no closed-form anti-derivative, so that all we have at our disposal with which to compute an approximation to $J(f)$ is n values of the integrand f. Let

$$N(f) = [f(t_1), f(t_2), \ldots, f(t_n)]$$

denote the *information pattern* available about the function f. To sprinkle a little analytic structure onto the situation, assume that $f \in C^r[0, 1]$ and that the derivatives of f are uniformly bounded by, say, 1. Thus, we let

$$F = \left\{ f \in C^r[0, 1] : |f^{(k)}(t)| \leq 1, \, k = 0, 1, \ldots r \right\}.$$

Our goal is to compute an ϵ-approximation $U(f)$ such that

$$|J(f) - U(f)| \leq \epsilon,$$

for all $f \in F$, where $\epsilon > 0$. Of course, the approximation $U(f)$ is generated by performing various mathematical operations on the information pattern $N(f)$, i.e., $U(f) = \phi(N(f))$, where ϕ is a function defining whatever arithmetic operations, comparison of real numbers, and evaluation of elementary functions needed to compute U for a given information pattern $N(f)$. We define the cost of computing $U(f)$ to be the cost of making the n function evaluations $N(f)$, together with the cost of the operations employed in computing ϕ.

Following in the footsteps of Joseph Traub and Henryk Woźniakowski, we can now define the computational complexity $\mathbf{c}(\epsilon)$ to be the minimal cost of computing an ϵ-approximation to $J(f)$ for all $f \in F$. Note that the quantity $\mathbf{c}(\epsilon)$ is an *intrinsic* property of computing an ϵ-approximation to a definite integral for the function class F. In particular, it does not depend on any specific method. It turns out that for definite integration, $\mathbf{c}(\epsilon) \sim (1/\epsilon)^{1/r}$. Over the past decade or so, Traub and his co-workers have used this line of argument to develop a whole theory of what they call *information-based complexity (IBC)*. Let's look at some of their results.

The three pillars underlying IBC are that information is incomplete, noisy and doesn't come for free. So, for instance, in the integration example, the information pattern is the set of function values $N(f)$, together with the knowledge that $f \in F$. Generally speaking, this is not enough information to pin down uniquely the function f. So the information at hand is only partial. Moreover, the numbers forming the set $N(f)$ are not computed exactly. Usually they come from some kind of numerical operations that contain round-off error or from measurements that are themselves imprecise. So the information is noisy. Finally, information, like everything else worth having in life, costs something to get. Depending on the situation, the cost may come from making the function values or from performing the combinatory operations represented by the function ϕ. But in all cases there is some cost associated with obtaining the information pattern. The primary question studied in IBC is how to obtain an approximate solution to a problem at minimal cost. Let's see how IBC theorists go about addressing this question of basic practical concern.

An extremely useful generalization of the problem of definite integration is to consider an operator

$$\mathcal{J} : \mathcal{F} \to \mathcal{G},$$

where $\mathcal{F}$ is a subset of a vector space of functions, while $\mathcal{G}$ is a normed vector space. Given $f \in \mathcal{F}$, we want to compute an approximation to $\mathcal{J}(f)$. To do this, we need an information pattern telling us something about f. We will assume this pattern is obtained in the following manner. Let Λ be a set of linear functionals. Then our information pattern will consist of what we know about the analytic properties of the function f, together with the function values

$$N(f) = [L_1(f), L_2(f), \ldots, L_n(f)],$$

where each $L_i \in \Lambda$. An idealized algorithm is then a mapping $\phi \colon N(f) \to \mathcal{G}$. Finally, we compute our approximation to be $\mathcal{U}(f) = \phi(N(f))$.

For worst-case analysis, we seek $\mathcal{U}(f)$ such that

$$\|\mathcal{J}(f) - \mathcal{U}(f)\| \le \epsilon,$$

for all $f \in \mathcal{F}$. On the other hand, if our interest is in the average-case performance as above, then we look for $\mathcal{U}(f)$ so that

$$\left(\int_{\mathcal{F}} \|\mathcal{J}(f) - \mathcal{U}(f)\|^2 \, d\mu(f) \right)^{1/2} \le \epsilon,$$

where μ is a probability measure on the function space $\mathcal{F}$.

Example: Definite Integration

To illustrate this more general setup, let's return to the integration example considered above. Here we have

$$\mathcal{J}(f) = \int_0^1 f(t) \, dt,$$

with $\mathcal{F} = \{f : f \in C^r[0, 1], \|f^{(r)}\|_\infty \le 1\}$ and $\mathcal{G} = \mathbb{R}$, the set of real numbers. For the functionals, we choose $L_i(f) = f(t_i)$. As for the algorithm ϕ, one possibility is the Riemann approximation considered earlier,

$$\phi(N(f)) = \frac{1}{n} \sum_{i=1}^n f(t_i).$$

In the average-case, we seek an approximation $\mathcal{U}(f)$ such that

$$\left(\int_{\mathcal{F}} |\mathcal{J}(f) - \mathcal{U}(f)|^2 \, d\mu(f) \right)^{1/2} \leq \epsilon,$$

where we could choose the probability measure μ to be a truncated Wiener measure on the rth derivatives of f.

In order to define what we mean by computational complexity, we have to have some model of the computational process. In this numerical-analytic setting, a reasonable set of postulates upon which to base such a model are:

1. For every $L \in \Lambda$ and for every $f \in \mathcal{F}$, the computation of $L(f)$ costs an amount $c > 0$;

2. Each combinatory operation associated with carrying out the algorithm ϕ can be performed without error and costs a unit amount. In other words, we can perform operations on real numbers without error, and each such operation costs us a unit amount.

So if we let **cost** $(\mathcal{U}, f)$ denote the total cost of computing the approximation $\mathcal{U}(f)$, then we have

$$\textbf{cost } (\mathcal{U}, f) = \textbf{cost } (N, f) + \textbf{cost } (\phi, N(f)),$$

where the first term on the right-hand side is the cost of obtaining the information $N(f)$), while the second term measures the cost of combining this information to form the approximation $\mathcal{U}$.

We can now define the computational complexity of a problem as

$$\textbf{c}(\epsilon) = \inf\{\textbf{cost } (\mathcal{U}) : \textbf{e}(\mathcal{U}) \leq \epsilon\},$$

where $\textbf{e}(\mathcal{U})$ is the error of the approximation $\mathcal{U}$. Note that here we adopt the convention that the cost is infinite if there are no ϵ-approximations. In the two settings we have been discussing, worst-case and average-case, the costs and error functions are given by

– Worst-Case Setting –

$$\textbf{e}(\mathcal{U}) = \sup_{f \in \mathcal{F}} \|\mathcal{J}(f) - \mathcal{U}(f)\|,$$

$$\textbf{cost } (\mathcal{U}) = \sup_{f \in \mathcal{F}} \textbf{cost } (\mathcal{U}, f).$$

– Average-Case Setting –

$$\mathbf{e}(\mathcal{U}) = \left(\int_{\mathcal{F}} \| \mathcal{J}(f) - \mathcal{U}(f) \|^2 \, d\mu(f) \right)^{1/2},$$

$$\mathbf{cost}\ (\mathcal{U}) = \int_{\mathcal{F}} \mathbf{cost}\ (\mathcal{U}, f) \, d\mu(f).$$

From these expressions we see that the complexity of a problem involves the intrinsic difficulty in solving that problem, and has nothing to do with the particular algorithm we employ. Rather, it depends on $\mathcal{J}$ and $\mathcal{F}$, as well as upon the setting and the set of allowable information construction operations Λ. Finally, it depends on the model of computation we use and, in the average-case setting, on the probability measure μ. So what do these complexities look like for various types of problems? Here are a couple of examples involving commonly-occurring types of problems.

Example 1: Numerical Quadrature

Our standard example has been the computation of the definite integral

$$J(f) = \int_0^1 f(t) \, dt,$$

for $f \in \mathcal{F}$. The information pattern, function class $\mathcal{F}$ and all the other factors needed to specify the problem are as in the earlier case. It turns out that for functions f of smoothness class r, we have

$$\mathbf{c}(\epsilon) \sim \left(\frac{1}{\epsilon} \right)^{1/r}.$$

From this expression, we can get a good handle on the effect of changing the smoothness and/or the error tolerance on the difficulty of calculating the integral of f. For instance, if we go from a once-differentiable function ($r = 1$) to twice-differentiable, then the complexity decreases by a factor of $\epsilon^{1/2}$. It's interesting to note that if we demand only that f be continuous ($r = 0$), then the complexity is infinite! This reflects the well-known fact that continuity doesn't really impose much by way of restricting how the function f can wiggle around. And it's the "wiggles" that cause difficulties in determining anything interesting about a function's nature (like its integral) from a finite amount of numerical information. In short, the less structure there is to exploit, the harder will be the computation.

Example 2: Nonlinear Optimization

Consider now the problem of minimizing the function $f_0(x)$, subject to the constraints $f_i(x) \le 0$, $i = 1, 2, \ldots m$. We assume all $f_i \in \mathcal{F} = C^r(X)$, where X is some Banach space of dimension d. So here the information pattern is just a collection of values of the functions $f_0, f_1, \ldots, f_m$. Note, in particular, that the only structure we impose at the outset is that the functions be r-times differentiable on X.

It turns out that under these circumstances the complexity of the optimization problem is

$$\mathbf{c}(\epsilon) \sim \left(\frac{1}{\epsilon}\right)^{d/r}.$$

So again we see that if the functions are only continuous, the cost is infinite. Note also that the complexity increases exponentially with the dimension d of the underlying Banach space X. This means that the problem is also practically uncomputable, i.e., intractable. We'll return to this point in a moment.

To see the effect of imposing additional exploitable structure on the situation, assume now that our space of functions $\mathcal{F} = \mathcal{F}_c$, the set of convex functions satisfying a uniform Lipschitz condition. In addition let the space X be a bounded, convex set. Adding this structure to the information pattern, the complexity of the optimization problem then becomes

$$\mathbf{c}(\epsilon) = k \log \frac{1}{\epsilon},$$

where the constant k depends on d and m in a polynomial fashion. Thus, adding the convexity conditions transforms the problem from one that's intractable to one that's solvable using a reasonable amount of computing resources.

Example 3: Large Linear Systems

Consider the linear algebraic system $Ax = b$, where A is a nonsingular $n \times n$ real matrix. A commonly-occurring case in which we do not have full information about the system comes about when n is very large, say on the order of several thousand. In such situations, either we may not know all the entries of A or the time involved in computing the solution by a direct method like Gaussian elimination may be prohibitive. In these circumstances we often resort to iterative methods, agreeing to settle for an approximate solution. So what we're looking for is a vector x such that

$$\|Ax - b\| \le \epsilon, \qquad \|b\| = 1.$$

Now what kind of partial information about A is it likely that we'll have available? In practice, the kind of information that's relatively easy to compute for large systems are vector-matrix products like Az, since they can be computed in a time linearly proportional to the dimension of the system n. So let's assume that the information pattern available is the set of vector-matrix products

$$N_k(A, b) = \{b, Az_1, Az_2, \ldots, Az_k\}.$$

In particular, if $z_1 = b$ and $z_i = Az_{i-1}$, then we have what in Chapter 6 was termed the reachability matrix for the system. In the world of linear algebraists, this set of quantities is called *Krylov information*. For this information pattern to be interesting, of course, we must require that $k \ll n$.

Quite often with such large systems, various symmetries in the underlying problem lead to a corresponding symmetry in A. Moreover, positivity conditions, especially in problems in physics and economics, naturally require that A be a positive-definite matrix, as well. Let's define the condition number of A to be $K_A = \|A\|/\|A^{-1}\|$. Then if we let $\mathcal{F}_1$ be the set of matrices $A > 0$ having condition numbers uniformly bounded by M, while denoting the set of symmetric A without any positivity assumption by $\mathcal{F}_2$, we find that the complexities are related as

$$\mathbf{c}_{\mathcal{F}_1} \sim M^{-1/2}\mathbf{c}_{\mathcal{F}_2},$$

when ϵ is small, M is large and $n > M \log(2/\epsilon)$. So if the spread of the characteristic values of A is great and we want relatively high accuracy for a large system, the positive-definiteness of A makes the problem easier by a factor of about $\sqrt{M}$.

There are many other results along these same lines relating to the interplay between the amount of information we have available and the intrinsic difficulty of solving a particular class of problems. But restrictions of space preclude our taking them up here. So we refer the reader to the stimulating discussions of these matters cited in the Notes and References. Now let's briefly compare the IBC ideas outlined in this section and the combinatorial complexity results sketched earlier in Section 8.

To begin with, both IBC theory and the combinatorial complexity theory share the feature that there is a model of computation stating what kinds of operations are allowed and how much they cost. The two theories also both define complexity to be the minimal computational cost involved in solving a particular problem. But there are important differences, too.

First of all, the model of computation in the two theories differs. For IBC, the model involves exact operations with real numbers, assuming a unit cost per operation; combinatorial complexity assumes a computational scheme involving calculations with n-bit integers, the cost of computation being proportional to the length of the operands (a so-called "bit model" of computation). And of even greater importance is the assumption the two theories make about the nature of the available information. Combinatorial complexity assumes that the information given about each problem is both *complete* and *exact*. So, for example, in the traveling-salesman problem the distance between each pair of cities is known (completeness) with total precision (exactness). IBC, on the other hand, assumes information is partial and contaminated by various kinds of noise. So we see that the two theories are in many ways complementary. To follow up this point, let's return for a moment to the notion of computability.

Given an error tolerance $\epsilon > 0$, we saw earlier that the computational complexity of many problems in the deterministic worst-case setting is of order $(1/\epsilon)^{d/r}$, where r measures the smoothness of the functions appearing in the problem, while d is the number of variables in these functions. Often we can take d to be a measure of the size of the problem. The nonlinear optimization example considered earlier is a good illustration of this sort of situation. So, with IBC we see a proven exponential increase in problem complexity as a function of the size of the problem; hence, nonlinear optimization is an *intractable* problem unless we can bring more information (like convexity) to bear on the situation. On the other hand, for NP-complete problems this kind of intractability has only been *conjectured,* but never proved.

IBC also gives us another perspective on the issue of computability. We say that a problem is *noncomputable* if it's impossible to calculate an ϵ-approximation at finite cost. More formally, if there exists an $\epsilon^* > 0$ such that $c(\epsilon) = \infty$ for all $\epsilon \leq \epsilon^*$, then in IBC terms the problem is not computable. This definition means that there is no algorithm for computing even an approximate solution in a finite number of steps. We have already seen a simple case of a noncomputable problem in our numerical integration example. If the function family $\mathcal{F}$ consists of the uniformly-bounded continuous functions, then the problem is not computable. This is only a special case of a much more general situation: If the smoothness $r = 0$, then for many classes of problems the complexity is infinite. So wherever you find continuous functions, noncomputability is usually lurking somewhere in the neighborhood.

Before closing this topic, it's important to note that these noncomputability results apply only to the *deterministic, worst-case* setting. This is one end of the complexity scale. The situation is a lot more encouraging if we stop being so risk-averse and live it up a little. So, for example, if we consider the average-case or if we randomly choose the functionals, their numbers and/or the algorithms, then noncomputable and/or intractable problems suddenly become solvable. However, this reduction in complexity doesn't come for free, as randomization reduces the assurances we can give about the error ϵ. For example, in the worst-case numerical integration problem with randomization of function values, it can be shown that $c_{ran} \sim 1/\epsilon^2$. Thus, the problem is both tractable and computable. Details of these matters are available in the books and papers cited in the Notes and References. With these rather specific ideas about the limits and possibilities of computation under our belts, let's now see how these ideas make contact with our earlier discussion of systems—dynamical and formal—and what we can expect them to do for us.

Exercises

1. Consider the linear algebraic system $Ax = b$, having input error δb. Define the following quantities:

$$\text{input precision} = -\log \|\delta b\|,$$

$$\text{relative input precision} = -\log \left(\|\delta b\|/\|b\|\right),$$

$$\text{relative output precision} = -\log \left(\|\delta b\|/\|x\|\right).$$

Here $\| \cdot \|$ is any convenient vector norm. (a) Show that if we define the overall loss of precision to be the difference between the relative input and output precisions, that the loss of precision is

$$\log \left(\frac{\|A^{-1}(\delta b)\|}{\|\delta b\|} \cdot \frac{\|Ax\|}{\|x\|} \right).$$

(b) Recalling that the condition number of the matrix A is defined to be the quantity $K_A = \|A\|/\|A^{-1}\|$, prove that the greatest precision loss that the system can have for fixed A is $\log K_A$.

2. Extend the IBC discussion given in the text for numerical quadrature to the case of multiple integrals.

3. Suppose you want to solve numerically the scalar Riccati differential equation $\dot{x} = 1 - x^2$, $x(0) = 0$. Calculate the information complexity associated with this problem. (Hint: Remember that solving a differential equation is essentially the same thing as performing an integration.)

10. Chaos, Gödel and Truth

Our ultimate goal in any modeling exercise is to make contact somehow with real-world "truths" through the medium of a mathematical representation of that world. But the theorems of Gödel and Chaitin showed us that no amount of mathematical finery is ever going to get at **all** the truth. There will always be something happening 'out there' that our mathematical models won't account for. Some have argued that the phenomena of chaos shows an inherent restriction even in mathematics itself on our ability to get at the scheme of things by following a set of rules. In this section we exploit the Dynamical System-Formal System dictionary in Table 9.5

to conclude that in a world without chaos, there wouldn't be very many truths—mathematical or otherwise—to make contact with anyway!

The most convenient type of dynamical system to focus our discussion upon is a 1-dimensional cellular automaton of the sort considered in Chapter 3. For simplicity, we consider such a system for which the state manifold is the infinite line $M = \mathbf{Z}^\infty/2$. Thus, at each moment in time $t = 0, 1, \ldots,$ the system state is simply an infinite sequence, each of whose entries is either 0 or 1. The vector field f then just a prescription telling us how each element of the sequence transforms to its value at the next moment of time. The system's trajectory is then given by a succession of such sequences. In Chapter 3 we saw that extensive computer experimentation and analytic work on such 1-dimensional cellular automata has shown that the attractor set of such a system can be one of the four characteristic types shown in Fig. 9.8, giving rise to the categorization of cellular automata by the not very imaginative labels Types A, B, C, and D.

Recalling the earlier discussion of such cellular automata, the top line of each diagram in Fig. 9.8 represents the initial state of the system at time $t = 0$, where each location on the line is either black (if the cell has the value 1) or white (if the cell has the value 0). The successive states on the trajectory at times $t = 1, 2, \ldots$ are then represented by the successive lines moving down the diagram from top to bottom. Thus, with the Type A system shown in the figure, we see that after a short transient period of 8 or 9 steps, the system settles into a fixed state which persists thereafter, i.e., it moves to a fixed point. With this picture in mind, we can identify the attractor types of a dynamical system with the four types of cellular automata as follows:

Fixed point	$\Longleftrightarrow$	Type A
Limit cycle	$\Longleftrightarrow$	Type B
Strange attractor	$\Longleftrightarrow$	Type C
Quasiperiodic orbit	$\Longleftrightarrow$	Type D

Now let's see what these matchups have to do with the undecidability results of Gödel, Chaitin & Co.

If there's any message at all for humankind in the results of Gödel, Turing, and Chaitin, it's that there is a forever unbridgeable gap between what's true and what can be proved. So where do chaos and strange attractors fit into the overall scheme? On the basis of the dictionaries linking dynamical systems, formal systems and programs, it seems natural to assert that the existence of a rich variety of mathematical and, hence, real-world truths depends in an essential way upon the existence of strange attractors. Here's why.

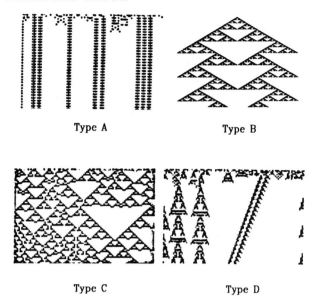

Type A   Type B

Type C   Type D

Figure 9.8. Cellular Automata Attractors

We have already seen that the idea of a provable real-world truth coincides with the decoding of a theorem in a formal system. Therefore, let's use S to represent the universe of true statements, while T denotes the set of theorems provable in some formal system. Speaking informally, Gödel's Theorem then just states that $T \subset S$.

From the discussion on complexity, it's clear that there exist computable numbers of arbitrary complexity. But these computable quantities correspond to the attractor of some cellular automaton. Thus, since there are an infinite number of such computable strings, there must exist cellular automata whose attractor set is infinite. But fixed points and limit cycles are both finite attractors. Hence, there must exist something "larger." But Chaitin's Theorem tells us that the attractor set must be smaller than the whole state manifold M, since it asserts that most strings can never be computed. In short, Chaitin's Theorem implies the existence of some kind of cellular automaton attractor beyond Types A and B. In short, there must exist strange attractors.

Now let's assume that such strange attractors exist. Since they don't fill up the entire manifold M, there must exist states that cannot be attained from any given initial state. But from the equivalence of formal systems and cellular automata, this is just another way of saying that $T \subset S$, i.e., Gödel's Theorem. Putting these two sets of arguments together, we arrive at the implications

Chaitin's Theorem $\implies$ strange attractors $\implies$ Gödel's Theorem

As the *pièce de résistance,* we come to the perhaps surprising conclusion that chaos implies truth, in the sense that a world without chaos would be very impoverished in the possible number of mathematical theorems that could be proved, even in principle. And this fact, in turn, would imply that whatever real-world truths there might be, the overwhelming majority of them could not be matched up to a mathematical theorem anyway because there just wouldn't be very many theorems. Of course, from this standpoint we might already be living in such a world. But at least the existence of strange attractors allows us to hold out hope for an eternity of surprises. Here's another mathematical reason why.

In the arguments above, we have exploited the equivalence between formal logical systems and dynamical systems, focusing attention on the implications of chaotic phenomena within the realm of formal systems. But in Table 9.3 we saw also that there is no real difference that matters between a dynamical system and a universal Turing machine. So it's of considerable interest to ask about chaos again, but this time within the world of Turing machines. Cristopher Moore looked into just this question in 1990 while still a graduate student at Cornell. What he discovered is that there exists a complex form of chaos showing that even if we knew the initial state of such a chaotic process *exactly,* it would still be almost impossible to answer any interesting questions about the system's behavior. Let's take a quick look at this remarkable result.

For the construction of complex chaos, consider the transformation of a square as shown in Fig. 9.9. Here the square is divided into eight sections, which Moore transforms separately. Usually, the boundaries between segments would violate continuity conditions. But Moore cleverly gets around these kinds of discontinuities by removing an infinitely thin net of points, leaving behind a Cantor set of the type we considered in Chapter 4 when discussing fractals. As the transformation is iterated, what gets left behind turns into a scrambled mess. But this technical virtuosity was only a pitstop on the road to complex chaos.

Moore's key insight was to recognize that the effect of this transformation on any point of the square is equivalent to a single step in a Turing machine program. Now having moved the dynamical process of transforming the square to the setting of Turing machines (by, in effect, invoking our dictionary), Moore shows that predicting the future behavior of the Turing machine comes down to the same thing as predicting whether or not the machine will halt. But we already know that the Halting Problem is unsolvable, i.e., undecidable. So by transferring this undecidability result to the dynamical process of mapping the square, we can only conclude that predicting the future behavior of any starting point in the square is also undecidable. So even if we know the coordinates of the point exactly, we can't tell what's going to happen to it under Moore's transformation.

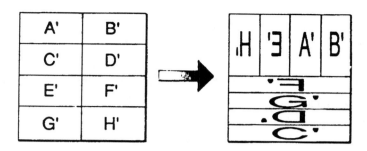

Figure 9.9. Transformation of a Square and Complex Chaos

What Moore has done, in effect, is to embed a UTM in a continuous dynamical system. The change of setting from computers to dynamical processes suggests many tantalizing questions. For example, is the famous three-body problem, involving calculation of the orbits of three gravitating bodies, computationally undecidable? Unfortunately, it can be shown that Moore's simple system corresponds to the motion of only a single classical particle in a three-dimensional box containing a number of plane and parabolic mirrors. But looking at things more positively, the result also shows that such a particle can be equivalent to a Turing machine and, hence, capable of universal computation.

But even if there may be such complex chaotic processes whose properties cannot be known, it does not mean that there is no structure underlying the chaos. So now let's look at some recent results illustrating the point that even in chaotic systems some degree of order may exist beneath the surface of chaos.

Consider a discrete-time system composed of N interacting particles whose dynamics are described by

$$x_i(t + 1) = (1 - \epsilon)f(x_i(t)) + \frac{\epsilon}{N}\sum_{j=1}^{N}f(x_j(t)), \qquad i = 1, 2, \ldots, N.$$

For simplicity, let's take the coupling function to have the logistic form $f(x) = 1 - ax^2$, where a is a parameter in the range $1.4 \leq a \leq 2$ and ϵ is a real number between 0 and 0.4. It's crucial to note, however, that the results reported below are not dependent on this particular map. Many others will do just as well.

Work by Kunihiko Kaneko has shown that the attractor set of this system has a very interesting "clustering" structure. In fact, there seem to be four quite distinct phases:

• *Coherent Phase:* In this phase, almost all initial conditions lead to a simple attractor in which $x_i = x_j$ for all i, j. Thus, the overall behavior is governed by just the single logistic map.

• *Ordered Phase:* Here the variables split into k disjoint clusters, with $k \ll N$. In each such cluster, we have $x_i = x_j$. In this ordered phase, these small clusters take up almost the entire space of initial conditions.

• *Partially-Ordered Phase:* Here the attractor consists of both small clusters of size $k \ll N$ and large clusters of size $k \approx N$. Both of these kinds of clusters occupy measurable volumes in the space of initial conditions.

• *Turbulent Phase:* Here almost all initial conditions are attracted to a single cluster of size $k \approx N$.

The overall picture of what happens as a function of the parameters ϵ and a is shown in Fig. 9.10.

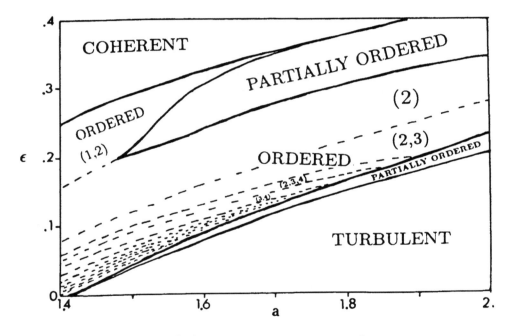

Figure 9.10. Structure of the Attractor Set

What these results strongly suggest is that order is somehow curiously immune to the onset of chaos. What Kaneko has shown is that coherent oscillation of the system from one phase to another is common, but that appropriate values of ϵ and a cause it to break up into clusters that oscillate out of phase with each other. What seems to be happening is that each component of the system is subject to two competing forces: (i) its individual tendency toward chaotic oscillation, and (ii) its tendency to conformity arising from the averaging effect of the system as a whole. The "phase diagram" of Fig. 9.10 shows that the greater the nonlinearity (the parameter a), the more widespread is the disorder; the greater the averaging effect (the parameter ϵ), the more the overall coherence. So the bottom line here seems to be that the chaotic states of a system may contain a measure of persistent coherence within themselves.

One of the implications of Kaneko's work is that we now have a quantitative way of supporting the intuition behind the by now time-worn phrase, "order in chaos." Even more importantly, perhaps, these coherencies buried deep within the chaotic states may be exactly what's needed to account for things like the persistence of neural memory in a disordered neural network or, more generally, the emergence of pattern from a collection of disordered individual agents.

We have now come to the end of our deliberations on algorithms, computation, logic, dynamics and truth. In the final section we summarize what we've learned about the possibilities and limitations of using mechanism as a way of getting at the way the world works.

Exercises

1. The Halting Problem for Turing machine programs is translatable into the formal system result we know as Gödel's Theorem. (a) What is the dynamical system interpretation of the Halting Problem? (b) How would you interpret a strange attractor in formal system terms?

2. Distinguish the difference between unpredictability of the type seen in chaotic processes, which is due to uncertainty in the initial conditions, and the unpredictability of a Turing machine of the sort guaranteed by the unsolvability of the Halting Problem. In particular, assume that you can faithfully code a Turing machine as a dynamical system. In that case, what does it mean to say that the Halting Problem is undecidable? Or that the long-term motion of the dynamical system is undecidable?

3. Consider the coherent phase of the Kaneko model given in the text. (a) Prove that the stability of the attractor can be calculated from the products of $N \times N$ Jacobi matrices associated with the mapping $f'(x_t)[\epsilon/N + (1-\epsilon)\delta_{i,j}]$, where $f(x) = 1 - ax^2$. (b) Show that for stability of the coherent attractor we must have $\lambda + \log(1-\epsilon) < 0$, where λ is the Lyapunov exponent for the logistic map (see Chapter 4 for details of this exponent).

11. The Rules of the Game

As we used the term in Chapter 1, a *scientific* law is a rule or procedure for determining how a system behaves. In other words, a law is the answer to a question about the behavior of a natural or human process. But if there's one message that we can now feel confident in trumpeting to the world, it's that to display such a law is tantamount to producing a computer program. So we find ourselves in the curious position of claiming that what we now mean by an "answer" is a program. To mathematicians schooled in an earlier, less computationally intensive era, this claim must surely come as a bit of a shock. For them, the answer to a mathematical question was a theorem or, perhaps, a formula. But certainly not a computer program (unless, of course, they were followers of L. E. J. Brouwer and the constructivist school of mathematical thought). However, our dictionaries establishing the equivalence of dynamical systems, programs and logical systems show that while fashions change, the underlying object we call "the answer" doesn't. So whether you're a believer in bits and bytes or axioms and attractors, in the final analysis it all comes down to the same thing: following a set of rules.

This fact raises the deep question of the degree to which *every* observable phenomena in the natural world is the outcome of some rule-based process. Or, put another way, are there physical processes that cannot even in principle be encapsulated in a set of rules? Recalling the Turing-Church Thesis from Section 3, this question boils down to asking: Is the Turing-Church Thesis true for physical processes? By assumption, the Thesis holds for computational processes. What we're asking here is whether it can be extended to all natural processes—physical or human.

The Oxford quantum physicist Roger Penrose has asserted that nature produces noncomputable processes—but at the quantum level only. He goes on to speculate that human intelligence is an example of just this sort of process. If this indeed turns out to be the case, then we would be forced to conclude that the Turing-Church Thesis is false for physical processes. But a recent thought experiment by N. da Costa and F. Doria casts doubt on Penrose's claim. Here's a brief sketch of why.

Suppose we construct a sequence of functions $\{f_m(x)\}$ whose values are each either identically 0 or identically 1, but it's undecidable which of these alternatives is the case. Now define a family of classical (i.e., nonquantum) dynamical systems, one for each m, with each system being either a free particle or a simple harmonic oscillator, depending on whether f_m is 0 or 1. Since the value of f_m is undecidable, there is no formal way to compute which of these two possibilities is actually the case. However, one can imagine having an analog computer that simulates the dynamics for any given m. It will then be apparent to an observer whether or not the free motion is simple harmonic motion or free—in one case the particle oscillates along an interval, while in the other it wanders off to infinity. So the analog machine can effectively decide the undecidable. Thus, a classical system computes something that is algorithmically undecidable, contrary to Penrose's claim. What this thought experiment shows is that if Penrose's claim is to be valid, then it must be posed for *real* systems, not mathematical ones. This is an idea that's difficult to rigorously formulate in a world that's really quantum.

What's at issue here is the dichotomy embodied in the Modeling Relation discussed in Chapter 1 and revisited in Section 4 of this chapter, a dichotomy centering on our discussion of the mathematical world and syntax vis-à-vis the real world and semantics. The key question is whether the laws of nature can themselves be formulated in purely syntactic terms. Or do they possess an inherent semantic component that cannot be formalized in syntax alone? All the results we've seen about undecidable propositions, uncomputable functions and the like are mathematical results. They live on the syntactic side of the divide between the worlds of mathematics and man. Whether or not they can be mapped in a faithful manner to the real-world side of the street depends on the nature of translation operations we called "encoding" and "decoding" in Chapter 1. For instance, if we could find even one example of a natural law that maps in a one-to-one fashion onto a noncomputable function, then the matter would be settled: the Turing-Church Thesis would be false for physical systems. And following from this would be the fact, not speculation, that there is indeed more to life than following rules. This distinction between syntax and semantics shows up in somewhat more transparent form when we dig a little deeper into the difference between a model and a simulation.

In naïve modeling exercises, a typical sequence of events involves the construction of some sort of formal mathematical model. The model is then used to "postdict" the results of past experiments, as well as to "predict" the outcomes of future observations. In the event these post- and predictions are in close enough agreement with what is actually observed, the formal system (program, dynamical system) is then declared to be a *model* of the process under investigation. Is it really? Is this all it takes to be deemed a model of some phenomenon? Shouldn't there be something more than just agreement with the data? After all, Ptolemy and his epicycles provided at least as accurate agreement with observed planetary positions as the ellipses of Copernicus and Kepler. Yet no one today seriously contends that the shaky edifice of epicycle piled upon epicycle is a *model* of the solar system. At best, such a structure provides a *simulation* of planetary motion. But surely not a model. So what's the difference between a simulation and a model? And if there is a difference, can we produce any procedures for separating one from the other?

The perceptive reader will recognize this problem as the same one we faced in Chapter 6, where we considered the input/output pattern of a black box, considering how to construct an internal model that matched this behavior. This internal model turns out to be nothing more than a simulation of the actual system, as we shall endeavor to show below. This simulation-versus-model problem is so ubiquitous in the applied modeling literature that it's well worth our taking some time to examine the differences between the two. We'll also take this as an opportunity to put forward a few ideas upon which answers to the above questions can be based. Perhaps the simplest way to outline the difference between a model and simulation is to follow an example presented originally by Robert Rosen.

Consider an abstract neural network described by a set of $nm + r$ formal neurons. Arrange the neurons into an $n \times m$ rectangular array below a single row of r output neurons. Then it's well known that we can represent this neural net by a finite-state machine $\mathcal{M} = (X, A, B, \lambda, \delta)$, where X is the state set, A is the set of inputs, B is the set of outputs and $\lambda\colon X \times A \to X$, $\delta\colon X \times A \to B$ are the state-transition and output maps, respectively. Here we assume the sets X, A and B have cardinalities n, m and r, respectively. By suitably *restricting* the inputs to the neural net, the behavior of the net can mirror the behavior of $\mathcal{M}$. Looked at from the perspective of the machine, we say the machine $\mathcal{M}$ constitutes a *realization*, or a *model*, of some particular behavior of the net.

Note the following key aspects of the above situation:

- There is a clear and direct correspondence between the states of the machine and the states of the net, as well as direct correspondences between the inputs (outputs) of $\mathcal{M}$ and the inputs (outputs) of the net.

- The machine $\mathcal{M}$ is *simpler* than the net it models, in that the net is capable of many behaviors that the machine is not. So given a machine $\mathcal{M}$, there exists a neural net that mirrors the behavior of $\mathcal{M}$ if we impose *restrictions* on the inputs and outputs of the net.

These properties are characteristic of the situation in which one system *models* another and doesn't just *simulate* it.

Now consider the case in which we want to use $\mathcal{M}$ to *simulate* another machine $\mathcal{A} = (\mathcal{S}, \mathcal{U}, \mathcal{Y}, \mu, \nu)$. For simplicity, assume that $\mathcal{A}$ is just a state-transition machine so that the behavior of $\mathcal{A}$ is characterized solely by the map $\mu\colon \mathcal{S} \to \mathcal{S}$, i.e., $\mathcal{U} = $ singleton, $\mathcal{Y} = \mathcal{S}$, $\nu = $ identity. In order to simulate $\mathcal{A}$ by $\mathcal{M}$, we must *encode* the states of $\mathcal{A}$ into the input strings A^* of $\mathcal{M}$, and we must *decode* the output strings of $\mathcal{M}$ into the states of $\mathcal{A}$. This means we must find maps

$$g\colon \mathcal{S} \to A^*, \qquad h\colon B^* \to \mathcal{S},$$

such that the following diagram commutes:

$$
\begin{array}{ccc}
\mathcal{S} & \xrightarrow{\;\mu\;} & \mathcal{S} \\
{\scriptstyle g}\big\downarrow & & \big\uparrow{\scriptstyle h} \\
A^* & \xrightarrow[\;\psi\;]{} & B^*
\end{array}
$$

Here ψ is determined by the initial state of the machine $\mathcal{M}$, as well as by the maps λ and δ.

Let's take a look at the relevant aspects of this setup for our simulation-versus-model comparison.

(1) The relation between $\mathcal{A}$ and its simulator $\mathcal{M}$ does not preserve structure. Roughly speaking, the "hardware" of $\mathcal{A}$ is mapped into the "software" of $\mathcal{M}$, and the hardware of $\mathcal{M}$ has no relation whatsoever to that of $\mathcal{A}$. In fact, if we try to describe the hardware of $\mathcal{M}$ as *hardware*, $\mathcal{A}$ will disappear completely.

(2) In general $\mathcal{M}$ is *bigger* than $\mathcal{A}$, usually much bigger; the state-set X of $\mathcal{M}$ must usually be much larger than S if $\mathcal{M}$ is to simulate $\mathcal{A}$. This is in stark contrast to our earlier situation when $\mathcal{M}$ served as a model of the neural network.

Returning now to the question of whether it matters that a simulation is not structure-preserving, assume that in the natural system N there is some property P of concern and that we have a simulation $\mathcal{M}$ of N. Then if P is preserved by $\mathcal{M}$, fine; if not, then we're in the kind of trouble that would never have arisen if $\mathcal{M}$ were a model of N rather than just a simulation. For example, if we're interested in the case when P is the motion of an arm, then a computer simulation of arm movement will certainly not possess the property P. But a robot constructed to mimic human arm movement will be a legitimate *model* of arm motion and, hence, will possess property P. Thus the robot represents a model of arm motion, whereas the computer program is only a simulation. The degree to which this very obvious distinction matters in practice remains open and depends to some extent on the particular situation. Nevertheless, the distinction is clear, mathematically testable, and bears heavily upon the question of model credibility. The problem here is that computation involves only simulation, which does not allow us to establish any congruence between causal processes in the real world and inferential processes in the simulator. These considerations argue for the issue to receive somewhat more attention than it has up to now in the applied modeling literature. Now let's return to the question of computability—in either a simulation or a model—looking at the implications of Gödel's and Chaitin's Theorems for getting a handle on the limits to mechanism.

Gödel's Theorem has many profound implications, both for science and for philosophy. It's of considerable interest for us here to pause to consider what Gödel's conclusions have to say about the limits of human reasoning. When all the mathematical smoke clears away, Gödel's message is that mankind will never know the final secret of the universe by rational thought alone. It's impossible for human beings to ever formulate a complete description of the natural numbers. There will always be arithmetic truths that escape our ability to fence them in using the tools, tricks and subterfuges of rational analysis. As logician and science-fiction author Rudy Rucker has expressed it, Gödel's Theorem leaves scientists in a position similar to that of Joseph K. in Kafka's novel *The Castle*. We scurry about running up and down endless corridors, buttonholing people, going in and out of offices and, in general, conducting investigations. But we will never achieve ultimate success; there is no final exit in the castle of science leading to absolute truth. However, Rucker notes, "to understand the labyrinthine nature of the castle is, somehow, to be free of it." And there's no understanding of the castle of science that digs deeper into its foundations than the understanding given by Gödel's Theorem.

It's thought-provoking to consider the degree to which Chaitin's result imposes limitations on our knowledge about the world. Consider the following back-of-the-envelope estimate: Suppose $\mathcal{K}$ represents our best present-day knowledge, while $\mathcal{M}$ denotes a UTM whose reasoning powers equal those of the smartest and cleverest of human beings. Then Rudy Rucker estimates the number t in Chaitin's Theorem as

$$t = \text{complexity } \mathcal{K} + \text{complexity } \mathcal{M} + 1 \text{ billion}$$

where the last term is thrown in to account for the overhead associated with the program of the machine $\mathcal{M}$. Plugging some plausible numbers for $\mathcal{K}$ and $\mathcal{M}$ into this expression, Rucker concludes that t is less than 16 billion. The bottom line then is that if any worldly phenomenon generates observational data having complexity greater than around 16 billion, no such machine $\mathcal{M}$ (read: human) will be able to prove that there is some short program (i.e., theory) explaining that phenomenon. Thus, recalling René Thom's idea of scientific theories as arbitrariness-reducing tools, Chaitin's work says that our scientific theories are basically powerless to say anything about phenomena whose complexity is much greater than about 16 billion. This does *not* mean that there's no simple explanation for these phenomena. Rather, it says that we will never understand this "simple" explanation— it's too complex for us! Complexity 16 billion represents the outer limits to the powers of human reasoning. Beyond that we enter the "twilight zone," where reason and systematic analysis give way to intuition, insight, feelings, hunches, and just plain dumb luck.

But despite occasional hiccups of this sort, most scientists continue to believe (or at least **act** as if they believe) that nature can be simulated by a Turing machine to an unlimited level of complexity and detail. According to this credo, there is no real difference between what a computer does and what nature does. It is taken for granted that simple computable operations, which include arithmetic and algebra, apply to the real world of apples and doughnuts. But the fact that they are computable depends, for one thing, on the stability of electrons. If electrons behaved as wildly as hurricanes or multiplied as uncontrollably as rabbits, then it would be impossible to build the sort of on-off circuits needed for a machine to do arithmetic. We have already seen that there are far more noncomputable functions than there are computable ones. It's lucky that the ones needed to make contact with most of life, the universe and everything else seem to be computable—as far as we know.

Exercises

1. The modeling relation discussed in this section is transitive. This means that if X is a model for Y and Y is a model for Z, then X is a model for Z. (a) Show that the simulation relation is not transitive? (b) What are the implications of this fact for modeling real-world processes as opposed to simulating them?

2. If an oracle told us that the Turing-Church Thesis were indeed true for physical systems, then it would entail the fact that every mathematical model of a material system must be simulable, i.e. expressible in terms of pure syntax. Discuss how this fact, in turn, would lead us immediately to conclude that reductionism as discussed in Chapter 1 is a universally valid method for analyzing physical processes.

3. In Chapter 7 we discussed the concept of an anticipatory system as being one that contained an internal model of itself. The key to Gödel's proof lies in employing number theory as a way of talking about numbers themselves, in effect using number theory as a model of itself. Consider this fact in assessing the claim that if anticipatory systems exist, then the Turing-Church Thesis is false for physical systems.

Discussion Questions

1. In connection with the distinction made in the text between a model of a natural system N and a simulation, do you think the following representations are models or simulations:

a) A linear system $\Sigma = (F, G, H)$ describing an electrical circuit?

b) A scaled-down version of an airplane wing in a wind tunnel?

c) A robot designed to do weldings on an automobile assembly line?

d) The Navier-Stokes equation describing the laminar flow of a fluid?

e) A linear regression curve describing the fluctuations of the Dow-Jones Industrial Averages on the New York Stock Exchange?

2. In the text we discussed the idea of simulating any dynamical process by use of an analog computer. This simulation would, in principle, enable us to "decide the undecidable." Discuss the contention that what makes this experiment work is that classical mechanics imposes no upper bound upon velocities. So it is possible to accelerate time in such equations in such a way that an infinite amount of real time passes within a finite amount of fake time. Thus, we have a system of classical equations that correspond to an infinitely fast computer. Such a machine can, in principle, compute functions that, in the usual sense of mathematical logic, are uncomputable. For example, it can solve the Halting Problem by running a computation for infinite real time and throwing a particular switch if and only if the program stops. In fake time this entire procedure is over within a fixed period. One then need only inspect the switch to see if it's been thrown or not. Do you see any flaws, theoretical and/or practical, in this argument?

3. Consider the following statements about the number π:

(A) "The $10^{1,000}$th digit of π is a 4."

(B) "The decimal expansion of π contains a run of $10^{1,000}$ 4's."

(C) "The expansion of π contains an infinite number of 4's."

In your opinion, which of these statements is decidable, at least in principle? That is, for which of the statements (A)–(C) can we construct a computable criterion for deciding the truth or falsity of the statement?

4. Computer scientist A. K. Dewdney has identified several variants of the Busy Beaver:

● *The Civil-Servant Beaver:* This enterprising creature seeks to advance itself as far as possible without producing anything.

● *The Scientific Beaver:* This animal seeks to maximize its total activity, again without producing anything.

● *The Dizzy Beaver:* This beaver produces nothing and goes nowhere, but in the process generates a maximum amount of activity.

How would you characterize these three behavioral types in precise Turing-machine terms? In other words, how would you describe the activities of these types of beavers in terms of machine state transitions, steps taken away from the starting square and number of 1's written on the tape before the corresponding beaver stops working? Write a program that illustrates each of the types.

5. The *tiling problem* involves covering large areas using square tiles with colored edges, such that adjacent edges have the same color. Here by a "tile" we mean a unit square that's divided into four by its two diagonals,

each quarter of the square being colored with some color. Moreover, we assume that the tiles have a fixed orientation and cannot be rotated.

The algorithmic problem is as follows: We are given a finite set T of tile types, and ask whether any area of finite size can be covered using only tiles from the set T. Of course, we assume that an unlimited number of tiles of any particular pattern are available.

a) Can you see why there is no algorithm for solving this problem? That is, can you see why there will always be input sets T upon which any candidate algorithm will either run forever or terminate with the wrong answer?

b) What's wrong with the following algorithmic "solution" to the tiling problem:

    i. If the types in T can tile any area, print YES and stop;

    ii. Otherwise, print NO and stop.

6. We have laid great emphasis on the equivalence between a computer program and a formal logical system.

a) Consider in this context the difference between an error in a proof done by a computer (i.e., an error in the program) and an error in a mathematical proof written by a mathematician.

b) On odd occasions one sees self-congratulatory statements in the mathematics literature to the effect that "the glory of mathematics is that existing methods of proof are essentially error-free." Do such statements make any sense to you? Can it really be the case that mathematical proofs are without error of any kind?

7. Virtually all mathematical theorems are assertions about the existence or nonexistence of certain entities. For example, theorems assert the existence of a solution to a differential equation, a root of a polynomial, or the nonexistence of an algorithm for the Halting Problem. A *platonist* is one who believes that these objects enjoy a **real** existence in some mystical realm beyond space and time. To such a person, a mathematician is like an explorer who discovers already existing things. On the other hand, a *formalist* is one who feels we construct these objects by our rules of logical inference, and that until we actually produce a chain of reasoning leading to one of these objects they have no meaningful existence, at all. Consider the pros and cons of these contradictory positions. Can you think of any third (or fourth) view of the nature of mathematical reality, one that would strike an intermediate stance between the platonist and the formalist?

8. Consider the following procedure for constructing a function $f(n)$: Build a machine that can pick up a pair of dice, shake them, roll them out, observe the two faces that come up and print out the sum of the points

on the two faces. For each number n, the quantity $f(n)$ is taken to be the number arising from the nth repetition of this process. Is this way of computing the function f a "mechanism"? Is it also an algorithm?

9. To the majority of theoreticians, the most significant piece of evidence in favor of the Turing-Church Thesis is that the many attempts to formalize the notion of computability by what initially appeared to be radically different means have all turned out to be equivalent. A platonist (see Discussion Question 7) would explain this fact by saying that the class of computable functions was there all along, and we have only managed to "see" it finally. Others argue that the Turing-Church Thesis is really about human capabilities, and is a consequence of the structure of our bodies and the way our brains happened to be wired up. Still others, like the logician Georg Kreisel, say that the support for the Thesis does **not** consist in the equivalence of different characterizations, since there could be some systematic error leading to the equivalence. Discuss the pros and cons of these arguments. In particular, can you think of any test that might be performed that would verify or refute the Turing-Church Thesis?

10. In the on-going debate over whether or not computers can duplicate human thought processes, there are two basic schools of pro-AI thought: Bottom-Up and Top-Down. The first group thinks that the physical hardware of the brain plays a crucial role in the emergence of human intelligence. Consequently, efforts to duplicate this intelligence in a machine should respect this fact. Such researchers are nowadays called "connectionists," with much of their activity being devoted to showing how to put together neural networks and other types of connectionist hardware that will display intelligent behavior of the human variety. Marvin Minsky, David Rumelhart and Douglas Hofstadter are representative practitioners of this Bottom-Up approach to AI.

Top-Downers, on the other hand, argue that what's important in human thought is symbol manipulation, and that the particular physical substrate in which these symbols move about is more-or-less irrelevant. Basically, this group claims that what's important are the symbols and the rules by which these symbols are created, destroyed and combined. In short, intelligence can be "skimmed off" the brain without having to pay any special attention to what's going on down at the level of the individual neurons and the physical hardware. The doyens of the Top-Down approach to strong AI are Herbert Simon and Alan Newell, as well as Roger Schank.

a) Consider the pluses and minuses of these competing views.

b) The kinds of algorithms employed by the two groups of researchers differ substantially in many ways, especially in the degree to which the details of the programs are specified in advance rather than emerging via learning as the problem-solving task unfolds. Which path do you think nature took in developing the human brain via the process of evolution?

c) Suppose we were to discover an intelligent life form on a planet orbiting the star Tau Ceti. To what degree do you think it's likely that this Cetacean intelligence would bear any resemblance to that of human beings? How do you think Bottom-Upers and Top-Downers would respond to this question?

d) Instead of an extraterrestrial intelligence, consider the same question as in part (c), except now for an earthly cetacean like a dolphin or a whale.

11. If nature can be viewed as a computational process, then the form of the laws of physics might be constrained by what in principle can be computed.

a) If something cannot be computed by the entire universe during the age of the universe, in what sense can it be said to be computable?

b) Does this picture of the computational limits of the cosmos imply that the laws of physics might somehow "fade away" as one goes back towards the initial singularity of the Big Bang, due to the fact that the computational power of the universe tends to zero as $t \to 0$?

12. Comment on the following statement of Oxford physicist David Deutsch: "The reason why we can perform mental arithmetic cannot be found in mathematics or logic. The reason is that the laws of physics 'happen' to permit the existence of physical models for the operations of arithmetic such as addition, subtraction and multiplication. If they did not, these familiar operations would be non-computable functions." In particular, consider the possibility that as we discover new laws of physics, functions that we now regard as noncomputable may become computable. Also, comment on the logical loop arising from the fact that the laws of physics generate the very mathematics that makes those laws both computable and simple.

13. Here are a list of candidates that have been proposed as formal measures of a system's level of complexity:

• *Life-Like Properties:* the ability of the system to grow, reproduce, and adapt.

• *Thermodynamic Potential:* the capacity of the system for irreversible change.

• *Computational Universality:* the ability of the system to be programmed through its initial conditions to simulate any digital computation.

• *Algorithmic Information:* the size of the smallest computer program needed to generate or describe the object in question.

• *Long-Range Order:* the existence of statistical correlations between arbitrarily remote parts of the system.

• *Long-Range Mutual Information:* the amount by which the joint entropy of two remote parts of a system exceeds the sum of their individual entropies.

• *Logical Depth:* the time required to generate the object by a minimal computer program.

• *Thermodynamic Depth:* the amount of entropy produced during a system's actual evolution.

While agreeing with intuitive notions of complexity for some classes of systems, none of these measures has won universal approval as the "right" way to characterize complexity. Can you pinpoint some of the flaws in each proposed complexity candidate? In particular, for each complexity candidate give an example of a system that is complex on intuitive grounds, yet fails to display the characteristics associated with that measure of complexity.

14. The theory of *functionalism* holds that a computer program duplicating the human brain must be comparable to the brain in every important aspect, including having consciousness (to the extent that that term has any objective meaning at all).

Now suppose that we learned about the sensation of pain in every relevant detail. Then if the functionalists are right, we could build a giant robot capable of feeling pain. The inside of our robot's head would be somewhat akin to a huge office. But instead of integrated circuits, you would see a lot of people sitting around at desks, each worker in the office having been trained to duplicate the function of a neuron. Moreover, each desk would have a number of telephones, the phone network simulating the neuronal connections in a brain capable of feeling pain.

Let's further imagine that right now the pattern of phone calls in the office is that which has been identified with excruciating pain. So according to functionalism, our poor robot is in agony. But where is this pain? All we see in the office is a group of disinterested, placid bureaucrats sitting around gossiping, flirting and talking on the phone. Discuss the connection of this functionalist argument with the model of the mind considered in Chapter 6.

15. Consider a jigsaw puzzle consisting of a finite number pieces that have no picture or pattern painted on them, but rather are all painted a single color. Thus, the only thing that distinguishes the pieces are their shape. In the worst case, how many distinct matching operations would you have to perform in order to put this puzzle together? Is this problem in class P, NP, both or neither?

16. To a large degree, science as an intellectual activity concerns itself with hypotheses that may be verified or refuted easily. Relate this general observation to the satisfiability problem. In particular, what does this suggest to you about the limitations of science as a method for getting at the "truth" by means of logical operations?

17. Suppose, like Descartes, you try to make a list of things you believe. Since you don't want to hold to any contradictory views, before any belief is added to your list it's first checked against the beliefs already on the list to ensure that it doesn't introduce a contradiction.

a) Show by example that you cannot detect contradictions merely by running through the list one item after another, making sure that the proposed new belief doesn't directly contradict any belief that's already been accepted.

b) In general, if you already have n beliefs on your list, you will have to perform $2^n - 1$ consistency checks before you can feel confident in adding the next proposed belief. Suppose you currently have 100 beliefs on your list. In addition, assume you have a "computer as big as the universe" and thus can make 10^{23} logical consistency check every second per processor, with your computer having 10^{24} processors. Calculate how long it would take (in years) to expand your list of beliefs from 100 to 300.

c) Using the same line of argument, show that if you started using your computer at the time of the Big Bang, your list could be no longer than 558 items. Does this mean that you can know at most 558 things? Why not?

18. In his book *The Idea of Economic Complexity,* David Warsh argues that the perceived "complexity" of the economy is simply the differentiation of its many parts, together with the manner of their interconnection. How does this view of complexity fit in with those considered earlier in Discussion Question 13? How does this view of economic complexity help to explain why five dollars went a long way in 1939 and doesn't go very far at all today? The standard answer is "inflation"—too much money chasing too few goods. Compare this answer with the alternative: "It's due to greater complexity."

19. Since every n-state Turing machine can be described by a program of finite length, it's clearly possible to draw up a list of every possible Turing machine simply by listing all the 1-state programs, then all the 2-state programs and so forth. Thus, the set of Turing machines is countable. Define the quantity $F_k(n)$ to be the value computed by the kth Turing machine when the input tape contains the number n. Now define the function

$$G(n) \doteq \begin{cases} 1, & \text{if } F_n(n) \text{ is undefined}, \\ 1 + F_n(n), & \text{otherwise}. \end{cases}$$

a) Is the function $G(n)$ computable? Why?

b) Is the value $G(n)$ an integer? Why?

c) What does the function $G(n)$ have to do with the Halting Problem?

20. Gödel's Theorem is normally stated in the language of formal systems, while Chaitin's Theorem is expressed in terms of programs.

a) Use Tables 9.3, 9.5-9.6 to translate these two results into the language of dynamical systems.

b) A large number of questions in dynamical system theory have been shown to be undecidable. These include whether the dynamics are chaotic, whether a trajectory starting from a given initial point eventually passes through some specific region of state space, and whether the equations are integrable, i.e., have explicit solutions in terms of elementary functions like polynomials, trigonometric functions and radicals. Translate these questions from dynamical system theory into equivalent statements about programs and formal logical systems.

21. We generally understand a model to be a symbolic representation of a set of objects, their relationships and their allowable dynamical motions. Models are generally used in three ways:

- *Description:* The model describes how the system works.
- *Computation:* The model is used to calculate the behavior of the system and/or to calculate controlling actions for the system.
- *Prediction:* We use the model to predict future states of the system.

Discuss the hierarchical relationship linking these different ways of using a model. Does this hierarchy suggest any connections between the complexity of a model and its reliability?

22. The following objections have been raised to the notion that the universe is just one big computer:

- *Space and Time:* This objection centers upon the argument that computers are far too limited to model anything other than a few simplified features of the universe.

- *Irreversibility:* The microscopic laws of physics are time-reversible. On the other hand, all computers that have been physically constructed are irreversible in their operation, a fact that can be traced to the irreversibility of the physical components used to build the logic gates that make up a computer's central processing unit.

- *Human Consciousness:* If the universe is a computer and if all computers are functionally equivalent, computers must be able to simulate every feature and event in the universe. Thus, computers must have the power to simulate conscious rational thought.

Consider the pros and cons of these arguments. (Note: Some of these matters have already been considered in another form in Chapter 3.)

23. System scientist and biologist Robert Rosen has noted the following compact way of stating Gödel's Theorem: "One cannot forget that number theory is about numbers." What do you think he meant by this pithy phrasing of the meaning of Gödel's monumental result? Discuss the claim made by some that the relation of number theory to any formalization of it is the same as the relation of the complex to the simple. Consider this whole circle of ideas within the underlying assertion of Hilbert's Program for the formalization of mathematics. Namely, that semantic truth can always be effectively replaced by syntactic rules.

24. Logician Martin Davis has raised the following intriguing query:

> ... how can we ever exclude the possibility of our being presented some day (perhaps by some extraterrestrial visitors), with a (perhaps extremely complex) device or "oracle" that 'computes' a noncomputable function? However, there are fairly convincing reasons for believing that this will never happen.

What do you think these convincing reasons are? Do *you* find them convincing?

25. If real-world experimental procedures are repeatable, then the Turing-Church Thesis must mean that any input/output function f is computable. Seen in this light, the T-C Thesis is an attempt to draw inferences about hardware (physics) from software (algorithms). Consider this claim in the context of von Neumann's attempt to learn about the *material* process of self-reproduction from a *formal* theory of computation, as we described in Chapter 3.

26. Consider the following sentence: "A is the smallest number that cannot be expressed in fewer than fifteen words." This sentence seems to single out a definite number A in fourteen words. Yet A is supposed to be the first number that *cannot* be expressed in less than fifteen words! This is the so-called *Berry Paradox,* which we considered briefly in the text. Discuss how you might resolve it. In particular, consider the Berry Paradox in the context of Gödel's and Chaitin's formalizations of truth and denotation, respectively.

27. An interesting corollary of Matyasevich's proof of the unsolvability of Hilbert's Tenth Problem is that for any listable set of natural numbers S, there exists a polynomial $P_S(y_1, \dots, y_n)$ with integer coefficients such that as the variables $y_1, \dots, y_n$ range over the nonnegative integers, the positive values of the polynomial are exactly the set S. To illustrate this result, here

is a polynomial equation in the 26 letters of the alphabet, whose positive values are the set of prime numbers:

$$P(a,b,\ldots,z) = (k+2)\{1 - [wz+h+j-q]^2$$
$$- [(gk+2g+k+1)(h+j)+h-z]^2$$
$$- [2n+p+q+z-e]^2$$
$$- [16(k+1)^3(k+2)(n+1)^2+1-f^2]^2$$
$$- [e^3(e+2)(a+1)^2+1-o^2]^2$$
$$- [(a^2-1)y^2+1-x^2]^2 - [16r^2y^4(a^2-1)+1-u^2]^2$$
$$- [((a+u^2(u^2-a))^2-1)(n+4dy)^2+1-(x+cu)^2]^2$$
$$- [n+l+v-y]^2 - [(a^2-1)l^2+1-m^2]^2$$
$$- [ai+k+1-l-i]^2$$
$$- [p+l(a-n-1)+b(2an+2a-n^2-2n-2)-m]^2$$
$$- [q+y(a-p-1)+s(2ap+2a-p^2-2p-2)-x]^2$$
$$- [z+pl(a-p)+t(2ap-p^2-1)-pm]^2\}$$

As the letters a through z run through all the integers, the polynomial P takes on positive and negative integer values. The positive values coincide with the set of prime numbers; the negative values may or may not be the negatives of primes.

a) This polynomial is expressed as the product of two factors. Yet, by definition, prime numbers have no factors other than 1 and themselves. Is there any contradiction here?

b) In Problem 3 of Chapter 6 we saw that the set of prime numbers has no finite-dimensional realization as a linear machine. That is, there is no finite-dimensional linear system whose output is the set of prime numbers. Yet here we see a polynomial formula for primes, which certainly **is** a finitely-describable "machine" whose output is this set of numbers. How can you explain this state of affairs?

28. One possible way to break intractability or noncomputability is randomization. For example, the famous Monte Carlo method for evaluating integrals uses randomized information to make the number of function evaluations needed to compute an ϵ-approximation to a multiple integral independent of dimension. It's clear that the complexity of a problem in the worst-case using randomization is never greater than when we use deterministic information, since deterministic information is just a special case of randomization. Multiple integration is one situation in which the complexity using randomization is *much less* than when using deterministic information. Can you think of others? Conversely, can you think of cases in which the two complexities are the same? Do matters change much if we shift from the worst-case setting to the average case?

29. The December 20, 1988 issue of *The New York Times,* in an article titled "Is a Math Proof a Proof If No One Can Check It?", reported the results of a computer search for an object called a finite projective plane of order 10. The search, which was initiated to settle a long-standing mathematical conjecture, consumed several thousand hours of time on a Cray supercomputer, eventually rendering the verdict that there are no such planes, thus confirming the conjecture. What C. W. H. Lam, the author of this search, did not tell the media was that the Cray is reported to have undetected errors at the rate of approximately one per thousand hours of operation. So it would certainly be reasonable to expect a few errors during the course of the computer's work on this search. As Lam remarked, "Imagine the expanded headline, 'Is a Math Proof a Proof If No One Can Check It and It Contains Several Errors?'" So when it comes to giving a "good" mathematical proof, how good is good enough? In particular, consider methods that might be employed to increase the confidence in a computer-generated proof like this one.

30. In classical quantum theory, the process of making a measurement of, say, an object's position causes a "collapse" of the Schrödinger wave function that describes the object and the experimental situation. This collapse of the wave function has the character of a fundamental incompleteness within quantum theory, similar to the incompleteness of axiomatic systems in mathematics as exemplified in Gödel's Theorem. Consider the claim that the difficulty of describing the measurement process by the Schrödinger equation reflects a limitation in formal language and, hence, quantum theory may thus require a formalism consisting of two levels of description—one for the dynamics, one for the measurement. What does this multi-level description correspond to in the mathematical setting?

31. The problem of *program correctness* can be stated as follows: Given an algorithm $\mathcal{A}$, a program P and a machine M, can we find a proof that P run on M carries out $\mathcal{A}$? James Fetzer has put forward the following chain of reasoning leading him to conclude that such a proof of program correctness is, in principle, impossible:

  i. The purpose of program verification is to provide a mathematical method for ensuring the performance of a program.

  ii. This is possible for algorithms, which cannot be executed by a machine, but not possible for programs that can be executed by a real, physical machine.

  iii. There is little to be gained and much to be lost through fruitless efforts to guarantee the reliability of programs when no guarantees are to be had.

The heart of Fetzer's claim is that computers are complex causal systems whose behavior, in principle, can only be known with the uncertainty that attends empirical knowledge as opposed to the certainty that attends specific kinds of mathematical demonstrations. In short, real machines are physical devices about which we can never have the same degree of certain knowledge that we can have about mathematical objects like Turing machines and algorithms. Therefore, program verification, being concerned with real programs running on real machines, will never cross the gap from the physical world to the universe of mathematical objects; hence, there can be no such thing as a proof of P's correctness. So there is an ambiguity within the program verification business: Are the theorems about a mathematical model of computation? Or are they about the real thing?

What do you think of Fetzer's arguments? Is it even theoretically feasible to give a mathematical proof that a given program P, when run on a machine M, will behave properly by being an embodiment of an algorithm $\mathcal{A}$?

32. Gödel's Theorem has been used by many people to argue that we cannot *write* a computer program equivalent in power to human mathematical intuition. But some (like Gödel, himself) claim that this does not mean that such a program could not exist, although it would be humanly incomprehensible. Discuss how such a program might arise. (Hint: Consider, for instance, a program arising as the outcome of an evolutionary process.)

33. Consider the three patterns of dots shown below:

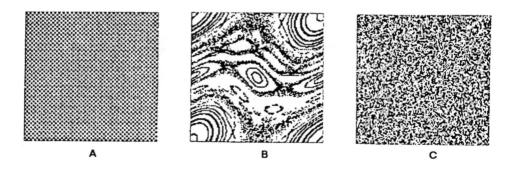

A   B   C

Which of the three patterns (A)–(C) appears the most "complex" to you? How would you rank these three patterns in terms of their algorithmic complexity? Their entropy?

Problems

1. Prove that the Busy Beaver function $BB(n)$ is uncomputable. (Hint: Show that the value of $BB(n)$ grows faster than that of any computable function $f(n)$ as n gets large.)

2. Prove that there is no program that can both test its input for the presence of a computer virus and simultaneously be guaranteed not to spread the virus itself. For the sake of definiteness, assume the following definition of a "safe program."

DEFINITION. *A program P spreads a virus on input x if running P under operating system OS on input x alters OS. Otherwise, P is safe on input x. A program is safe if it is safe for all inputs.*

(Hint: Assume there exists a safe program and derive a contradiction.)

3. Use a computer to check how many 1's the following 5-state Turing machine prints before it halts. (Remark: The answer you obtain will be a lower bound for the Busy Beaver function value $BB(5)$.)

State	Symbol Read	
	0	1
A	1, R, B	1, L, C
B	0, L, A	0, L, D
C	1, L, A	1, L, STOP
D	1, L, B	1, R, E
E	0, R, D	0, R, B

4. Consider the following problem: There are n persons $p_1, \ldots, p_n$ who must find the maximum of n numbers $a_1, \ldots, a_n$.

a) Show that if the people cannot communicate with each other, then any program that solves this problem will take at least $O(n)$ steps. That is, each person will have to examine all the numbers.

b) Now suppose that each person can communicate with every other by means of a shared memory. Assume, for simplicity, that $n = 2^k$ for some integer k, and that each person can find the maximum of two numbers in one step. Produce an algorithm that solves the problem in $O(\log n)$ steps.

(Remark: This Problem shows that there are some problems—but not all—that will run much faster on a parallel computer with several independent processors than on a serial machine.)

5. Suppose we are searching for a name in a telephone book containing N names. Show that binary search is optimal. That is, in the worst case we cannot find a name in less than $\log_2 N$ comparisons.

6. Prove that the N-disk Tower of Hanoi problem cannot be solved in fewer than $2^N - 1$ moves, thus showing that the Tower of Hanoi problem is not in class P.

7. The figure below shows two different sets (A) and (B) of three tiles. The tiles from a given set are to be used to tile a rectangular area so that the adjacent edges of any two tiles have the same color. Assume that the tiles cannot be rotated but must be used in the orientation shown in the figure. One of these patterns can tile an area of any size; the other cannot even tile the smallest area. Which is which?

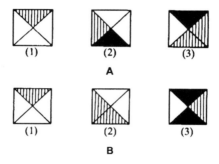

Two Tile Patterns

8. Consider the linear programming problem: Find $x \in R^n$, subject to the constraints, $x \geq 0$, $Ax \leq b$ such that x maximizes the sum $\sum_{i=1}^{n} c_i x_i$, where $b \in R^m$, $c \in R^n$. The standard algorithm for solving this problem is the famous "simplex method," due to George Dantzig. Prove that for a randomly-chosen A, b and c with m fixed, the number of iterations needed for the simplex method to converge grows linearly in the number of problem variables n.

9. In the ★-✠-ℰℬ-system considered in Section 4, show that the following rules constitutes a decision procedure for the theoremhood of a string:

    i. The string must begin with a ✠.

    ii. The number of ℰℬ's in the string must not be divisible by 3.

10. Define $SH(n)$ to be the maximum number of shifts that an n-state Turing machine can perform in the Busy Beaver game when starting in state A on a blank tape. Analogously, define $SC(n)$ to be the maximum

number of tape squares that an n-state machine scans before halting. Finally, let $H(n)$ be the number of different n-state Turing machines that halt when started in state A on a blank tape.

a) Prove that

$$BB(n) \leq SC(n) \leq SH(n) \qquad \text{and} \qquad H(n) < (4n+4)^{2n}.$$

b) Establish the inequalities

$$SC(n) < BB(3n) \qquad \text{and} \qquad SH(n) \leq nSC(n)2^{SC(n)}.$$

c) Consider a Turing machine that's randomly programmed by, say, a chimpanzee. Give arguments showing that the probability of such a machine halting is

$$\lim_{n \to \infty} \frac{H(n)}{(4n+4)^{2n}}.$$

11. Let $p \colon \mathbb{R}^n \to \mathbb{R}$ be a real polynomial of degree 4. The *4-Feasibility Problem over* $\mathbb{R}$ asks for an algorithm to decide whether or not p has a real zero.

a) Prove that the 4-Feasibility Problem over $\mathbb{Z}$ is unsolvable, i.e., there is no algorithm for determining whether p has an integer root.

b) Show that the 4-Feasibility Problem is **not** in NP over $\mathbb{Z}$.

12. Suppose f is a quadratic polynomial with distinct roots, and define the Newton operator

$$N_f(z) = z - \frac{f(z)}{f'(z)}.$$

a) Prove that there exists a linear fractional transformation $g(z) = (z - \beta)/(z - \alpha) = \omega$ such that $g(N_f) = \omega^2$.

b) Use the results of part (a) to show that, in general, Newton's method is not convergent.

13. Show that the "size" of a polynomial of degree n can be expressed in $O(\log n)$ bits of information. (Note: Here the "size" is a measure of the polynomial's degree and the bits needed to express its coefficients.) Use this result to prove that the problem of factoring a polynomial is in class P.

14. The natural way of multiplying two n-digit binary numbers requires on the order of n^2 steps. Show by example that there exist methods requiring only $kn \log n \log \log n$ steps.

15. Suppose you have n items of varying sizes and you want to find the item of median size, i.e., half the items are larger, half smaller. Show that this item can be found by performing $O(n)$ pairwise comparisons, thus proving that this problem is in class P.

16. Show that the Busy Beaver function value $BB(5) \geq 4,098$ by providing an explicit program for a 5-state Turing machine that writes this many 1's on a blank tape before halting.

17. Show that a Turing machine that can read n tapes cannot compute anything that cannot already be computed by the standard single-tape machine (but it may compute it a lot *faster*).

Notes and References

§1–2. An introductory account of the matters of these sections can be found in Chapter Six of

Casti, J., *Searching for Certainty: What Scientists Can Know About the Future*, Morrow, New York, 1991.

In this same introductory vein, see also

Hoffman, P., *Archimedes' Revenge*, Norton, New York, 1988,

Rucker, R., *Mind Tools*, Houghton-Mifflin, Boston, 1987.

More technical accounts of Turing machines and their connections with not only decision problems, but also formal languages, are given in the volumes

Davis, M. and El Weyuker, *Computability, Complexity, and Languages*, Academic Press, Orlando, FL, 1983,

Harel, D., *Algorithmics*, Addison-Wesley, Reading, MA, 1987,

Davis, M., *Computability and Unsolvability*, McGraw-Hill, New York, 1958 (reprint edition: Dover, New York, 1982),

Boolos, G. and R. Jeffrey, *Computability and Logic*, 3rd edition, Cambridge University Press, Cambridge, 1989,

Epstein, R. and W. Carnielli, *Computability: Computable Functions, Logic and the Foundations of Mathematics*, Wadsworth & Brooks/Cole, Pacific Grove, CA, 1989.

A stimulating collection of essays reviewing current knowledge about Turing machines and their many implications and ramifications in other areas is presented in

The Universal Turing Machine: A Half-Century Survey, R. Herken, ed., Oxford University Press, Oxford, 1988.

§3. For more details on the construction and operation of a UTM, the reader should see the Rucker book *Mind Tools* referenced above.

The Turing-Church Thesis lies at the heart of the currently fashionable artificial intelligence debate, which revolves about the question of whether or not a computer can think like a human being. *If* all human thought processes can be shown to be "effective," and *if* the Turing-Church Thesis is correct, then it necessarily follows that there is no barrier, in principle, between the "thought processes" of machines and those of humans. But both of these "ifs" are very big if's indeed, and no one has yet been able to give a knockdown argument resolving either half of this conundrum. For an account of the current state of play, as well as for an extensive bibliography on the whole issue, see Chapter Five of

Casti, J. *Paradigms Lost: Images of Man in the Mirror of Science.* Morrow, New York, 1989 (paperback edition: Avon, New York, 1990).

A much more technical and philosophically-oriented approach to the implications of the Turing-Church Thesis for both psychology and the philosophy of mathematics is presented in

Webb, J., *Mechanism, Mentalism, and Metamathematics,* Reidel, Dordrecht, Holland, 1980.

In this same connection, see also

Arbib, M., *Brains, Machines, and Mathematics,* 2nd ed., Springer, New York, 1987,

Penrose, R., *The Emperor's New Mind,* Oxford University Press, Oxford, 1989.

The Busy Beaver Game was dreamed up by Tibor Rado of Ohio State University in the early 1960s. A compact, introductory discussion of what's currently known about this problem and about the Busy Beaver function can be found in the articles

Dewdney, A., "Busy Beavers" in *The Armchair Universe,* Freeman, New York, 1988, pp. 160–171,

Brady, A., "The Busy Beaver Game and the Meaning of Life," in *The Universal Turing Machine,* R. Herken, ed., Oxford University Press, Oxford, 1988, pp. 259–277.

A relatively technical update of the current state of play of the Busy Beaver is given in the article

Machlin, R. and Q. Stout, "The Complex Behavior of Simple Machines," in *Emergent Computation,* S. Forrest, ed., MIT Press, Cambridge, MA, 1991, pp. 85–98.

The Turing Machine Game is taken from the very illuminating survey article

Jones, J., "Recursive Undecidability: An Exposition," *Amer. Math. Monthly,* September 1974, pp. 724–738.

§4. A simple, easy-to-understand introduction to formal systems is given in

Levine, H. and H. Rheingold, *The Cognitive Connection,* Prentice Hall Press, New York, 1987.

For a more technical account, emphasizing the connections between formal systems and languages, see the book

Moll, R., M. Arbib, and A. Kfoury, *An Introduction to Formal Language Theory,* Springer, New York, 1988.

A good account of formal systems and Hilbert's Program for the formalization of mathematics is found in the Epstein and Carnielli text noted under §3. For a more leisurely, semi-layman's account, see the Casti volume cited under §1–2.

The ★-✠-ℰℬ-system used in the text example is a rewrite of the MIU-system described in

Hofstadter, D., *Gödel, Escher, Bach: An Eternal Golden Braid,* Basic Books, New York, 1979.

It goes without saying that this Pulitzer-prize winning volume is also a primary reference for all the other material of this section, as well.

§5. To the best of my knowledge, the first account of Gödel's Theorem written expressly for the general reader, and still one of the best, is the short volume

Nagel, E. and J. R. Newman, *Gödel's Proof,* New York University Press, New York, 1958.

An English translation of Gödel's pioneering paper, as well an enlightening biographical account of his life, can be found in the first volume of Gödel's collected works:

Kurt Gödel: Collected Works, Volume 1, S. Feferman, et al., eds., Oxford University Press, New York, 1986.

An assessment of Gödel's Theorem from both a philosophical as well as mathematical point of view is contained in the collection of reprints

Gödel's Theorem in Focus, S. Shanker, ed., Croom and Helm, London, 1988.

People often wonder whether or not long-standing, seemingly impenetrable mathematical questions like Goldbach's Conjecture (every even number is the sum of two primes) are undecidable in the same way that Cantor's Continuum Hypothesis turned out to be undecidable. Musings of this sort give rise to the consideration of whether or not Gödel's results really matter to mathematics, in the sense that there are important mathematical questions that are truly undecidable. With the recent work of Chaitin and others, the comforting belief that there are no such problems seems a lot less comforting than it used to. For a discussion of some other "real" mathematical queries that are genuinely undecidable, see

Kolata, G., "Does Gödel's Theorem Matter to Mathematics?" *Science,* 218, (19 November 1982), 779–780.

Many details of Gödel's personality, views on life and philosophy, as well as an assessment of both his mathematical and philosophical work, are found in the following book written by the well-known mathematical logician Hao Wang, who was a long-time acquaintance of Gödel:

Wang, H., *Reflections on Kurt Gödel,* MIT Press, Cambridge, MA, 1987.

Additional information about Gödel's life is given in

Dawson, J., "Kurt Gödel in Sharper Focus," *Mathematical Intelligencer,* 6, No. 4 (1984), 9–17,

Kreisel, G., "Kurt Gödel: 1906–1978," *Biographical Memoirs of Fellows of the Royal Society,* 26 (1980), 148–224.

The text discussion of "mirroring" and Gödel numbering follows that given in the Nagel and Newman book noted above. Hofstadter's switching-yard metaphor for Gödel numbering and transformations in a formal system can be found in the expository paper

Hofstadter, D., "Analogies and Metaphors to Explain Gödel's Theorem," *College Mathematics Journal,* 13 (March 1982), 98–114.

The discussion by Rudy Rucker likening Gödel's Theorem to the plight of Joseph K. in his wanderings through Kafka's *The Castle,* may be found in the very enlightening (but slightly technical) book

Rucker, R., *Infinity and the Mind,* Birkhäuser, Boston, 1982.

In connection with Gödel as a person, this book is especially recommended for its account of several meetings that Rucker had with Gödel in the years shortly before Gödel's death in January 1978.

§6. The citation for Solomonoff's original paper on the complexity of scientific theories is

Solomonoff, R., "A Formal Theory of Inductive Inference," *Information and Control,* 7 (1964), 224–254.

The complete story of Chaitin's independent discovery of algorithmic complexity and its connection with randomness is contained in his collection of papers

Chaitin, G., *Information, Randomness, and Incompleteness,* 2nd ed., World Scientific, Singapore, 1990.

Quite independently of both Chaitin and Solomonoff, the famous Russian mathematician Andrei Kolmogorov also hit upon the idea of defining the randomness of a number by the length of the shortest computer program required to calculate it. His ideas were presented in

Kolmogorov, A., "Three Approaches to the Quantitative Definition of Information," *Problems in Information Transmission,* 1 (1965), 3–11.

The original formulation of Berry's Paradox involved a statement like: "The smallest number that cannot be expressed in fewer than thirteen words." Since the preceding phrase contains twelve words, the paradox follows for exactly the same reasons as given for the more general phrase used in the text. A fairly complete account of the Berry Paradox and its relationship to complexity and Gödelian logic is available in the Rucker book *Infinity and the Mind* already noted.

§7. A very easy-to-understand, illuminating discussion of Hilbert's Tenth Problem is given in Chapter Six of

Devlin, K., *Mathematics: The New Golden Age,* Penguin, London, 1988.

A somewhat more technical account is presented in the article

Davis, M., "What is a Computation?", in *Mathematics Today: Twelve Informal Essays,* L. A. Steen, ed., Springer, New York, 1978, pp. 241–267,

as well as in the volume

Salomaa, A., *Computation and Automata,* Cambridge University Press, Cambridge, 1985.

Each of these sources also gives a good account of Matyasevich's resolution of the problem.

An introductory account of Chaitin's fabulous Diophantine equation straight from the horse's mouth, so to speak, is found in

Chaitin, G., "Randomness in Arithmetic," *Scientific American,* 259 (July 1988), 80–85.

Creation of Chaitin's "monster" equation followed the flowchart below, involving the creation of a sequence of machine-language and LISP programs:

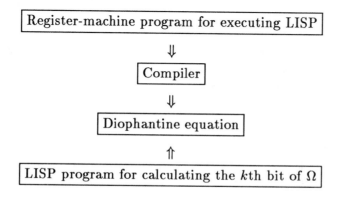

The technical details surrounding this monumental intellectual and programming effort are given in the book

Chaitin, G., *Algorithmic Information Theory,* Cambridge University Press, Cambridge, 1987.

§8. The definitive general reference on computational tractability, NP-completeness and other matters of this ilk is

Garey, M. and D. Johnson, *Computers and Intractability,* Freeman, San Francisco, 1979.

Introductory accounts of NP-completeness are given in Chapters 31, 38, 42 and 50 of

Dewdney, A., *The Turing Omnibus,* Computer Science Press, Rockville, MD, 1989.

See also the following references for further details

Poundstone, W., *Labyrinths of Reason,* Doubleday, New York, 1988,

Kleitman, D., "Algorithms," *Advances in Math.,* 16 (1975), 233–245,

Håstad, J., "Lower Bounds in Computational Complexity Theory," *Notices Amer. Math. Soc.,* 35 (1988), 677–683.

An introductory account of Cook's Theorem is found in Chapter 42 of the Dewdney book cited above.

For a discussion of Yao's work, see the article

Kolata, G., "Must 'Hard Problems' Be Hard?", *Science,* 228 (1985), 479–481.

§9. The famous mathematician and Fields Medalist Stephen Smale has been one of the leaders in calling for a rethinking of numerical analysis to emphasize the complexity of algorithms. An account of his work is given in the review articles

Smale, S., "On the Efficiency of Algorithms of Analysis," *Bull. Amer. Math. Soc.,* 13 (1985), 87–121,

Smale, S., "Some Remarks on the Foundations of Numerical Analysis," *SIAM Review,* 32 (1990), 211–220.

When we shift attention from methods to problems, the investigation of the intrinsic complexity of problems of analysis was instituted by Joseph Traub and his co-workers some years ago. An account of their "information-based complexity" is found in

Traub, J. and H. Woźniakowski, "Information-Based Complexity: New Questions for Mathematicians," *Math. Intelligencer,* 13 (1991), 34–43,

Packel, E. and H. Woźniakowski, "Recent Developments in Information-Based Complexity," *Bull. Amer. Math. Soc.,* 17 (1987), 9–36,

Traub, J., G. Wasilkowski and H. Woźniakowski, *Information-Based Complexity,* Academic Press, Orlando, FL, 1988.

§10. The relationship between Chaitin's Theorem, Gödel's Theorem and chaos is taken from

Casti, J., "Chaos, Gödel and Truth," in *Beyond Belief: Randomness, Explanation and Prediction in Science,* J. Casti and A. Karlqvist, eds., CRC Press, Boca Raton, FL, 1991, pp. 280–327.

Moore's ingenious use of the dynamical system-program dictionary to create a complex chaotic system whose behavior cannot be predicted even if we know the initial conditions *exactly* is first reported in

Moore, C., "Unpredictability and Undecidability in Dynamical Systems," *Physical Rev. Lett.,* 64 (1990), 2354–2357.

A summary for the scientifically-literate layman is

Bennett, C., "Undecidable Dynamics," *Nature,* 346 (1990), 606–607.

Kaneko's results showing that the attractors of dynamical systems with global coupling often contain lots of structure—even when the dynamics is formally chaotic—are given in

Kaneko, K., "Chaotic But Regular Posi-Nega Switch Among Coded Attractors by Cluster-Size Variation," *Physical Rev. Lett.,* 63 (1989), 219–223,

Kaneko, K., "Globally Couples Chaos Violates the Law of Large Numbers But Not the Central-Limit Theorem," *Physical Rev. Lett.,* 65 (1990), 1391–1394.

§11. The modeling vs. simulation arguments using neural nets and mathematical machines is examined further in

Rosen, R., "Causal Structures in Brains and Machines," *Int. J. Gen. Syst.,* 12 (1986), 107–126.

DQ #3. These matters are discussed in much greater detail in the article

Gale, D., "The Truth and Nothing But the Truth," *Mathematical Intelligencer,* 11, No. 3 (1989), 62–67.

DQ #4. For more information about Dewdney's zoo of beavers, see

Dewdney, A., *The Armchair Universe,* Freeman, New York, 1988.

DQ #6–7. Fuller discussions of these matters are available in

Goodman, N., "Mathematics as an Objective Science," *American Math. Monthly,* 86 (1979), 540–551,

Snapper, E., "What Do We Do When We Do Mathematics?," *Mathematical Intelligencer,* 10, No. 4 (1988), 53–58.

DQ #10. This whole Top-Down vs. Bottom-Up approach to strong AI is recounted in detail in Chapter Five of

Casti, J., *Paradigms Lost: Images of Man in the Mirror of Science,* Morrow, New York, 1989 (paperback edition: Avon, New York, 1990).

DQ #13. Consideration of the assets and liabilities of these competing measures of complexity is reported in

Bennett, C., "How to Define Complexity in Physics, and Why," in *Complexity, Entropy and the Physics of Information,* W. Zurek, ed., Addison-Wesley, Redwood City, CA, 1990, pp., 137–148.

DQ #18. For the details surrounding Warsh's ideas, see his book

Warsh, D., *The Idea of Economic Complexity,* Viking, New York, 1984.

DQ #29. For a personal recounting of Lam's work on the projective planes, along with his views on computer proofs in general, see

Lam, C. W. H., "How Reliable Is a Computer-Based Proof?," *Mathematical Intelligencer,* 12 (1990), 8–12.

DQ #31. The program correctness issue is discussed in some detail in

Barwise, J., "Mathematical Proofs of Computer System Correctness," *Notices Amer. Math. Soc.,* 36 (September 1989), 844–851.

DQ #33. More consideration of this aspect of the complexity problem is given in

Grassberger, P., "Toward a Quantitative Theory of Self-Generated Complexity," *Int. J. Theor. Physics,* 25 (1986), 907–938.

PR #7. A much fuller discussion of this tiling problem, which exemplifies a large class of solvability problems is given in the Harel book cited under §1–2.

PR #9. The proof that this is indeed a decision procedure for the ★-✠-꒰꒱-system is given in

Swanson, L. and R. McEliece, "A Simple Decision Procedure for Hofstadter's *MIU*-System," *Mathematical Intelligencer,* 10, No. 2 (1988), 48–49.

PR #11. This problem arises during the course of developing a theory of computation and complexity over the real numbers instead of just the integers. As one might suspect when moving to a broader class of numbers, a number of questions that seem puzzling over $\mathbb{Z}$ dissolve, or at least become a whole lot easier, when looked at over a bigger field of numbers. For details, see the article

Blum, L., M. Shub and S. Smale, "On a Theory of Computation and Complexity over the Real Numbers: NP-Completeness, Recursive Functions and Universal Machines," *Bull. Amer. Math. Soc.,* 21 (1989), 1–46.

CHAPTER TEN

How Do We Know?: Myths, Models and Paradigms in the Creation of Beliefs

1. Ordering the Cosmos

In his *Metaphysics,* Aristotle remarked that "all men by nature desire to know." This desire springs from both a psychological and a practical need to organize our observations of nature somehow, as well as our desire to structure the events and activities of daily life into a coherent pattern that we call "reality." Each domain of knowledge has its own observables, and we generally can't construct the observables in one domain from knowledge of those in another. And so arises the need to create many realities in order to make a lawful ordering of the cosmos. In this volume our reality-generating mechanism has been that handmaiden of theoretical science, the mathematical model. But it would be remiss in a work of this type to leave the reader with the impression that science as we know it today is the only—or even the best—tool for this reality-generation task. In fact, Western science is a johnny-come-lately in the reality-generation game, having arisen historically in the Middle Ages as a response to the inability of the competition to offer a satisfactory explanation of the Black Death. So in this final chapter we shall delve a bit more deeply into some of the philosophical issues underlying the relationship between realities and models, and at the same time take a critical look at the degree to which the concepts, methods and techniques of science are any better at playing the Reality Game than the competition.

At this juncture, it's worth reflecting for a moment on the idea of an "external" reality as we've been using that term. In everyday language, the notion of an external reality comes down to a belief that the objects in the world around us enjoys an independent existence. In other words, lamps, tables, TVs, cars and atoms are all 'out there' whether we observe them or not. According to this belief, the universe is defined to be the totality of such independently existing "things." Loosely speaking, this commonsense belief in the nature of the universe constitutes what's generally thought of as "objective reality." It forms one of the pillars upon which modern science is precariously perched. As we've noted in earlier chapters, recent results in quantum theory cast some doubt on this view. Let's summarize why.

According to the now-famous Bell's inequalities, one may either retain the idea of an objective reality or obey the universal speed limit of no faster-than-light signaling—but not both. Recent experiments by Alain Aspect have confirmed the theoretical predictions of Bell's result, leaving us in the uncomfortable position of either having to throw away objective reality or being forced to discard the key assumption underlying Einstein's theory of relativity, thereby leaving us open to a host of time travel and causality paradoxes. Most practicing physicists are loathe to dispense with the idea of an objective reality; nevertheless, this seems to be the only viable alternative left to us by nature if we want to avoid the ambiguities and puzzles of acausal effects. Fortunately for most practical day-to-day matters, there is no harm done in clinging to the illusion of an objective reality, and the practice of science at the meso- and macro-level goes on much as before. However, when we move into the microworld, it's well to bear in mind that it is not the *fact* that there may be no objective reality "out there" that is important. Rather, what counts is that there may be as many "realities" as you want. This point will surface again in our subsequent discussion.

The traditional means for structuring experiences in virtually every culture is the *myth*. In my dictionary, the term "myth" is traced back to the Greek *mythos* meaning "word," in the sense that it is the decisive or definitive statement on the subject. A myth presents itself as an authoritative account of the facts which is not to be questioned, however strange it may seem. According to the mythologist Joseph Campbell, myths serve several quite distinct purposes:

• *Metaphysical*—Myths awaken and maintain an "experience of awe, humility and respect" in recognition of the ultimate mysteries of life and the universe.

• *Cosmological*—Myths provide an image of the universe and explanations for how it works.

• *Social*—Myths validate and help maintain an established social order.

• *Psychological*—Myths support the "centering and harmonization of the individual."

It will become clear as we proceed that the idea of "objective" science is one of Western man's most widely accepted myths. Of course, myths need be neither true nor false, but only useful fictions. However, we should not confuse myths with the kind of ordinary fiction that has entertainment value alone and does not pretend to be true. The opposite side of the coin of *mythos* is *logos,* the Greek term for an account whose truth can be demonstrated and debated. This duality should be kept in mind as we compare the role of science as myth with its various competitors like religion, art, humanism and mysticism.

Throughout this book we have sought answers from science, not mythology. Thus we claim there is indeed something more to science as it is practiced than just a story taken to manifest some aspect of cosmic order. Basically our position is that science involves myths of a special type, namely, myths that are *predictive, empirically testable and cumulative*. In the following pages we shall elaborate upon these conditions, trying to explain a few of the reasons for the success of modern science, as well as point out its glaring deficiencies as a reality-structuring mechanism. But for now it's sufficient just to think of the connection between science and myth as being

$$\text{science} = \text{myth} + \text{discipline}.$$

2. Models and Theories

In works on the philosophy of science, many different meanings are attached to the term "model." There are *physical* models such as the well-known billiard ball model for the behavior of an enclosed gas; there are *logical* models involving a collection of symbolic entities satisfying a particular set of axioms leading via logical deduction to theorems; there are *mathematical* models consisting of symbolic representations of quantities appearing in physical, social, and behavioral systems. Our focus has been solely upon the latter class of models, and we have been distinctly pragmatic in the sense that we have considered a model of a natural system N to be the representation of the observables and equations of state of N using the elements of a formal mathematical system F. To a certain degree this view edges dangerously close to that held by the now-defunct school of *logical positivists*, who claimed that the only meaningful statements that could be made about the world were those that could be translated into observables, and that the actual meaning (semantic content) of a statement was identical with a rule for verifying it. Although we are sympathetic to some aspects of the positivist program, we hold no brief for the extreme view that theoretical terms can (and must) be translated into equivalent observational terms. Thus, for us models are tools for reality organization, i.e., a tool for ordering experiences rather than a *description* of reality.

If models are only a means for ordering our experiences, it follows that there can be many different models of the same experiences; hence, many alternative realities—at least to the degree that we see reality in our models. This position is made strikingly clear in the linguistic work of Edward Sapir and Benjamin Whorf, who considered the role of natural—rather than mathematical—language as a tool for ordering and describing experiences. The Sapir-Whorf hypothesis asserts that the way one sees the world depends upon the structure of one's language. If true, such a contention would imply that there is no correct way of seeing the world. Alternately, different languages give rise to different worlds—literally.

As support for their claims, Whorf spent considerable time studying the languages of several American Indian tribes and found, for instance, that the Hopi language has one noun that covers everything that flies with the exception of birds, which as a class are covered by another noun. Thus, the only distinction that can be made between flying objects in Hopi is the distinction between birds and nonbirds. Similarly, Eskimo languages have different words for expressing various gradations of snow—wet, dry, powdery, icy—distinctions that are be made in everyday English only with difficulty or by using complicated constructions. Consequently, it can be argued that the snow "seen" by an Eskimo is not the *same* snow that I see today on the street as I look out my window in Vienna.

There are many examples of this kind, and they form the basis for the claim that language is the prime force in shaping our conception of the world. For science, this means that no individual is free to describe nature with absolute objectivity, but is constrained to certain modes of interpretation even when he thinks himself most free. Of course, we can argue that the "language" of mathematics provides a universal *lingua franca* enabling us to describe nature in as impartial and universal way as can be imagined— and surely far better than any natural language. There is some merit to this argument. But the same principle that underlies the Sapir-Whorf hypothesis for natural languages also applies to mathematical descriptions, since every formal system contains its own symbols, axioms and provable truths. But we know from Gödel's Theorem that there are other truths that cannot be expressed within that system. So in the final analysis we must recognize that just as it is with objective reality, so it is too with objective science, and there is no possibility of attaining the nirvana of objectivity untainted by the "prejudices" built into our method of describing nature, be it linguistic or mathematical. The best we can hope for is to illuminate various facets of natural systems using different—and inequivalent—formal descriptions.

The traditional philosophy of science literature distinguishes carefully between a scientific *theory* and a *model.* Roughly speaking, a theory is an interpreted mathematical proposition (in the terms of Chapter 1, a *decoding*), whereas a model is generally thought of as something like the billiard ball model of a gas mentioned above. Throughout the book we have blurred this distinction, primarily because we have held steadfastly to the position that the only types of models we're interested in are mathematical models. With this bias regarding our model class, there is very little remaining to separate a theory from a model, at least at the practical level that we are considering in this book. For us a model is the mathematical manifestation of a particular theory, and we are happy to leave the debate about exactly what constitutes that theory to the philosophical literature. Nevertheless, it's worth dipping into that literature briefly to examine some of the issues surrounding the criteria for a good theory and the standards that a theory must meet before it can be accepted by the scientific community.

In what follows we shall take the semantic view and regard a theory as being synonymous with a *family* of related models, where the term model has our usual interpretation as being a formal mathematical system. Theories like relativity, quantum mechanics, gene translation, natural selection and continental drift that have come to have a high level of universal acceptance, and by common agreement constitute what are usually termed "good" theories, generally have a number of identifying fingerprints:

• *Consistency*—Good theories contain no self-contradictory statements. So, for example, a good theory of gravitation has no statements asserting that a ball rolls both up and down an inclined plane under seemingly identical experimental circumstances. Similarly, a good theory of cellular development would not lead to situations in which identical genotypes give rise to radically different phenotypical forms.

• *Noncircular*—The conclusions of a good theory are not subtly buried in its premises. In other words, good theories contain no circular arguments.

• *Cumulative*—A good theory is somehow "better" than previous theories in the sense that it contains most—but not necessarily all—of the results of the earlier theories yet still adds something new. It is in this way that the Special Theory of Relativity is better than Newtonian gravitation, since all the claims of the older theory are contained as special cases of Einstein's extension. So a good theory always encompasses all the observational evidence of its predecessor and still manages to add something new.

• *Testable*—Good theories always make predictions or claims that are testable by experiment, at least in principle. This criterion of *falsifiability* has been emphasized by the philosopher Karl Popper as the line separating scientific from nonscientific theories, and has been used to dismiss the claims of astrologers, Marxists, and creationists that their brand of medicine qualifies to be labeled "scientific."

Despite these rather severe conditions, the scientific graveyard is littered with the corpses of "good" theories that never managed to gain acceptance. This leads us to wonder whether there are any criteria or standards beyond those imposed above for the acceptance of a scientific theory. In his pioneering work on the ways of science and scientists, Thomas Kuhn identified at least two additional conditions that a theory must meet before it can be accepted by the scientific community.

One of Kuhn's conditions is that the new theory must answer a significant number of questions answered by the old theory, while at the same time providing insight into issues that the old theory cannot or does not address. Note that this doesn't mean that the new theory must answer *all* the questions answered by the old theory. As we shall see in a moment, it's perfectly possible for the new theory and the old theory to have only substantial points of intersection, rather than having the new theory completely encompass the old. For example, in the theory of ideal gases Robert Boyle originally thought of the gas molecules as little coiled springs that exerted a strong repulsive force when compressed. Newton improved upon this picture by imagining the gas particles as being perfect little spheres that were mutually repelled by an inverse square law. Both of these theories are inherently *static,* for they regard the gas as a sort of sponge that shows an elastic expansion when compression is removed. A radically different theory was put forth by Bernoulli, who thought of the now familiar billiard ball model of molecular motion in which the gas molecules careen about in the container, bouncing off one another in much the same way as a collection of randomly moving balls on a billiard table. This is a *dynamic* theory of gas behavior.

The Bernoulli theory clearly answers many of the same questions about the pressure-temperature-volume relationship of gas as the older theories, but it also contains the possibility for addressing substantial new issues by virtue of its dynamic view of the origin of the observed behavior. However, there are questions that the dynamic theory cannot address, questions that are within the purview of the earlier theories such as the elasticity constants of the "springs" in the Boyle view or the "gravitational" constant of the spheres in Newton's picture. But since such questions aren't important for addressing the major issues surrounding the behavior of gases, they don't work to the disadvantage of the new theory insofar as its acceptance by the scientific community is concerned. What's important are the questions that are answered and the new questions that are generated, not those that are ignored. This observation leads to Kuhn's second criterion for theory acceptance.

New theories, like their predecessors, always generate new questions as well as answer old ones. An important component of the success of a new theory—or at least its acceptance—is that it provide enough new puzzles to keep scientific research alive. In other words, a new theory can't be **too** successful in answering the old questions or it won't leave enough material for the next generation of scientists to keep themselves busy. The lifeblood of science is unsolved problems. And if the current theory doesn't leave enough problems to keep scientific activity alive (and the research-granting machinery humming), it won't be accepted by the scientific community.

This situation is radically different from the way science is usually portrayed in the media and cinema, where the popular vision is of the scientist as a seeker after all truths, large and small. Nevertheless, science is done by scientists, and scientists are no more immune to the psychological and sociological pressures and constraints of their culture than are butchers, bakers or candlestick makers. So with this point in mind, it's perhaps not surprising to see such subjective factors entering in to the acceptability of a scientific theory.

These social and psychological considerations led Kuhn to postulate his now famous idea of a "paradigm" to explain the workings of science and, most importantly, the manner in which the scientist's view of the world undergoes discontinuous periodic shifts. Let's take a deeper look into the heart of the "paradigm" concept.

3. Paradigms, Revolutions and Normal Science

A scientific paradigm is basically a theory plus a set of presuppositions about physical or mental models, together with a symbolism and the other social and technical trappings of the time. To use a somewhat fanciful metaphor, think of science as the *terra incognita* of the ancient geographers and mapmakers. In this context, a paradigm may be thought of as a crude sort of map in which territories are outlined—but not too accurately—with only major landmarks like large rivers and prominent mountains initially appearing on the map. From time to time explorers venture into this ill-defined territory and come back with accounts of native villages, desert regions, minor rivers, and so forth, which are then dutifully entered onto the map. Often such new information is inconsistent with what was reported from earlier expeditions, so it's periodically necessary to redraw the map in accordance with the current best estimate of how things stand in the unknown region. Furthermore, there is not just one mapmaker but many, each with his own set of sources and data regarding the lie of the land.

As a result of this process of map-making, there are a number of competing maps of the same region. So the adventurous explorer has to make a choice of which map s/he will believe before embarking upon an expedition to the "New World." Generally, the explorer will choose the old, reliable firm of mapmakers, at least until gossip and reports from the Explorers Club show too many discrepancies between the standard maps and what has actually been observed. As these discrepancies accumulate, the explorers eventually shift their allegiance to a new firm of mapmakers whose picture of the territory seems more in line with the observational evidence.

This exploration fable gives a fair picture of the birth and death of a scientific paradigm. Kuhn realized that revolutionary changes overturning old theories in science are not in fact part of the normal process of science. Nor do theories start small and grow more and more general as claimed by Bacon. Nor can they ever be axiomatized as asserted by Newton and Descartes. Rather, for most scientists major paradigms are like a pair of spectacles that they put on in order to solve puzzles. Occasionally a paradigm shift occurs when the spectacles get smashed, forcing the scientists to put on a new pair that turns everything around into new shapes, sizes and colors. Once this shift takes place, a new generation of scientists is brought up wearing the new glasses and accepting the new vision as "true." Through these new glasses scientists then see a whole new set of puzzles.

The paradigms have great practical value for the scientist, just as maps have value for the explorer; without them, the scientist wouldn't know where to look or how to plan an experiment (expedition) and collect data. This observation brings out the crucial point that there is no such thing as an "empirical" observation or fact; we always see by interpretation. And the interpretation we use is given by the prevailing paradigm of the moment. In other words, the observations and experiments of science are made on the basis of theories and hypotheses contained within the prevailing paradigm. As Einstein put it, "the theory tells you what you can observe." According to Kuhn's paradigmatic view of scientific activity, the job of "normal" science is to fill-in the gaps in the map given by the current paradigm, and it's only seldom, and with great difficulty, that the map gets redrawn when the normal scientists (explorers) turn up so much data that doesn't fit into the old map that the map begins to collapse of its own weight. But what happens during these times of a paradigm crisis?

Suppose we are at the initial stages of such a crisis, where the old paradigm can't account for certain anomalies, strange observations, and the like. Two new theories emerge that offer different explanations for these aberrations. These theories represent different maps, or sets of spectacles, i.e., different realities. After a period of competition, one of these theories begins to gain the acceptance of the scientific community. The reasons may not be objective, but may revolve about matters like simplicity, elegance, the social position of its adherents or government science policies. This support leads to experiments that then "corroborate" the theory; the more evidence that accumulates, the more supporters the theory gathers, especially among the "young Turks" in the scientific community. Soon "reality" begins to take on the look of the new theory, and scientists begin to universally see and test for certain features of this reality and ignore others.

But what if the community had given its initial support to the competing theory? According to Kuhn, in that event "reality" would have taken quite a different turn, and the scientific view of the world would have been seen through that pair of spectacles rather than the first. This means that there is no such thing as scientific "progress," at least not in the sense that one paradigm builds upon its predecessor. Rather, the new paradigm turns in an entirely different direction, and as much knowledge is lost from the old paradigm as is gained from the new. Instead, we "know" a *different* universe.

If Kuhn's thesis is true, then it also destroys one of the main pillars of the scientific method, since the whole idea of a scientific experiment rests upon the assumption that the observer can be essentially separate from the experimental apparatus that tests the theory. Kuhn contends that the observer, his theory and his equipment are all essentially an expression of a point of view. So the results of the experimental test must be an expression of that point of view, as well. This means, in essence, that science is not *objective*. But at the same time we know that science is not totally subjective

either, since paradigms are eventually overthrown. So what is the relationship of the scientist to the universe he observes? This question leads us to
a consideration of what it means to say a model constitutes an explanation
of nature as opposed to just being a description.

4. Explanations versus Descriptions

Many would argue that the basic business of science is to be able to say "because." Planets travel around the Sun in elliptical orbits *because* they obey
the inverse square law of gravitation; diamonds are the hardest substance
known *because* their crystalline structure resists compression more strongly
than any other; it's raining today *because* at the current temperature and
pressure the outside air cannot hold all the water vapor present. It's in the
nature of an *explanation* to answer the question: "Why?" On the other
hand, a *description* just gives an account of "what is," without entering
into any underlying reasons for the "why of things." For the most part we
have been concerned with descriptions, and our modeling philosophy has
been directed primarily at providing formal structures that describe what
is without being too concerned with whether they are in any way "true" in
the explanatory sense of that. However, it's patently clear that an explanation is far preferable to a description, in just the same way that a formal
proof is much better than a heuristic plausibility argument. Consequently,
it's worth taking some time to consider a few of the differences between an
explanation and a description.

   In common parlance, an explanation involves giving an account of an
already *known* fact on the basis of logical conclusions drawn from well-established general theories. Often there is some sort of hierarchical structure
of explanations, with the fact to be "explained" residing at a higher level
than the components of the explanation. Such a hierarchy, of course, lies
at the heart of the reductionist program for scientific explanation in which
all phenomena are ultimately explainable by some kind of elementary or
rock-bottom level of facts or entities. So, the essence of an explanation is to
answer the question "Why?" at one level with a "Because" constructed from
entities and logical operations at a lower level. This is the reductionistic
pattern we see, for example, in the explanation of the temperature of an
enclosed gas in terms of the kinetic behavior of its constituent molecules,
the explanation of gross national product in terms of the contributions of
individual industries and in numerous other situations.

   From the above considerations it's evident that there's an intimate connection between giving an explanation for a process and providing a causal
description of it; in fact, the two are virtually synonymous. For this reason
it's of interest to note that the idea of a causal description is at least as old
as ancient Greek science, in which we find a number of attempts to provide
a causal explanation of events. The dominant causal structure for over two

millennia was put forth by Aristotle, who identified a set of "causes" in his treatment of physical phenomena. According to Aristotle, we can answer "because" in four mutually exclusive and collectively exhaustive ways:

• *material cause*—Things are as they are because of the material elements from which the system is constructed.

• *efficient cause*—Things are as they are because of the physical work or energy that went into making them as they are.

• *formal cause*—Things are as they are because of the plan or design according to which they were created.

• *final cause*—Things are as they are because of the desire of an external agent to have occur as they do.

In Aristotle's epistemology, the highest place was reserved for the last category, final causation, which, amusingly enough, was banished from polite scientific discourse by the paradigm of Newtonian mechanics that finally overthrew Aristotle's view of the world. Since it's of some interest to see how this elimination of final cause was orchestrated by Newton, let's look at the situation in a bit more detail.

In Newton's world of particles and forces, every possible "why" question is answerable in terms of two aspects of the system: (1) the positions and velocities of all the particles as measured from some appropriate reference frame, and (2) the external forces imposed upon the particles. In mathematical terms, the state $x(t)$ representing the positions and velocities of all particles of the system at time t is given by

$$x(t) = x_0 + \int_0^t g_\alpha(x(s), f(s)) \, ds,$$

where x_0 is the initial state, $f(\cdot)$ represents the external forces acting upon the system, and $g_\alpha(\cdot, \cdot)$ is the law of motion of the system, including the constitutive parameters α, which represent the force of gravity, particle masses and so on. In Aristotelian terms, we can use the above framework to answer the question "Why is the system in the state $x(t)$?" in the following inequivalent ways:

• *material cause*—The system is in the state $x(t)$ because it was in the state x_0 at time $t = 0$.

• *formal cause*—The particles are in the state $x(t)$ because the system's constitutive parameters assumed the value α.

• *efficient cause*—The state is now $x(t)$ because the integral operator $\int_0^t g_\alpha(\cdot) \, ds$ acted upon previous states to bring the system to the state $x(t)$.

Note that there is no need here—and no room, either—for any sort of final causation in Newton's epistemology. It's for this reason that the idea of final cause has been turned into an unfashionable, even crankish notion in modern scientific thought. Here we see a perfect example of the "different worlds" of competing paradigms in the Kuhnian sense described earlier.

Classical science regards the banishment of final cause as a significant step forward and a selling point for the Newtonian paradigm. Our view is distinctly bleaker. Although holding no brief for final cause as teleology, in the sense of the future generating the present, we do believe there is a type of quasi-teleological principle at work in all systems involving living organisms. This principle forms the root cause of the difference between the physical sciences, on the one hand, and the life, social and behavioral sciences on the other. By banishing final cause from scientific discussion, Newton in effect threw out the baby with the bathwater, eliminating all notions of anticipation and self-reference from respectable scientific consideration. We have given extended treatment in Chapter 7 as to how we think it's possible to incorporate this most important of Aristotelian causes within a major extension of the Newtonian *Weltanschauung.* Now let's turn to the matter of predictions and the use of models as tools for data correlation.

5. *"Good" Models and "Bad"*

Earlier we examined some criteria for a "good" theory, noting that from a semantic point of view we could consider a theory to be a collection of related models, where the term "model" here denotes the kind of formal system that we have centered attention on in this book. Now we want to investigate the related matter of what constitutes a "good" model, together with the delicate issue of how to validate or confirm a proposed model of a specific natural process. To serve as a focal point for these deliberations, let's return to that most ancient of modeling problems, planetary motion, and consider the two competing models put forth by Ptolemy and Copernicus.

Ptolemy's model of the solar system had the planets moving in orbits that were described by a collection of superimposed epicycles, the Earth being at the center of the system. From a predictive standpoint, this structure gave remarkably good accounts of where the planets would be at any time, and it was successfully used by many generations to predict eclipses, comets and, most importantly, lunar positions influencing the flooding of the Nile. The discovery of new solar bodies and their influence on the motion of others could always be accommodated in this model by piling on another epicycle or two to make the behavior of the model agree with the observed data. In this sense, the model was somewhat akin to the process of fitting a polynomial through a set of data points. We know that the fit can be made exact if we're willing to take a polynomial of high enough degree.

The major change introduced by Copernicus was to assume that the Sun rather than the Earth lies at the center of the solar system. This assumption, later translated into mathematical form by Kepler and Newton, led to the now familiar picture of the planets moving about the Sun in elliptical orbits, and has generated a highly accurate set of predictions about future planetary positions. However, it's of the greatest importance to note that from a predictive standpoint the Copernican model is no better than the Ptolemaic. In fact, if anything the epicycles of Ptolemy provided a mathematical framework that could make *more accurate* predictions than the ellipses of Kepler and Newton. Again this is analogous to the situation with curve-fitting by polynomials, in that you can't do better than to match the data *exactly*. And the Ptolemaic model was capable of doing a better job of this than the heliocentric Copernican picture. So why was the Ptolemaic view discarded in favor of Copernicus? The obvious answer is that the Copernican picture is *simpler:* it contains fewer *ad hoc* assumptions and leads to a less cumbersome, more straightforward mathematical formalization. Fundamentally, Ptolemy's model of planetary motion was cast aside on aesthetic grounds, rather than because the Copernican model was in better agreement with the observations or gave better predictions about future positions of the planets. This kind of subjective criterion is one that we shall see more of in a later section when we come to consider competition between models.

The above scenario for choosing one model over another has been formalized in the principle now termed Occam's Razor, which asserts that one should not multiply hypotheses beyond what is required to account for the data. This principle leads to the corollary that everything else being equal (which, of course, it never really is), take the simplest model that agrees with the observations. This leads immediately to another issue that we have touched upon a number of times in this volume—the problem of complexity. Given two models, by what criterion do we say one is simpler (or more complex) than the other? So *simplicity,* however you choose to define it, is certainly one of the principal attributes of a "good" model. But there are others.

In the competition between Copernicus and Ptolemy, we saw another important criterion for a "good" model: it must agree to a reasonable degree of accuracy with *most* of the observed data. Here we emphasize the term "most," since it's important in this connection to bear in mind our earlier dictum that the theory tells you what you can observe. And for us theory is a collection of models. Consequently, if the aesthetic reasons for choosing a model are strong enough, then it's only necessary for the model to agree with the majority of the observations.

As an example of the foregoing precept, consider Newton's model of gravitation as embodied in the inverse-square law and his Third Law, $F = ma$. This led to predictions about the orbit of Mercury that were not in agreement with observation. Nevertheless, this model of gravitation is probably the most successful model in the history of science, and it needed only some fine-tuning by Einstein to account for this Mercurial anomaly. In fact, one might say it's the hallmark of a "good" model when the data that it **doesn't** agree with lead to the next scientific revolution. Thus, *good, but not necessarily perfect,* data agreement is a second benchmark test for a "good" model.

The third element in our Holy Trinity of good modeling is *explanatory power.* A good model should be in a modeling relation to the natural system it represents and not be just a simulation. This means that there should be some interpretable connections between the entities of the formal system comprising the model and the physical entities characterizing the natural system under study. Since we have already spent some time elaborating our view of the differences between an explanation and a description, it's not necessary to repeat those arguments here, other than to note that the notion of explanatory power forces us to confront head-on the problem of how we would go about validating a proposed model. To address this issue, we again return to matters central to twentieth-century philosophy of science.

6. Validation and Falsification

The problem of the validation of a scientific theory or a formal model was brought into sharp focus by the extreme position taken by the logical positivist movement in the 1930s. According to the positivists, a scientific theory is not a representation of the world but a shorthand calculational device for summarizing sensory data. The positivists asserted that the scientist should use only concepts for which he can give "operational definitions" in terms of observations. According to this position, length and time are not attributes of things in the world, but are relationships defined by specifying experimental procedures. The culmination of the positivist position was the notorious *verification principle,* which demanded that only empirical statements verifiable by sense experience have meaning (formal definitions and tautologies are also meaningful but convey no factual information). Consequently, by this criterion most statements in philosophy, and all those in metaphysics, ethics and theology, are neither true nor false, but meaningless; i.e., they state nothing and merely express the speaker's emotions or feelings. The difficulties with this position are by now well chronicled so we won't enter into them here. In our quest for insight into the model validation question, the importance of the positivist position is the implied claim that a model is valid only if it can be tested (verified) by experimental procedures; otherwise, it's not only invalid, it's nonscientific.

A major outgrowth of the verification thesis was put forward by Karl Popper in his view that scientists use models not just as explanations of natural phenomena, but as probes to provoke nature into behavior which the scientist can recognize as an observation. This observation can then be used to: (1) provide further inductive evidence in favor of the model, or

(2) provide a counterexample to the claims of the model. It is the second case that Popper termed *falsifying* the model.

Popper's basic position is that science is not in the business of validating models at all, but rather should be trying to falsify them. His claim is that real scientific models are set forth in a way that spells out observations and predictions that can be tested experimentally. If the prediction fails, then the model is falsified and must be abandoned or completely re-thought. But if a model passes its crucial test, it's not validated but only "corroborated." In this case the process of testing must continue.

An interesting and important consequence of Popper's theory is that controversies about the merits of competing models shouldn't really exist. When faced with two competing models, each representing a given situation, scientists would simply choose the "better" one—the one that could survive the toughest tests. As we have already noted, Kuhn realized that in practice this is impossible. The parties in paradigm debates speak such different languages and wear such different glasses that even if they look at the same clock, they won't agree on which way it's running.

Kuhn concluded that Popper's falsification standard for scientific validation was a myth. In fact, he claimed that most scientists hardly ever see an anomaly in the observations as a challenge to the paradigm that lies behind it. Just as astrologers claim that when their predictions fail it's due to factors like too many variables in the system, too much uncertainty, and so forth, viewing the failure as in no way disproving their cosmic model, so it is also with meteorologists, who make the same excuses when their prediction of a sunny weekend turns into a torrential downpour.

So, although the theory of falsifiability of models looks logically attractive, it seems to have little to do with how science actually operates. And it certainly doesn't appear to form the basis for any objective procedure for model validation. When all is said and done, we are left in much the same position we found ourselves in earlier when considering standards for theory acceptance. There are no purely objective criteria by which we can select one model over another. In practice, as often as not the choice hinges upon psychological and sociological factors having little or nothing to do with the empirical content of the model. So let's take a longer look at the way real-life science functions in separating the wheat from the chaff in picking what model to back.

7. Competition among Models

Scientific theories and their models prove their mettle in a forum of debate. Most of us are familiar with the kind of school-day debates in which one party argues the affirmative of an issue while another supports the opposite position. But what is the nature of the debate by which models are accepted or rejected in science? There are at least three basic ground rules.

First of all, each of the competing models must be an available alternative. By this we mean that the model must be known to the scientific community, and it must be developed to the extent that its advantages are explicit. A model or theory must be a serious contender in its own time, which means that it must be able to explain the facts in question, as well as either fit into the current scientific framework or else have something going for it that makes it worth considering even though it violates some basic assumptions. It should also be noted here that competing models, unlike school-day debates, are usually not contradictory in that one is not the denial of the other. Thus, in contrast to the familiar debating situation where all it takes to show one side right is to show the other side wrong, in the modeling derby it could very well be that all sides are wrong.

The second basic ground rule of scientific debate is that the arguments and evidence offered in support of a model must be submitted to evaluation according to the standards of the scientific community. What sorts of reasons do the scientific community accept? It depends to a large extent upon the particular field. It's somewhat easier to say what kinds of reasons are *not* acceptable, a detailed treatment of which we shall defer to the next section.

Our final ground rule is that the winning model must prove itself better than its competitors. It's not enough that there are reasons that seem to favor it. And it's not enough that there is solid evidence that seems to confirm it. It has to be shown that the model has more going for it than the other candidate models. What are the identifying features of such a model? Well, just those features we outlined earlier in our discussion of model credibility: simplicity, agreement with the known facts, explanatory power and predictive capability.

Although the primary filtering mechanism of modern science is the debate outlined above, we shouldn't discount the social and psychological factors that also enter into a model's or a theory's acceptance or rejection. To begin with, the deck is always stacked in favor of the existing orthodoxy since this is the paradigm by which we see theoretical constructs. In addition, the existing power structure in the scientific community generally has a large professional, psychological and financial stake in protecting the *status quo*. After all, who wants to see a lifetime's worth of work on developing a chemical analysis technique or a delicate asymptotic approximation get wiped out overnight by some young hotshot's computer-based substitute? This ego factor is only exacerbated by the system of research grants administered by federal agencies, which more often than not reward Kuhn's "normal" science and strongly penalize adventuresome revolutionaries. This process is further reinforced, of course, by the promotion and tenure policies in the universities, where advancement is contingent upon publication of lots of "potboiler" articles and the hustling of research grant money as much as by painstaking scholarly activity. All these factors taken together introduce an enormous bias into the model acceptance process, a bias strongly leaning toward preservation of the existing orthodoxy. This situation calls to mind Max Born's famous aphorism that, "New theories are never accepted, their opponents just die off." This has never been more true than in today's world of science.

Nevertheless, there is no shortage of theories about how the world is ordered, many of them at least clothed in the trappings of science. A glance at the shelves in your favorite bookshop will turn up an almost endless array of volumes on UFOs, hollow-earth theories, creationism, astral projection, extraterrestrials, telekinesis, numerology, astrology, spoon-bending, pyramidology and the like, many of which lay claim to being legitimate alternatives to mainstream science. This is not to mention the various types of religions such as Scientology, Christian Science, or even more exotic forms given to beliefs involving speaking in tongues, handling of poisonous snakes, infant damnation and other practices. We shall deal with religion later. For now, and by way of contrast between science and the "pseudosciences," let's take a harder look at some of the features of these alternative belief systems.

8. Science and Pseudoscience

Many working scientists (and certainly every editor of a science journal) has had the experience of receiving a bulky package in the mail plastered with stamps and addressed in a semi-legible scrawl, containing the results of a lifetime's worth of work claiming to square the circle, create perpetual motion, refute the theory of relativity or definitively resolve in some way another outstanding issue in science. Generally such efforts come accompanied with rambling accounts of why the scientific community has misunderstood the basic issues involved, and how the author has seen the way clear to set the record straight using what he claims is the scientific method of gathering data, generating hypotheses and doing experiments. After a time, the experienced eye learns to recognize the telltale signs of such cranks and pseudoscientists, and learns to never, ever, under any circumstances enter into any sort of dialogue with such people over the merits of their "discoveries." Our aims in this section are not to lay out the complexities of the scientific information-dissemination process, but rather to point out that there are ways of operating that are not found in science but that are routinely employed in pseudoscience. These ways have to do with the reasons for proposing alternative hypotheses, with what are accepted as facts to be explained, with what counts as supporting evidence and with what counts as a theory.

Here is a checklist of criteria for identifying pseudoscience when you see it. Anyone who meets even one of these conditions is practicing in pseudoscience.

• *Anachronistic Thinking*—Cranks and pseudoscientists often revert to outmoded theories that were discarded by the scientific community years, or even centuries, earlier as being inadequate. This is in contrast to the usual notion of crackpot theories as being novel, original, offbeat, daring and inventive. Many of them actually represent a return to a world view that was dismissed by the scientific world as being too simplistic or just plain wrong years earlier. Good purveyors of this kind of crankishness are the creationists who link their objections to evolution to catastrophism, claiming that geological evidence supports the catastrophic rather than uniformitarian view of geological activity they associate with evolution. The argument is anachronistic insofar as it presents the uniformitarianism-catastrophism dichotomy as if it were still a live debate.

• *Seeking Mysteries*—Scientists do not set out in their work looking for anomalies. Max Planck wasn't looking for trouble when he carried out his radiation emission experiments, and Michelson and Morley certainly were not expecting problems when they devised their experiment to test for the luminiferous ether. Furthermore, scientists do not reject one theory in favor of another solely because the new theory explains the anomalous event. On the other hand, there are entire schools of pseudoscience devoted to enigmas and mysteries be they the Bermuda Triangle, UFOs, Yetis, spontaneous combustion or other even more offbeat phenomena. The basic principle underlying such searches seems to be that "there are more things on heaven and earth than are dreamed of in your philosophies," coupled with the methodological principle that anything that can be seen as a mystery ought to be seen as one.

• *Appeals to Myths*—Cranks often use the following pattern of reasoning: Start with a myth from ancient times and take it as an account of actual occurrences; devise a hypothesis that explains the events by postulating conditions that obtained at that time but which no longer hold; consider the myth as providing evidence for support of the hypothesis; argue that the hypothesis is *confirmed* by the myth as well as by geological, paleontological or archaeological evidence. This is a pattern of reasoning that is absent from the procedures of science.

• *Casual Approach to Evidence*—Pseudoscientists often have the attitude that sheer quantity of evidence makes up for any deficiency in the quality of the individual pieces. Further, pseudoscientists are loathe to ever weed out their evidence. And even when an experiment or study has been shown to be questionable, it is never dropped from their list of confirming evidence.

• *Irrefutable Hypotheses*—Given any hypothesis, we can always ask what would it take to produce evidence against it. If nothing conceivable could speak against the hypothesis, then it has no claim to be labeled "scientific." Pseudoscience is riddled with hypotheses of this sort. The prime example of such a hypothesis is creationism; it is just plain not possible to falsify the creationist model of the world.

• *Spurious Similarities*—Cranks often argue that the principles that underpin their theories are already part of legitimate science, and see themselves not so much as revolutionaries but more as poor cousins of science. For example, the study of biorhythms tries to piggyback upon legitimate scientific studies carried out on circadian rhythms and other chemical and electrical oscillators known to be present in the human body. The basic pseudoscience claim in this area is that there is a similarity between the claim of the biorhythm theorists and the claims of the biological researchers. Therefore, biorhythms are consistent with current biological thought.

• *Explanation by Scenario*—It's commonplace in science to offer scenarios for explanation of certain phenomena, such as the origin of life or the extinction of the dinosaurs, when we don't have a complete set of data to construct the exact circumstances of the process. However, in science such scenarios must be consistent with known laws and principles, at least implicitly. Pseudoscience engages in explanation by scenario *alone,* i.e., by mere scenario without proper backing from known laws. A prime offender in this regard was Immanuel Velikovsky, who stated that Venus' near collision with the Earth caused the Earth to flip over and reverse its magnetic poles. Velikovsky offers no mechanism by which this cosmic event could have taken place, and the basic principle of deducing consequences from general principles is totally ignored in his "explanation" of such phenomena.

• *Research by Literary Interpretation*—Pseudoscientists frequently reveal themselves by their handling of the scientific literature. They regard any statement made by any scientist as being open to interpretation, just as in literature and the arts, thinking that such statements can then be used against other scientists. They focus upon the words, not on the underlying facts and reasons for the statements that appear in the scientific literature. In this regard, the pseudoscientists act more like lawyers gathering precedents and using these as arguments rather than attending to what has actually been claimed.

• *Refusal to Revise*—Cranks and crackpots pride themselves on never having been shown to be wrong. It's for this reason that the experienced scientific hand never, under any circumstances, enters into dialogue with a pseudoscientist. But immunity to criticism is no prescription for success in science, as there are many ways to avoid criticism: write only vacuous material replete with tautologies; make sure your statements are so vague that criticism can never get a foothold; simply refuse to acknowledge whatever criticism you do receive. A variant of this last ploy is a favorite technique of the pseudoscientist: he always replies to criticism but never revises his position in light of it. They see scientific debate not as a mechanism for scientific progress but as a rhetorical contest. Again the creationists serve as sterling examples of the power of this principle.

The major defense of pseudoscience is summed up in the credo: "Anything is possible." Earlier in this chapter we considered the question of competition between models and theories and drew up a few ground rules by which the competition is generally carried out in legitimate scientific circles. Let's look at how the pseudoscientist, hiding behind his "anything is possible" shield, enters into such competition.

In the competition among theories, the pseudoscientist makes the following claim: "Our theories ought to be allowed into the competition because they may become available alternatives in the future. Scientists have been known to change their minds on the matter of what is and is not impossible, and they are likely to do so again. So who is to say what tomorrow's available alternatives may be?" In other words, anything is possible! This argument clearly violates our first rule of scientific debate. The fact that a theory may become an available alternative in the future does not constitute a reason for entering it into the competition today. Every competitor now must be an available alternative now. The pseudoscientist suggests that we may as well throw away the current scientific framework since it will eventually have to be replaced anyhow.

By referring to a future but as yet unknown state of science, the cranks are in effect refusing to participate in the competition. This would be all right if they didn't at the same time insist on entering the race. It's as if one entered the Monaco Grand Prix with a jet-propelled car and insisted on being allowed to compete because, after all, someday the rules may be changed to a jet car race!

The pseudoscientist also worms his way into the competition by putting the burden of proof on the other side. He declares to the scientific community that it is up to them to prove his theory wrong, and if they cannot do so he then states that his theory must be taken seriously as a possibility. The obvious logical flaw is the assumption that not proving a theory impossible is the same thing as proving it is possible. Although the principle of innocent till proved guilty may be the norm in a court of law, scientific debate is not such a court. The reason why the pseudoscientist thinks he can put the burden of proof on the scientists can be traced to his mistaken notion of what constitutes a legitimate entry in the competition. The pseudoscientists think that the scientific method places a duty on the scientific community to consider *all* proposed ideas that are not logically self-contradictory. So, to ignore any idea is to be prejudiced.

Finally, we note that the pseudoscientists often act as if the arguments supporting their theory are peripheral to the theory. Thus, they fail to see that what makes a theory a serious contender is not just the theory, but the theory plus the arguments that support it. Cranks think that somehow the theory stands on its own, and that the only measure of its merits for entering the competition is its degree of outlandishness (i.e., "imagination"). Hence, they think the scientific community has only two choices: admit their theory into the competition or else prove it to be impossible. So we conclude that when it comes to defending a theory or model in scientific debate, there is just no room for the "anything is possible" school of pseudoscience. No room, that is, unless the theories are supported by scientifically-based arguments.

Belief systems outside science come in many forms, some of them covered by the general umbrella of pseudoscience. However, by far the most interesting and important alternative to a scientific ordering of the world is that provided by the principles and tenets of organized religion. From the beginnings of Western science in the Middle Ages, there has been a sort of undeclared guerrilla war being waged between the church and the scientific community on the matter of who is the keeper of true knowledge about the nature of the cosmos. In the next section we examine this conflict as our final statement about the alternate realities by which we shape our existence.

9. Science, Religion and the Nature of Belief Systems

In the Reality Game, religion has always been science's toughest opponent, perhaps because there are so many surface similarities between the actual practice of science and the practice of most major religions. Let's take mathematics as an example. Here we have a field that emphasizes detachment from worldly objects, a secret language comprehensible only to the initiates, a lengthy period of preparation for the "priesthood," holy missions (famous unsolved problems) to which members of the faith devote their entire lives, a rigid and somewhat arbitrary code to which all practitioners swear their allegiance, and so on. These features are present in most other sciences as well, and bear a striking similarity to the surface appearances of many religions. In terms of similarities, both scientific and religious models of the world direct attention to particular patterns in events and restructure how one sees the world. But at a deeper level there are substantial differences between the religious view and that of science.

Let's consider some of the major areas in which science and religion differ:

• *Language*—The language of science is primarily directed toward prediction and control; religion, on the other hand, is an expression of commitment, ethical dedication and existential life orientation. So even though we have superficial surface similarities at the syntactic level, the semantic content of scientific and religious languages are poles apart.

- *Reality*—In religion, beliefs concerning the nature of reality are presupposed. This is just the opposite of the traditional view of science, which is directed toward *discovering* reality. Thus religion must give up any claims to truth, at least with respect to any facts external to one's own commitment. In this regard, the reality content of most religious beliefs is much the same as in the myths considered earlier in the chapter. Fundamentally, what we have in science is a basic belief in the comprehensibility of the universe, a belief that is not necessarily shared by all religions

- *Models*—Although both scientific and religious models are analogical and used as organizing images for interpreting life experiences, religious models also serve to express and evoke distinctive attitudes, and thus encourage allegiance to a way of life and adherence to policies of action. The imagery of religious models elicits self-commitment and a measure of ethical dedication. These are features completely anathema to the role of models in science. In religion, the motto is "live by these rules, think our way and you'll see that it works." The contrast with science is clear.

- *Paradigms*—In the discussion of paradigms we saw that scientific paradigms were subject to a variety of constraints like simplicity, falsification and influence of theory on observation. All of these features are absent in religious paradigms.

- *Methods*—In science there is a method to get at the scheme of things: observation, hypothesis and experiment, i.e., the scientific method. In religion there is a method, too: divine enlightenment. However, the religious method is not repeatable, nor is it necessarily accessible to every interested investigator—even in principle.

So both pseudoscience and religion provide alternate reality-structuring procedures that are radically different in character from those employed in science. It's of interest to ponder the point as to why there is such a diversity of brands of nonscientific knowledge, especially in light of the claims of virtually every sect that its own brand of medicine is uniformly most powerful.

Our view on this matter is quite simply that neither science nor religion nor pseudoscience offer a product that is satisfactory to all customers. The wares are just not attractive enough. In some cases the beliefs are not useful in the way that people want to use them. For example, many people have a deep-seated psychological need for security and turn to conventional religion for myths of all-powerful and beneficent Beings who will attend to these needs for protection. Science with its mysterious and potentially threatening pronouncements about black holes, the "heat death" of the universe, evolution from lower beings, nuclear holocausts, and the like, offers anything but comfort to such primal needs and, as a result, loses customers to the competition. Basically, beliefs thrive because they are useful. And the plain fact is that there is more than one kind of usefulness.

To the practicing scientist, the foregoing observations may come as a sobering, if not enervating, conclusion because they seem to put into jeopardy the conventional wisdom that the road to real truth lies in the "objective" tools of science, not in the subjective, romantic ideas of believers and crusaders. But if we accept the existence of alternative and equally useful belief systems, we are inexorably led full circle back to the position that there are many alternate realities not just within science itself, but outside as well. And the particular brand of reality we select is dictated as much by psychological needs of the moment as by any sort of rational choice.

In the final analysis, there are no complete answers but only more questions, with science providing procedures for addressing an important subset of such questions. We conclude with the admonition that the only rule in the Reality Game is to avoid falling into that most common of all human delusions, the delusion of a single reality—our own!

Discussion Questions

1. If one accepts the Kuhnian picture of science as a succession of paradigms, it's difficult to imagine such a thing as scientific "progress," since successive paradigms don't usually encompass all the observations accounted for by their predecessors. How can you reconcile this notion of no progress with the obvious advances in medicine, communication, and transportation over the past decades?

2. Consider the claims of some religions such as Scientology or the Christian Scientists that their brand of faith is based on "scientific" principles. Also consider the same claim made by political ideologies such as Marxism. Is there any interpretation, other than honorific, by which you could term these belief systems "scientific" in the sense we have used that term?

3. It has been noted that the training and practice of mathematics as a profession has much in common with that of many religions. For example, there are obsessions with the abstract, together with a professed detachment from worldly matters, a long and arduous training period, during which the acolytes learn a special language that's incomprehensible to the uninitiated, claims that the work makes contact with deep mysteries of how the universe is ordered, sacred missions (i.e., famous unsolved problems) to which some members of the priesthood devote their entire lives, and so forth. Does this analogy make any sense to you? (As an aside, one of the corollaries of this analogy is that as it seems somewhat less fashionable nowadays than in earlier times to enter the religious life, many people of a basically theological

turn of mind have found the more rarefied corners of mathematics, philosophy and theoretical physics to be a socially acceptable way of satisfying their essentially religious leanings and yearnings.)

4. In quantum theory, at least in the classical Copenhagen interpretation, there are two contradictory models for describing the behavior of an object: a *particle* model by which the object displays behavior characteristic of point particles, and a *wave* model by which the object displays properties like interference that are associated with waves. Is there any way you could see either of these models as providing an "explanation" for the behavior of an object? Do either of these models of the quantum object provide the basis for describing an external reality or are they only tools for predicting the results of measurements?

5. Give an account of why each of the following beliefs might be considered pseudoscience: astrology, graphology, numerology, dowsing, orgone therapy, ESP, the prophecies of Nostradamus, Adam and Eve, theosophy, psychoanalysis, psychokinesis, Noah's Ark, creationism.

6. We have given great weight to the Law of Parsimony, or Occam's Razor, as a guiding principle in the selection of models for natural phenomena. How can you reconcile this principle with the increasingly complex paradigms of modern particle physics, evolutionary biology, paleontology and other branches of science? For example, in particle physics it was once thought that the only fundamental particles were protons, electrons and neutrons. Today the prevailing orthodoxy involves a host of quarks with different "colors," "flavors" and "orientations." This view is clearly not as simple as the earlier model. Give reasons why, despite this added complexity, the quark model has replaced the earlier picture.

7. In science we often see intense competition to obtain credit for a discovery. Prominent examples are Charles Darwin's rushing to print to achieve priority over Alfred Wallace, James Watson's and Francis Crick's race with Linus Pauling to unravel the genetic code and the competition between Luc Montagnier and Robert Gallo for priority in discovery of the AIDS virus. To what degree do you feel these psycho-sociological factors enter into the acceptance of one paradigm and the rejection of another?

8. Modern Western culture has traditionally based its view of reality upon two pillars: *materialism* (there is an external reality composed of "things") and *cartesian duality* (there is a separation between these things of the world and the things of the observing mind). Thus, in cartesian terms reality is divided into the *res extensa* and the *res cogitans.* To what degree have the discoveries of modern physics (Bell's Theorem, for example), neurophysiology, computer science and linguistics altered this vision of reality?

Is there any way we can interpret or identify the altered states of consciousness of the mystic with the alternative realities of the scientist?

9. The position that we have presented on the question of how to mirror nature in a formal system is rather close to the philosophical view termed "instrumentalism." Discuss the following objections that have been raised against this approach to modeling nature and man:

a) There is too much prominence given to *formal* analogies and not enough attention to *substantive* ones, i.e., analogies between observables in different natural systems N.

b) The approach is unable to provide for the extension of theories; i.e., it can't generate new encoding/decoding operations or extend observables.

c) The approach neglects the importance of physical models because it's not concerned with the process of discovery.

d) Instrumentalism asserts that explanation is tantamount to prediction, i.e., they are equivalent concepts.

Problems

Dear Reader:

If you've worked even a small fraction of the Exercises and Problems in the preceding chapters, you're entitled to a well-deserved rest; if not, please go back and do so now! The Exercises, Discussion Questions and Problems are an integral part of the message of this book. So if you want to hear the message, you'll have to make an active effort by trying your hand at doing, not just listening and reading. Fortune favors the prepared mind. And the only preparation for the art and science of modeling is practice. So please don't fall into the trap of believing that the skills of good modeling practice can be acquired without the effort of thinking—and doing.

Notes and References

A reasonably short, general introduction to material of this chapter is available in Chapter One of

Casti, J., *Paradigms Lost: Images of Man in the Mirror of Science,* Morrow, New York, 1989 (paperback edition: Avon, New York, 1990).

§1. According to Bertrand Russell, "It is not what the man of science believes that distinguishes him, but *how* and *why* he believes it." It's the "how" and "why" that set science apart from other belief systems like religion, mysticism and pseudoscience. For introductory discussions of these and other matters pertaining to the role of science as a belief system, see

Stableford, B., *The Mysteries of Modern Science,* Routledge and Kegan Paul, London, 1977,

McCain, G., and E. Segal, *The Game of Science,* 4th ed., Brooks and Cole, Monterey, CA, 1982,

Spradlin, W., and P. Porterfield, *The Search for Certainty,* Springer, New York, 1984.

An authoritative account of myths and their place in Western civilization can be found in

Campbell, J., *The Masks of God: Occidental Mythology,* Viking, New York, 1964.

§2. Good accounts of the types of models used by science and the way in which theories rise and fall are given in

Oldroyd, D., *The Arch of Knowledge,* Methuen, London, 1986,

Barbour, I., *Myths, Models and Paradigms,* Harper & Row, New York, 1974,

Goodman, N., *Ways of Worldmaking,* Hackett, Indianapolis, IN, 1978.

Benjamin Whorf, whose daytime occupation was as a fire inspector for a large Hartford insurance company, was living testimony to the claim that an amateur can play ball on the same field as the professionals—if he is dedicated and talented enough. His views on the relationship between human thought and language can be summed up in the postulates: (1) all higher levels of thought are dependent on language, and (2) the structure of the language that one habitually uses conditions the manner in which one understands his environment. An account of the origin and evolution of his views is found in

Whorf, B., *Language, Thought and Reality,* MIT Press, Cambridge, MA, 1956.

An easily-readable introduction to the basics of Whorf's work is found in Chapter Four of the Casti volume cited under General References. For an entertaining view of the folk-tale concerning the many Eskimo words for snow, see

Pullum, G., *The Great Eskimo Vocabulary Hoax,* University of Chicago Press, Chicago, 1991.

§3. Thomas Kuhn's classic account of the origin of scientific paradigms, as well as a response to some of the criticisms of his ideas, is found in

Kuhn, T., *The Structure of Scientific Revolutions,* 2d ed., University of Chicago Press, Chicago, 1970.

§4. An excellent account of the tenuous connections between science and philosophy is given in

Gjertsen, D., *Science and Philosophy: Past and Present,* Penguin, London, 1989.

The resurrection of Aristotelian causation as a basis for questioning the Newtonian paradigm, especially in the context of biological systems, is taken up in the work

Rosen, R., "On Information and Complexity," in *Complexity, Language and Life: Mathematical Approaches,* J. Casti and A. Karlqvist, eds., Springer, Heidelberg, 1986, pp. 174–196.

§5. The story of the rise of the heliocentric view of Copernicus and its supplanting of the geocentric model of Ptolemy has been told in many places. One of the best is

Kuhn, T., *The Copernican Revolution,* Harvard University Press, Cambridge, MA, 1957.

Copernicus' life and times are presented in graphic detail in the novel

Banville, J., *Dr. Copernicus,* Martin Secker & Warburg, Ltd., London, 1976.

§6. The position of the positivists and the many objections raised against their view of scientific knowledge and how to get it are chronicled in the Oldroyd volume referred to under §2.

Sir Karl Popper has been a key figure in the twentieth-century philosophy of science, and his views on the aims of science and the means for attaining those aims have been put forth in many forums. Two excellent sources for the ideas, as well as for information about the man himself are

Popper, K., *Unended Quest,* Fontana, London, 1976,
Magee, B., *Popper,* 2d ed., Fontana, London, 1982.

For a more detailed account of Popper's views in his own words, see

Miller, D., ed., *A Pocket Popper,* Fontana, London, 1983.

§7. The way in which competing models of a phenomenon are judged in a forum of debate is most properly a topic within the sociology, rather than the philosophy, of science. A good account of this process is found in

Merton, R. K., *The Sociology of Science: Theoretical and Empirical Investigations,* University of Chicago Press, Chicago, 1973.

Case studies of the sociology of science are given in the works

Fisher, C. S., "The Death of a Mathematical Theory," *Arch. Hist. Exact Sci.,* 3 (1966), 137–159,

Fisher, C. S., "The Last Invariant Theorists," *Arch. Euro. Soc.,* 8 (1967), 216–244,

Raup, D., *The Nemesis Affair,* Norton, New York, 1986,

Sheldrake, R., *A New Science of Life,* 2d ed., Anthony Blond, London, 1985.

Other accounts of the way in which some models survive and others die are found in the Kuhn volume listed under §3, as well as in the works noted under §2.

§8. Pseudoscience is now flourishing as never before, with the volume of material available seemingly only limited by the capacity of the printing presses to supply the public's insatiable appetite for easy explanations and quick "scientific fixes." Our treatment of this topic follows that in

Radner, D., and M. Radner, *Science and Unreason,* Wadsworth, Belmont, CA 1982.

Martin Gardner, former editor of the Mathematical Games column of *Scientific American,* has been a tireless crusader against the debasement of science by pseudoscience. Some of his work in this direction is reported in

Gardner, M., *Science: Good, Bad and Bogus,* Prometheus, Buffalo, NY, 1981,

Gardner, M., *Order and Surprise,* Prometheus, Buffalo, NY, 1983,

Gardner, M., *The New Age: Notes of a Fringe Watcher,* Prometheus, Buffalo, NY, 1988.

For an account of the history of the Velikovsky controversy, see

Bauer, H., *Beyond Velikovsky: The History of a Public Controversy,* University of Illinois Press, Urbana, IL, 1984.

Pseudoscience is closely related to the question of how rational people are in the course of forming their beliefs and taking actions. A good summary of this rationality issue is provided in the volume

The Limits of Rationality, K. Cook and M. Levi, eds., University of Chicago Press, Chicago, 1990.

§9. The contrast between religion and science as competing belief systems is brought out in sharp detail in the Stableford and Barbour books cited under §1 and §2 above. Other works along the same lines are

Peacocke, A., "A Christian Materialism?", in *How We Know,* Michael Shafto, ed., Harper and Row, San Francisco, 1985, pp. 146–168,

Hummel, C., *The Galileo Connection: Resolving Conflicts Between Science and the Bible,* InterVarsity Press, Downer's Grove, IL, 1986.

DQ # 4. The science bookshelves are filled with volumes purporting to explain the mysteries of the quantum world to the layman. Chapter Seven of the Casti book cited under General References gives an introduction to the issues and competing theories. Another recent account is given in

Peat, F. David, *Einstein's Moon,* Contemporary Books, Chicago, 1990.

DQ # 6. For a blow-by-blow account of the development of the current state of elementary particle physics, see

Crease, R., and C. Mann, *The Second Creation,* Macmillan, New York, 1986,

Building the Universe, C. Sutton, ed., Basil Blackwell, Oxford, 1985,

Superstrings: A Theory of Everything?, P. Davies and J. Brown, eds., Cambridge University Press, Cambridge, 1988,

Barrow, J., *Theories of Everything,* Oxford University Press, Oxford, 1991.

DQ #8. The relationship between the altered states of consciousness of the mystic and the realities of the modern physicist is explored in the popular volumes

Capra, F., *The Tao of Physics,* 2d ed., Shambhala, Boulder, CO, 1983,

Zukav, G., *The Dancing Wu Li Masters,* Morrow, New York, 1979.

Some of the same territory has also been scouted in the more recent works

Krishnamurti, J., and D. Bohm, *The Ending of Time,* Gollancz, London, 1985,

Wallace, B. Alan, *Choosing Reality: A Contemplative View of Physics and the Mind,* Shambhala, Boston, 1989,

Weber, R., *Dialogues with Scientists and Sages,* Routledge and Kegan Paul, London, 1986.

INDEX

Page numbers for *The Fundamentals* appear in roman type. Page numbers for *The Frontier* appear in *italic* type.

Page numbers for *The Fundamentals* appear in roman type. Page numbers for *The Frontier* appear in *italic* type.

Page numbers for *The Fundamentals* appear in roman type. Page numbers for *The Frontier* appear
in *italic* type.

Page numbers for *The Fundamentals* appear in roman type. Page numbers for *The Frontier* appear in *italic* type.

Page numbers for *The Fundamentals* appear in roman type. Page numbers for *The Frontier* appear in *italic* type.

Page numbers for *The Fundamentals* appear in roman type. Page numbers for *The Frontier* appear in *italic* type.

Page numbers for *The Fundamentals* appear in roman type. Page numbers for *The Frontier* appear in *italic* type.

Page numbers for *The Fundamentals* appear in roman type. Page numbers for *The Frontier* appear in *italic* type.

Page numbers for *The Fundamentals* appear in roman type. Page numbers for *The Frontier* appear in *italic* type.

Page numbers for *The Fundamentals* appear in roman type. Page numbers for *The Frontier* appear in *italic* type.